Commercializing High Technology: East and West

International Series on Technical Innovation and Entrepreneurship

General Editors: George Kozmetsky, David V. Gibson, and Robert S. Sullivan

This series provides the latest insights, experiences, and research on ways to promote and sustain technical innovation and entrepreneurship in a global context. Each volume seeks to integrate the insights of businesspersons with the findings of academics and the experiences of government policymakers. Sponsored by the IC^2 (Innovation, Creativity, and Capital) Institute at The University of Texas at Austin.

Technology Companies & Global Markets: Programs, Policies, and Strategies to Accelerate Innovation and Entrepreneurship
Edited by David V. Gibson, 1991.

University Spin-Off Companies: Economic Development, Faculty Entrepreneurs, and Technology Transfer
Edited by Alistair Brett, David V. Gibson, and Raymond W. Smilor, 1991.

Commercializing High Technology: East and West
Edited by Judith B. Sedaitis, 1997.

Commercializing High Technology:

East and West

Edited by

Judith B. Sedaitis

ROWMAN & LITTLEFIELD PUBLISHERS, INC.
Lanham • Boulder • New York • London

ROWMAN & LITTLEFIELD PUBLISHERS, INC.

Published in the United States of America
by Rowman & Littlefield Publishers, Inc.
4720 Boston Way, Lanham, Maryland 20706

3 Henrietta Street
London WC2E 8LU, England

British Cataloging in Publication Information Available

Library of Congress Cataloging-in-Publication Data

Commercializing high technology : East and West / edited by Judith B. Sedaitis.
p. cm. — (International series on technical innovation and entrepreneurship)
Includes bibliographical references and index.
ISBN 0–8476–8399–0 (alk. paper).
1. Economic conversion—Russia (Federation) 2. Research, Industrial—Russia (Federation) 3. High technology industries—Russia (Federation) 4. Economic conversion—United States. 5. Research, Industrial—United States. 6. High technology industries—United States. 7. Technology transfer—Russia (Federation) 8. Technology transfer—United States. I. Sedaitis, Judith B. II. Series.
HC340.12.Z9D4346 1997
338.4'762'000947—dc20 6–35862
CIP

ISBN 0–8476–8399-0 (cloth : alk. paper)

Printed in the United States of America

The paper used in this publication meets the minimum requirements of American National Standard for Information Sciences—Permanence of Paper for Printed Library Materials, ANSI Z39.48–1984.

Contents

Foreword

The transformation of research and development from the Cold War to the fragile "Cold Peace" which has followed has not been a smooth process. Nations on both sides of the ideological divide have attempted to exploit past investments in their national, and especially military, R&D sectors. Because research on the process of change has been primarily focused on Western industrialized countries, we are pleased to include a volume in the IC2 Institute's International Series on Technological Innovation and Entrepreneurship which begins to redress this gap in the literature by including comparative studies on technology commercialization in Russia while offering perspectives from China as well as lessons from the U.S. experience.

Financial pressure to commercialize research is daunting in the Russian Federation, where the transition away from state support threatens the very survival of the extensive Russian research sector. At the same time, military technologies represent a pinnacle of Russian technological achievement that in many areas is competitive with and even superior to Western developments.

Investigating the technology transfer issue in transitional economies is important for at least two reasons: it can reveal new sources of potential partners and invention for commercialization, and it can promote global development by strengthening the prospects for a new peaceful world order.

The disintegration of Russia's science and technology base would spell the loss of a potential source of future economic growth and job creation that could threaten the very foundations of peaceful reform and democratization in that country. Given the similarities between the U.S. and Russian defense sectors, the experience of technology transfer by U.S. national labs is especially salient for their Russian counterparts. Both systems were honed in a period of intense rivalry for military and technical superiority, and both developed similar patterns in isolation from the pressures of commercial markets that they are now attempting to transform.

The blueprint for the West during the Cold War period was to provide guns and butter for a free world through a world-class foundation of revolutionary scientific research and development while at the same time vanquishing the forces of communism. Today more than ever, we need to struggle with, debate, and then develop a vision and a road map for the future so that we can resolve the current political and ideological problems which beset us.

From a larger political and ideological point of view, two critical issues exist. First, we need to develop stronger linkages between academia, business, and government. We can no longer take for granted that excellence alone in each of the institutions that make up the three sectors will provide for the general interest and welfare of our citizens. Collaboration must replace confrontation and gridlock. More important, future leaders need to be more involved in coordination of the priorities between and within the three sectors.

Second, new efforts, institutions, and mechanisms must be put into place that mitigate and solve urban, suburban, and regional problems in parallel with each other. This means that we must become more individually active in our communities to help bring together academia, business, and government. Economic development, especially that which produces high-quality jobs, takes place within a community rather than in the legislative halls.

There seems to be emerging a globally competitive world that may engage in a yet-to-be-understood and yet-to-be-defined "economic warfare," even a "scientific warfare." A less dramatic way to state what is happening in terms of economic forces is that the very nature of capitalism is changing. The capitalism of the future will be a system that neither Adam Smith nor Karl Marx nor Lenin presaged.

This newer form of capitalism is based on science and technology, and it results in explosions of economic diversity. Twenty-first century science and technology will generate an economy based on diversity of product and service developments as a continuous stream of new product designs emerge to meet unique global market niches and human needs.

This new system will no longer be dependent on large-scale production and efficiency, but on technology, adaptability, and flexibility. Success in this new system requires a global shift in technology investment and development away from defense to civilian production. We hope this volume spurs future discussion and analyses of these trends by our former ideological rivals. In this process lies the key to future economic competition and technology-based wealth creation for shared prosperity at home and abroad.

George Kozmetsky, Chairman
IC2 Institute Advisory Board

Introduction

Judith B. Sedaitis

In the aftermath of the Cold War, global economic competition has come to play an increasingly important role in defining national security and the shape of the future world order. As international conflict shifts from military to economic competitiveness, many nations are now hoping to extract economic advantage from their investments in defense research and production. This volume brings together papers on several key aspects of defense commercialization and attempts to bridge the divide between research on conversion efforts in the United States and studies of transition in post-Communist economies.

In Section One of this volume, authors focus on U.S. and Russian state policy toward conversion and the implication of past U.S. conversion experience as Russian defense enterprises struggle with their new austerity. In the United States, policies have been driven by the dual pressures of economic competitiveness and structural adjustment. As John Alic points out in the leading chapter, the growing trade deficit and alarming decline in the U.S. share of global markets have increased pressure to find ways to utilize investments in defense R&D to increase the competitiveness of U.S. industry. Improving the interface between federal labs and research sites and private industry is seen as a crucial mechanism for increasing productivity, stimulating economic growth, and creating jobs in the reoriented defense R&D sector. Yet, Alic argues, the history of defense spending and the nature of developmental research hampers an easy transfer of defense technology to commercial use. To expect otherwise assumes a "pipeline" approach where innovation is neatly separate from product development, which belies concrete experience and may explain the rather dismal performance of attempts to spin off new businesses from the national laboratory system. Alic suggests that

U.S. policymakers should redefine current policies in light of the nature of developmental research to support the diffusion of knowledge as well as the generation of new knowledge, particularly in the form of critical infrastructural technologies. The latter are key to future U.S. development, yet historically have proven too costly for the private sector to bear. Linda Cohen and Roger Noll extend further a critical review of past U.S. commercialization policy, which they fault as motivated by political, and not economic, rationale. Instead, they agree with Alic that state dollars should go to support the strengths of our current national R&D program. In their view this entails a focus on basic research historically supported by federal dollars at universities and the national labs. In this way, research is integrated with education in the United States more so than in any other country. As a result, faster and more broadly based technical discoveries take place that ultimately permit a greater range of potential application in the long run.

The chapters by Jacques Gansler and Harley Balzer turn to the policies of conversion in post-Soviet Russia. Gansler succinctly critiques the three main approaches to past defense conversion attempts in the United States to highlight in particular the dangers of diversifying. While new products should be driven by market demand and not the technology at hand, converting firms should be careful not to lose their core competencies by producing below their technological competencies. He suggests that the Russian government stands to play an important role providing assistance for needed market research and capital investment, and by focusing on high-technology infrastructural needs, such as improving telecommunications, railways, and air traffic controls. As Harley Balzer writes, however, the Russian state is playing catch-up with its policies on science and technology. While it continues to place great stock in the value of a plan, its science and technology (S&T) policies are still in the making. Despite several new government committees focused on industrial and science policy, ministerial rivalries have hampered consensus on the future of state research. At the same time, Balzer finds the majority of enterprise directors expect industrial policy to be nothing short of a bailout of their unstructured enterprises.

At the heart of accessing the potential of defense R&D is the process of technology transfer, in which an idea or innovation becomes a concrete, marketable product. In Section Two, "The Problems and Prospects of Technology Transfer," authors focus on different aspects of that usually long and complicated process. Based on his research of MCC, the first R&D consortium in the United States, David Gibson theorizes on the different stages inherent to technology transfer and suggests several elements key for its success. He finds that participants in the process need to actively communicate with one another, which in turn is facilitated by their physical, strategic, and cultural proximity. The nature of the technology also affects the process: innovations of narrower breadth that are easy to grasp are also easier to apply. Ray Radosevich and Gary Smith further articulate the concrete steps and resources necessary for commercializing public sector technology in particular. Based on their experience in New Mexico, their model is generalizable to other areas of relatively low technology infrastructure such as Russia and envisions the building of local areas of high-tech growth based on former defense R&D.

The prospects of technology transfer in Russia are assessed from two very different angles. First, Jacques Sapir presents a rather bleak outlook on the possibility and even interest in conversion by defense enterprise managers. He spells out the disincentives of retooling the former Soviet military-industrial complex especially in light of the absence of domestic demand and suggests that arms sales will increasingly become the option of choice for financially strapped defense managers. Drawing on survey data from 100 former defense R&D enterprises, Judith Sedaitis paints a different picture of R&D managers. She finds that the majority are engaged in adapting their technologies for civilian production and are actively encouraging civilian commercial activity among their technologists. However, she finds that holdovers from Soviet R&D management still loom large in the process, including a preference for centralized organizational forms and a lack of tools for understanding or researching market demand. Instead of turning to new arms production in lieu of domestic markets, those enterprises most actively engaged in technology transfer practices are also those that focus on attracting foreign partners.

Lori Coakley and Linda Randall use case study data to hone in more precisely on the question of marketing strategies in the third section of the volume, "Managerial Strategies and Strategic Alliances." The authors colorfully demonstrate the uneven abilities of Russian high-tech managers to develop strategic plans for converting to new production lines. They also describe several cultural variables considered of particular importance, including the reliance on "technology push" of which Gansler warned earlier. The authors found that production decisions are driven by managerial commitment to an enterprise's technological abilities rather than by market demand. They were also struck by a clear difference between the Russian managers of conversion and their U.S. counterparts in the strong Russian aversion to downsizing their labor pool.

The importance of managerial perception and decision-making runs throughout this section. In his case studies of how defense managers are learning to learn about investment decisions, Yevgeny Kuznetsov highlights managerial perception as key to restructuring. He finds that managers who perceive potential value in their firm are more likely to accept the high risks of a long-term investment strategy. Those who do not create a self-fulfilling prophecy by pursuing short-term rent-seeking behaviors that simply exploit the firms' assets without building for the future. Drawing on survey data, Andrew Aldrin examines the impact of managerial leadership on enterprise restructuring as an indicator of positive adjustment. He finds, for instance—as Gibson theorizes—that the less complex and more asset specific a firm's core competencies, the more likely it is to reorganize and restructure. Like Sedaitis, he also finds cooperation with foreign firms a strong and positive indicator of adaptation to new market demands.

Industrialized nations have increasingly relied upon inter-organizational partnering to spread the risk of commercializing technologies and adapt to the fast pace of technologically driven changes. In his case study of AmTech, a nonprofit facilitator of technology partnerships, Stephen Gomes points up the importance of alliances for commercializing technologies from the public sector in particular. In addition to the difficulties generally of finding new product markets, public-private partnering faces a cultural divide in terms of internal procedures and pace which third party organizations can help bridge. With very little domestic demand currently, however, Russian managers need to turn to international partnerships.

In their chapter on international high-technology alliances, John Hagedoorn and Bert Sadowski indicate the tendency to engage in such alliances is clearly dominated by the triad of Japan, Western Europe, and the United States. Nonetheless, they find that Russia is the dominant country in alliances with Eastern Europe. After a decrease in high-tech alliances during the early reform years of 1990–1992, the number of alliances began to grow and are now strongest in the areas of telecommunications, aircraft production, and chemicals. The warring Russian ministries which Harley Balzer described would do well to mark these leading sectors as strategic to the further development of Russia's technology base. By all accounts in this volume, foreign investment is critical to that end. Because the Russian private sector is a relative latecomer to the arena of high-tech partnering, the potential for high-tech firms in this sector to continue to expand their global presence is high. The extent to which this potential is realized ultimately rests with the attractiveness of the individual Russian firm itself.

Several characteristics combine to define the general attractiveness of foreign firms to Western investors, including the investors' own past experience in global venturing. We focus on three key areas that consistently emerge in our experience with U.S. investors: Russian firm structure, investment pathways, and legal constraints. David Ellerman does a superb job in articulating the value of an increasingly popular organizational form, the "spin-off," used by R&D managers in particular to retain their teams of specialists by offering them entrepreneurial opportunities in addition to their traditional research. The foreign investors win as well, by contracting with a unit that is smaller, more flexible, more cost-effective, and often more efficient than old parent structures have generally been. Corinna-Barbara Francis adds a new dimension to our understanding of this structural fragmentation by her case study of spin-offs in China's "Silicon Valley." Despite what Western theory can classify as only "fuzzy" property rights at best, these new organizational forms prove an important intermediate between the private and state-owned sector that in large part accounts for the great success of China's economic reforms.

In addition to choosing the right organizational form for their needs, Western investors are generally keen on sourcing the widest array of investment funds available and on methods for recouping the investment made. Larry Schwartz addresses some of these concerns by outlining three trends that he feels will help increase venture capital activities in Russia. He also discusses limited exit strategies for Western partners in high-technology ventures, which is of special concern to those who invest in Russia. Emily Silliman and Edward Kayukov round out our discussion of business concerns by their historical overview of new company formation in Russia since 1987. They then focus on the practical aspects of new company formation, the advantages and disadvantages of forming an attached or daughter company, and the growing legislative attempts to stem what some officials see as the "cherry-picking" of high-technology assets by Westerners.

In the end, the major thrust of this volume is to confirm the validity and value of drawing on both the experience of and investment from the West to help access and commercialize the great potential for civilian production in the Russian defense sector and in the high-technology end of the sector in particular. Despite the exasperation of some in the Western business community and fears by some in the Russian state administration, partnering with Western firms has been linked in our chapters to more active technology transfer, restructuring, and new firm creation that in balance will strengthen the transition of defense activities to civilian production. We hope our volume sheds some light on that process and encourages continued mutual learning and cooperation.

SECTION ONE

Technology Policy, Defense Conversion, and Economic Competitiveness

1

Science, Technology, and Economic Competitiveness

John A. Alic

Three-quarters of the U.S. trade deficit in manufactured goods is created by motor vehicles and electronics. Japanese firms ship many of these products to the United States, from Japan or from manufacturing plants elsewhere in Asia. Americans buy them because they like their functional attributes, performance, quality, and price. Behind the design features and value of goods from Honda and Toyota, Sony and Matsushita, stand thousands of engineers and technicians, industrial designers and stylists, product planners and market researchers, factory managers, foremen, and blue-collar workers. Most importantly, the Japanese government has active and articulate industrial policies to sustain high levels of technology and hence bolster national competitiveness. To explore the link between national policy and national competitiveness, this chapter argues that the U.S. manufacturing sector in particular deserves technical support to improve its productivity, since it presently drives the U.S. trade deficit. The link is further developed by distinguishing technology from science, research from development, and by examining the limits to the increased productivity that we can hope to derive from commercializing military technologies. Nonetheless, a new type of proactive science and technology policy is necessary, principally to help promote the diffusion of existing knowledge.

Technology and Productivity

Manufacturing industries have special importance for the U.S. balance of trade. The table below, summarizing 1994 trade by broad categories, shows deficits in some technology-intensive goods and surpluses in others. In fact, only one of the goods categories—textiles, apparel, and shoes—is not technology-intensive in a fairly obvious way (and some textiles are made from high-technology synthetics, while garments are frequently cut with computer-controlled lasers).

An earlier version of this chapter was presented at the conference on Alexander Hamilton's Report on Manufactures, Hagley Museum and Library, Wilmington, Delaware, November 8, 1991.

Manufactured goods	$124 billion deficit
Motor vehicles and parts	$ 63 billion deficit
Textiles, apparel, shoes	$ 42 billion deficit
Electronics, electrical equipment	$ 30 billion deficit
Primary metals and materials[a]	$ 21 billion deficit
Aircraft, aircraft engines	$ 20 billion surplus
Chemicals, plastics, pharmaceuticals, fertilizers	$ 15 billion surplus
Other manufactures	$ 3 billion deficit
Oil, gas, energy	$ 46 billion deficit
Food, agricultural products	$ 18 billion surplus
Services	$ 58 billion surplus
Total goods and services	$ 94 billion deficit[b]

[a] Excluding plastics.
[b] Census basis, which differs from figures reported on a balance of payments basis.
Source: Aggregated from Bureau of the Census and Bureau of Economic Analysis tables in *International Trade Reporter*, 22 February 1995, pp. 380–383.

The dollar value of U.S. oil imports has not changed much in recent years. Nor has the value of the nation's agricultural exports. Although the surplus on trade in services has been increasing steadily—and is probably a good deal larger than the official statistics indicate (because of incomplete coverage and poor data)—it does not come close to counterbalancing the deficits in oil and manufacturing.[1] Nor is this likely to change much in the future, given that many service products are inherently non-tradable (because production takes place at the point of sale). Because the United States will continue to import oil, the nation will, in one way or another, be forced to increase its exports of manufactures, reduce its imports, or both.

There are many routes to adjustment in manufacturing trade, but it is in the interests of all Americans that the nation's competitive position improve through increases in productivity as a result of technology rather than through reductions in wages and living standards. First of all, more competitive firms can afford to pay higher wages (although this is not to say that they will). Second, the U.S. economy must support at some level all those living within the nation's borders, whether employed or not. A more competitive, more productive set of industries means more wealth, in principle, for all U.S. residents (though how that wealth is distributed brings another set of issues to the fore).

Productivity is the most important factor in determining competitive position, and productivity growth is a function almost exclusively of technology—broadly defined to include, for instance, individual and organizational skills and competencies. It is no surprise, then, to find that although the United States retains an aggregate lead in productivity by almost all indicators, in cases where U.S. productivity has fallen behind, the lags are generally in manufacturing industries (including autos and auto parts, consumer electronics, and steel).[2] Although the service sector has long since outstripped manufacturing as a creator of jobs and wealth, and indeed deserves far more attention than it has received from analysts and policymakers, manufacturing remains the principal creator and consumer of technology and therefore is central to any discussion of science and technology (S&T) policy.

Current U.S. Science and Technology Policy

U.S. science and technology policies have three primary elements. First, funding for research in basic science receives more than 20 percent of the federal R&D budget. Second, R&D in support of government missions focuses largely on defense, space, and health. Finally, a rather miscellaneous array of measures are intended to stimulate innovation and entrepreneurial activity. These include strong protection for intellectual property, which goes back to the founding of the republic, and R&D tax credits, which have been in place in one form or another since the early 1980s.

The federal R&D budget, which for the past few years has hovered around $70 billion, accounts for more than 40 percent of all U.S. R&D spending. Funds flow from more than a dozen agencies, with the Department of Defense (DoD) at the top of the list. Defense accounts for nearly 60 percent of the federal total. The money goes to industry, universities, and government laboratories. Over the postwar period, federal R&D grew enormously—the total was less than $1 billion in 1950—but the overall policy structure has hardly changed.

Although R&D spending decisions lie at the heart of "normal" U.S. S&T policy, the debates that began in the 1970s over industrial policy and competitiveness largely ignored federal R&D. President Reagan's Commission on Industrial Competitiveness helped shift attention to technology, but officials in both the Reagan and Bush administrations were vociferous in holding that private industry should be left to commercialize the results of basic scientific research and mission-oriented R&D, encouraged if necessary by tax and other indirect incentives.

R&D got more attention during the second half of the 1980s, when Democrats in Congress began to focus on technology policy, which came to be viewed in some circles as a slimmed-down version of industrial policy. The 1988 Omnibus Trade Act, for example, created the Commerce Department's Advanced Technology Program. In 1992, Congress added $1.4 billion in budget authority for civilian technology to President Bush's fiscal 1993 requests and authorized the multi-agency Technology Reinvestment Project for support of dual-use R&D. When President Clinton took office, Congress and the administration finally found themselves in general agreement, but the consensus lasted only two years. The Republicans who gained control of the 104th Congress, opposed to federal programs of nearly all kinds and bent on crafting a budget in which R&D accounts for nearly 15 percent of discretionary spending, immediately resurrected the debate over the legitimacy of government support for commercial technologies. In recent times, this debate began with President Carter's domestic policy review of Industrial Innovation, a concerted reaction to concern over the slowdown in productivity growth beginning around 1973. In the longer view, it is a reprise of Thomas Jefferson versus Alexander Hamilton, a debate over the legitimacy of government policies of all types designed to foster economic development.

From Science to Technology

It has been common in policy circles, at least since Vannevar Bush's *Science—The Endless Frontier* appeared in 1945, to view science and technology as a seamless web. After all, physics helped win the war (radar, the atomic bomb), Bush's report set the stage for decisions in the 1950s that continue to shape U.S. S&T policy, and it is scientific research that today underlies high-technology industries ranging from microchips and computers to biotechnology and satellite communications. As a result, the perception has grown that research and the policies affecting it form a kind of core activity that more or less automatically determines downstream processes of technology development and commercialization. In fact, this "pipeline" view, in which research leads naturally to development and then to the introduction of new products and processes, gives an incomplete and distorted impression of the technological enterprise.

The Two Meanings of Development

One reason science and technology are so often confused is that we give "development" rather different meanings depending on whether it is coupled with research or with design. As an extension of research, as in R&D, development implies reducing new knowledge to practice—verifying and validating experimental results and theoretical predictions, exploring specific cases, determining the accuracy and limits of mathematical models and the methods and techniques based on them. One example is the development of practical methods for predicting the growth of fatigue cracks in aircraft structures.[3] But when coupled with design, development implies the steady refinement of quite concretely conceived products, processes, and systems through an iterative sequence of conceptualization, preliminary design, analysis, testing, and redesign. This is the everyday technical work of industrial firms, often called product development or product engineering (or process development, software development, etc.). An example of this is the application of methods for predicting fatigue crack growth to determine the spacing of "crack stoppers" in an aircraft fuselage, followed by testing to verify the results.

It might be better to think of design and development as D&D, in contrast to R&D. The former is rightly associated with science (including "engineering science"), but should be viewed as only loosely, and sometimes not at all, linked with product/process design and development. D&D draws eclectically on the results of R&D. It is R&D that leads to high-temperature superconductors that may someday have practical applications. It is D&D that characterizes the processes by which companies arrive at the bundles of attributes that characterize particular products. It is these attributes—including, for a pickup truck or passenger car, cost, carrying capacity, reliability, fuel economy, safety, styling, and resale value—that customers evaluate and compare when they make their purchases. Product attributes depend on technology, much of it embodied in the skills, knowledge, and decisions of a company's employees. Most of this technology is well removed from science and the laboratory. Yet research does play a part, as illustrated by the ability of automakers to simultaneously, over the past two decades, increase fuel economy and reduce exhaust emissions—two fundamentally conflicting objectives. Managing the tradeoffs between fuel economy and emissions has been a matter of steady improvements in electronic engine controls, depending in turn on cheaper, more powerful chips and on a better understanding of the thermodynamics and chemical kinetics of combustion and post-combustion reactions. There are few if any secrets in D&D, but a great deal of know-how, much of it unwritten and some of it proprietary.

A pipeline view that sees new technologies as moving progressively and predictably from R&D to D&D is false. Only in rare cases does new science lead simply and directly to major technological innovations. Far more often, new products and processes are pieced together from a crazy quilt of old and new science and technology, liberally supplemented with tacit know-how, good and bad guesses, intuitions, and heuristics. The technological enterprise remains in substantial part an art and a craft, drawing on science and depending on it, but also, and not infrequently, going beyond hard scientific understanding.

The simple pipeline model has been repeatedly discredited.[4] It is true of neither past nor present, and has led to mistaken policies. Nevertheless, with the ever-accelerating pace of change in science and technology, many people have assumed that the two activities have, in some sense, converged. In fundamental ways, however, they have not. Technology is broader than science (it includes, for example, a great deal of unwritten tacit knowledge), and in many respects more complex and difficult (scientists can often simplify where technologists must find some means for dealing with nature's stubborn realities). Scientists and technologists define and approach problems in contrasting ways. They belong to separate though overlapping professional communities, and fill rather different kinds of jobs in the economy. Scientists are trained to do research; their job is to analyze—to break things down and understand them. Engineers are trained to undertake product and process design and development; their job is to synthesize—to put things together. Even though some scientists find themselves working as engineers, and some engineers engage in research, most people bear the imprint of their educational background throughout their working lives. At the same time, of course, technology depends on science, and more so than in the past, just as science depends on technology (with instruments such as microscopes, telescopes, and now missions into space the obvious examples). But the two activities have grown in parallel as much as they have grown together.

New technology sometimes builds on new science. Always it builds on old science, and on existing technology. The invention of the transistor followed the theoretical understanding of electrical conduction in crystal lattices developed during the 1930s, science preceding technology by perhaps two decades. No one could know that the original research would be put to this use. Today, the design of very large-scale integrated circuits (ICs) proceeds at a rapid pace despite theoretical foundations that are surprisingly weak.[5] There is great interest in using x-rays to fabricate very dense ICs; such applications could hardly be anticipated early in the century when x-rays were new tools for studying atomic structure. Even today, some innovations continue to emerge quite independently of scientific knowledge.[6] For example, the just-in-time or kanban production systems pioneered by Toyota, which have had impacts reaching far beyond the auto industry, evolved over more than two decades of shop-floor experience and trial-and-error development.

If the distinctions between technology and science are not as sharp today as they once were, they still exist. Scientists explore nature to understand it. Technologists design and develop tools, products, and systems with purposeful ends in view. And it is technology, more than science, that drives economic growth and determines competitive outcomes internationally, especially in the short and medium term.

Design, Development, and Commercialization

There are roughly 370,000 manufacturing firms in the United States. Fewer than 20 percent conduct R&D, while all except those that function purely as make-to-order subcontractors necessarily engage in D&D. During the 1970s, many U.S. manufacturers, large and small, fell behind their overseas competitors in process know-how. Their costs were too high, their quality not high enough. During the 1980s, even as attention swung to manufacturing, growing numbers fell behind in product development. Not only did American manufacturers have trouble providing the features customers wanted at a cost they were willing to pay, they found it sometimes took them twice as long as their rivals abroad to design, develop, and introduce new products.

When it comes to the manufacture of internationally competitive products, whether consumer goods like VCRs or capital goods like 777s, the single most critical set of design decisions are those made during the early stages, when overall parameters are set, performance criteria and cost targets are established, and product attributes are defined in at least a preliminary way.[7] This set of decisions will, in the normal course of events, have little to do with science (except as physical reality constrains what is possible and practical), and everything to do with perceptions of market demand, business risk, and the anticipated difficulty of downstream D&D.

Both ultimate customer appeal and total life-cycle costs depend in large measure on decisions made before the time-consuming, iterative process of design and analysis, redesign and refinement, and testing and modification even begins. Early and largely irreversible choices—whether a new car, for instance, will have front-wheel drive, or a transverse or longitudinal engine with four cylinders or six—come first. Often these follow marketing decisions (or, in the case of military systems, requirements set by defense planners). Subsequent decisions involve fundamental trade-offs between conflicting goals—acceleration versus fuel economy versus exhaust emissions, weight versus crash safety. Already, design choices have become heavily constrained. Although a great deal of fluidity may remain at the level of detail design, early choices largely determine the course of downstream development.

The story of the videocassette recorder illustrates in concrete fashion the significance for competitiveness of design and development.[8] Often portrayed as a case of pioneering U.S. innovation followed by imitative Japanese commercialization, it was nothing of the sort. RCA began its R&D on video recording in 1951. In 1956, Ampex, another U.S. firm, introduced a recorder intended for use by television stations; the machine was the size of a closet and cost $50,000. By this time, Japanese companies had started their own projects. Indeed, helical scanning, one of the key steps in the process of developing a compact recorder, had already been invented—in Japan. During the 1960s and 1970s, some fifteen companies—American, Japanese, and European—demonstrated at least nine different technical approaches to home video. Matsushita entered pilot production first, in 1973, but soon decided its design was not good enough and pulled its products from the shelves. Two years later, Sony's Betamax opened the consumer market. Including Sony's pioneering industrial models, the Betamax represented the seventh generation of the company's engineering development. Sony, of course, eventually lost out to a resurgent Matsushita, whose VHS standard found greater acceptance in the marketplace.

The basic technology of recording electromagnetic signals on flexible tape covered with ferric oxide depends on scientific understanding. But commercialization meant solving a long chain of tough engineering problems, so that reliable VCRs could be produced cheaply with features consumers wanted. This is where Japanese companies succeeded, while American and European firms failed. Once the VCR became a commercial reality, competition centered on cost reduction, image quality, and longer recording times. The competitive race—those many generations of product development and testing—turned on engineering skills and management commitment, not on scientific expertise. Similar stories could be told in many other product categories and industrial sectors.

The problem for Russia is more difficult. To continue its transformation to a market economy, Russia must produce a wider range of consumer and capital goods in growing volume. The country's future needs center on D&D more than R&D, on upgrading the skills and competencies of individuals and organizations so that salable goods can be produced for both local and export markets. The skills and know-how required depend only a little on research and science, and a great deal on product design and expertise in high-volume manufacturing. Neither the "big science" traditions inherited from the former Soviet Union, nor the organizational habits passed along from military design bureaus and the Soviet space program, would seem to have much to offer.[9]

Among other things, Russia needs inflows of know-how from the West on best practices in D&D and management of technology, and programs for retraining technicians, engineers, and scientists in commercially oriented product/process design and development methods.

From Military R&D to Civilian Technology

How does federal R&D affect design, development, productivity, and competitiveness? For fifty years, with agricultural and biomedical research the primary exceptions, war and the prospect of war dominated U.S. policy decisions in technology and science. The end of the Cold War led to a reassessment of U.S. national security policy. The need for reassessment of U.S. S&T policies may not be so obvious. But recall that in the United States federal agencies pay more than two-fifths of a national R&D bill totaling about $170 billion. Nearly three-fifths of the federal contribution goes for defense—about $42 billion—a percentage that has been considerably higher at various times in the past. In effect, U.S. S&T policy has been, and still is, subsidiary to national security policy.

In Japan, the picture is vastly different. Companies pay for more than 80 percent of all R&D, spending substantially more as a percentage of revenues than their American counterparts. Corporate funds in both countries go almost entirely for D&D—i.e., for new products and the manufacturing processes needed to produce them. Japan's government budgets only about $1 billion annually for military R&D. Some observers in the United States have been tempted to conclude from such contrasts that military R&D distorted the U.S. technological enterprise, attracting the best and the brightest among American engineers and scientists to work on stealthy airplanes and strategic weapons rather than consumer products and capital goods, while Japan and other countries (e.g., Germany) grew wealthy under the shelter provided by American military power.

Others draw quite different conclusions. Some argue that postwar U.S. prosperity stemmed largely from the technological fruits of defense-related R&D—a

view widely held in Europe during the 1960s, where it was popularized by J.J. Servan-Schreiber's *The American Challenge* (and can be heard even now on the continent). Indeed, supporters of President Reagan's Strategic Defense Initiative sometimes suggested that the program could be justified solely on the basis of "spin-offs" to the civilian economy.[10]

Both these extreme views of the consequences of defense-related R&D spending are wrong, however. Consider first an aspect of this question which critics of U.S. military spending have, over the years, made a central theme—diversion of technical talent from the civilian side of the economy. At the height of the Reagan defense buildup, between 15 and 20 percent of American engineers and scientists worked in defense.[11] These 15–20 percent may well have included a disproportionate share of the best and the brightest, although there is no way to know if this is true. More important, no one has ever demonstrated that the U.S. labor market for technical professionals fails to function reasonably well. If the labor market works—so that, for example, supply moves more or less in parallel with demand, with a time lag associated with vocational decisions by college students—it would be wrong to see engineers and scientists who work in defense as a drain on the rest of the economy. Rather, defense would simply add to total demand for, and hence eventually the supply of, technical professionals.[12]

The more difficult question concerns the impact of defense-related spending on technology itself.

Commercializing Military Technology

In the 1950s and 1960s, a number of U.S. industries got substantial boosts from defense spending. Defense agencies, bent on maintaining technological superiority in weapons systems and anxious to guard against technological surprise (e.g., a Soviet breakthrough in quiet submarines), sponsored R&D in a wide variety of fields. DoD contractors, many with good-sized commercial divisions, pursued R&D in their own interests and, with government funding, in the national interest. University research increased rapidly and steadily, fueled by federal funding.

For two or more decades, military technologies led civilian. American companies were able to take advantage of waves of innovation associated with aerospace technologies, computers, and microelectronics. Some of the commercial leads established over those years still persist, albeit in attenuated form.

With the wave of innovation spurred by World War II long past, the impacts of military spending on technology have become largely indirect. Most defense R&D is directed not toward research, but the design and development of weapons systems—work performed primarily by aerospace and electronics firms. Design features and performance requirements for these systems—command, control, and communications (C3I) networks; surveillance satellites; airplanes, tanks, ships, missiles—differ greatly from those for civilian products. This is true even when superficial resemblances exist, as they do for some kinds of software-intensive C3I systems.

On the other hand, differences at the level of the end product or system may hide similarities at the level of materials, components, subsystems, and manufacturing processes—and, most important, technical tools and methods. The field service experience necessary for widespread applications of fiber-reinforced composite materials, for example, began with military applications. Precision bearings and ICs found in ballistic missile guidance systems resemble commercial components, although they are rarely identical (and the ICs are likely to be several generations older). The same kinds of machine tools and fabrication processes are used to make military equipment and commercial products. Finally, many of the design techniques, analytical procedures, and experimental methods used by engineers and scientists are the same whether turned to the task of reducing the aerodynamic drag of a military jet or a commercial airliner—or, indeed, an automobile. It is therefore wrong to look, for instance, at a "stealthy" airplane like the B-2 bomber, designed to avoid radar detection, and conclude that, since civilian aircraft should be as visible as possible to radar, such a project has little or nothing to offer on the civilian side of the economy. What, if anything, it has to offer can only be determined through careful examination and evaluation of the technical knowledge gained during the course of the particular project.

To reiterate, much of this technical knowledge takes the form of methods for engineering design and analysis. Over many years, defense agencies have supported, both directly (through R&D and D&D) and indirectly (through production contracts) the improvement and extension of these techniques—structural integrity technology, software engineering, computational fluid dynamics, mathematical models for aircraft guidance and control, etc. Companies like Boeing and Airbus Industrie, to pursue the aircraft example, also contribute new knowledge as a consequence of their own spending. The end result is a set of engineering methods, in use worldwide, that individual companies supplement with proprietary know-how and standard practices (e.g., computer codes for certain kinds of design/analysis calculations). Many of the tools used in the design and development of pickup trucks and VCRs thus have their origins, directly or indirectly, in the military R&D of past years and past decades.

For half a century, U.S. defense spending provided much of the funding for the development and validation of techniques and methods used particularly in aerospace and electronics. Knowledge gained as a result of government spending targeted to the particular needs of mission agencies like DoD, the Energy Department (which is responsible for nuclear weapons), and NASA often turns out to be broadly useful. Today, automobile structures are designed to crush in controlled fashion during a collision using computer methods developed originally in military R&D on aircraft structures. Engineers use similar methods to design bridges, buildings, and mechanical systems of all kinds.

However, by the 1980s DoD's narrow focus on aerospace and electronics seemed an increasingly inadequate basis for support of a vast and variegated U.S. economy. And, defense spending is often not a very efficient way to generate technical knowledge. In the past, the slow, indirect migration of knowledge from the military to the civilian side of the economy seemed of little importance, so long as American firms remained well ahead of rivals elsewhere. Today, however, the migration is two-way, and U.S.-based firms face a wide array of highly competent rivals. Even in countries like Mexico, the best companies, often the subsidiaries of multinationals, exhibit world-class productivity and quality. It has become much more difficult for U.S.-based firms to create and maintain technologically based advantages than in past decades. Yet technology, in the broad sense, is the major and perhaps only sustainable source of long-term competitive advantage for the United States. The problem is how to capitalize on this advantage.

Today military technologies, with some exceptions, are as likely to lag as to lead commercial technologies. Microelectronics is the prototypical example. During its formative years, the U.S. microelectronics industry grew by selling to defense and aerospace firms that supplied end systems under contract to government; federal procurement was more important than federal R&D.[13] In the middle 1950s, the Air Force, seeking denser, more reliable solid-state electronic circuitry to be packed into airplanes and missiles, initiated an R&D program on three-dimensional "molecular electronics." Most of the contract funds for this ill-defined concept went to Westinghouse, then a leading manufacturer of discrete transistors as well as a major defense contractor. But neither Westinghouse nor other defense-oriented firms were able to meet the Air Force's requirements. Instead, the first integrated circuits emerged, without benefit of federal funding, from two small companies, Texas Instruments (TI) and Fairchild Semiconductor. The government's willingness to buy the new chips spurred the rapid growth of the merchant semiconductor industry. Demand created by the Minuteman II missile and the Apollo project pushed U.S. firms rapidly down design and production learning curves. TI got the lion's share of Minuteman contracts, Fairchild most of the orders for Apollo. Both companies became industry leaders, with Fairchild the early centerpiece of Silicon Valley.

As costs came down, computer manufacturers designed the new ICs into their products. Indirect demand, through purchases of general-purpose computers by defense contractors and government agencies, especially federal laboratories, helped fuel expansion. By the end of the 1960s, industrial and consumer applications of microelectronics were growing much more rapidly than DoD purchases. Defense accounted for 72 percent of IC sales in 1965, and for only 21 percent in 1970; the defense share has remained below 15 percent since the mid-1970s. During the period of transition in the late 1960s, the semiconductor firms that grew the fastest were those able to negotiate the shift from specialized military requirements to the needs of commercial firms. Nontechnical dimensions like price and delivery schedules were far more important to commercial customers than to defense contractors; as computer price/performance ratios plummeted, seemingly month by month, the pace in defense electronics seemed relaxed by contrast.

With no meaningful foreign challenge until the 1970s, competing U.S. semiconductor firms sought to anticipate and understand the needs of customers and be the first to satisfy them. Rapid increases in circuit density—the number of transistors on a chip—followed from improvements in design and fabrication technologies. Rising production volumes and learning-curve effects drove prices down. New chip designs and lower prices, in turn, opened additional markets, including consumer electronics applications which Japanese companies already dominated. By the early 1970s the U.S. lead over Japan especially in processing technology began to shrink, and during the 1980s the U.S. share of the world semiconductor market tumbled. Although the story is complicated, in large part it is a familiar one: as ICs, particularly memory chips, became more standardized, it became harder and harder for U.S.-based firms to maintain a competitive advantage (in part, because chip production was at the time relatively labor-intensive).

As the semiconductor case suggests, long before the Cold War began to wind down, the Pentagon's contribution to the U.S. and world technology base was declining—not because defense stagnated, but because commercial markets and commercial technologies were expanding more vigorously. In 1960, U.S. defense R&D accounted for one-third of the Western world's R&D; now it is less than one-seventh.[14] Even if defense R&D does not fall in the years ahead—and many have argued that the United States must maintain its technologically based advantages in military systems—U.S. S&T policies, established during the 1950s and relatively little changed since, cannot provide adequate support for an economy that is the largest and most diverse the world has ever seen.

Reshaping U.S. Science and Technology Policy

This paper has suggested that U.S. technology and science policies continue to reflect the position of the United States during the earlier years of the Cold War, when American technology and American companies were preeminent. Today, the United States needs to reconsider these policies.[15] On the defense side, the Pentagon must find ways to foster dual use at component and subsystem levels, and to design and develop more effective systems and produce them affordably. To accomplish this, DoD must avoid isolation from the flourishing world of commercial high technology.

On the commercial side, the United States will have to find ways to fill in the gaps that have been appearing in the national portfolio of technology investments as defense R&D shrinks in relative terms. Defense R&D and procurement earlier compensated, however partially and imperfectly, for the underinvestment by private firms in long-term R&D.

In brief, reshaping U.S. S&T policies means adding at least three distinct objectives to the nation's strategy:

1. First, there must be support for pathbreaking R&D. In the absence of a superpower rival, the DoD may mistakenly feel it has less reason to sponsor long-term, high-risk projects. The state should fund projects with longer time horizons and higher risks than industry is likely to fund on its own because the paybacks are too uncertain and/or too far in the future. Examples include fusion energy and high-temperature superconductivity.

2. Second, DoD spending should target more immediately useful technologies of two types:

Infrastructural or generic technologies. Industry tends to underinvest in these technologies because no one firm can hope to capture the rewards. By definition, such technologies have widespread applications, making it difficult for an investor to appropriate the full economic benefits. Examples include computer-aided software engineering methods, more efficient manufacturing processes, control of corrosion and wear, and improved highway pavements.

Strategic R&D. Government also needs to support areas where business risks and financial constraints combine to slow development of technologies having substantial importance for industrial competitiveness, national security, or both. SEMATECH, a program focusing on IC processing in which costs have been shared between industry and government, was an example of combined government-industry efforts to regain a strategically important market.

3. Finally, defense R&D spending could promote the diffusion of both new and existing technologies. The agenda here could range from education and training to industrial extension programs directed toward smaller companies that cannot expect to be technologically self-sufficient.

The last of these objectives, although not necessarily the most controversial, poses the most complicated set of problems and issues. Broadly speaking, this objective contemplates a shift in U.S. technology policy from a near-exclusive focus on the creation of new knowledge (through R&D) toward diffusion of both new and existing knowledge.

Promoting Knowledge Diffusion

The current ratio of federal spending on knowledge creation to knowledge diffusion probably exceeds 99:1. Even when diffusion is a normal outcome, as when federal dollars support graduate students in engineering and science, the money often comes from research contracts and grants. This near-exclusive focus on knowledge generation is one of the strongest legacies of the Cold War, when it seemed essential to stay ahead of the Soviet Union in each and every field of science and engineering, even those with the most tenuous links to national security. Given the many and various ways in which technical knowledge serves as an input to D&D and to production, it seems clear that a wider range of diffusion-oriented policies could help the United States take advantage of new and existing knowledge more quickly and more effectively.

At the same time, U.S. technology policy should not de-emphasize generation of new knowledge. That would be foolish. Rather, it should emphasize diffusion-related aspects of knowledge creation such as training, information dissemination, cooperative R&D (which includes technology transfers), and technical assistance.

Although the government funds many education and training programs, few of these support training for those already in the labor force.[16] Yet most private firms have few incentives to invest in training; companies that invest in their workers risk losing them to other employers, including free-riding competitors (another type of market failure). The needs of employed workers are threefold: basic skills (reading, writing, simple arithmetic); training in job-related technical skills; and better social and interpersonal skills (which take on greater significance in reorganized workplaces, including so-called high-performance work systems). At professional levels, the ongoing explosion of technical knowledge makes it difficult for engineers and scientists to keep up, especially for workers who are well along in their careers.

By the same logic that underlies support for research, government could also subsidize information dissemination since the social returns exceed private returns. At present, federal agencies undertake interpretation, validation, and dissemination mostly as necessary for particular missions—e.g., when research becomes the basis for drug approvals or regulation of environmental hazards. Dissemination could also include screening, interpretation, and validation of new knowledge, support for databases and computer/telecommunications networks, overseas technical missions, and translations from foreign languages. The same explosion of knowledge that makes it important to retrain engineers and scientists also makes it important to screen and evaluate technical information so that it can be utilized more quickly and effectively. There is a large unmet need for reviews of key technical subjects, development of engineering databases that codify information in forms useful to practitioners of D&D (rather than research), and evaluations of research results and the norms and practices of engineering in other countries.

The growing number of consortia and increasing cooperation among firms is also an important avenue for diffusion, even if the primary objective is R&D. Technology demonstrations also diffuse know-how, although they have the disadvantage of sometimes tempting federal agencies too far toward commercialization. Past U.S. experience includes many examples of demonstration programs that have suffered from an excess of technology push, in the absence of market pull that might help shape technical objectives and discipline decisions by program managers (and by Congress).[17] For more than a decade, the United States has also sought to foster exploitation of technologies resident in the federal laboratory system, or otherwise stemming from government-funded R&D, through revised technology transfer policies. As part of this effort, decision-makers continue to explore ways of more effectively integrating government laboratories, particularly those funded by the Energy Department, into the nation's S&T system. Technical assistance and industrial extension programs, finally, grew relatively quickly at both federal and state levels during the 1980s. Given that many of the hundreds of thousands of small and medium-sized U.S. manufacturing firms do not employ even a single trained engineer or scientist, the existing industrial extension programs have hardly begun to address the need.

Remarkably little is known about the actual processes by which know-how moves through the economy (and across national borders). Thus a broad range of possibilities, beyond the straightforward diffusion-oriented policies summarized above, awaits analysis and debate.

In theory, the problem for the United States is a simple one. Although aggregate levels of productivity remain the highest in the world, the United States is no longer a world leader in technology. Today, rough parity is about the best most American companies can expect. And, even this position of parity may be precarious. Absent strong continuing productivity growth to create wealth for wage earners, rising income inequality will aggravate the social strains already plaguing the nation. But productivity and competitiveness depend on technology. Indeed, the dependence is so strong that it is almost tautological to say that the post-1973 slowdown in productivity growth must in some sense reflect a decline in returns to investments in technology. The question is then how to increase those returns. This requires a fresh look at the nation's S&T system, recognizing that effective use of existing knowledge is just as important as the creation of new knowledge.

Notes

1. *Trade in Services: Exports and Foreign Revenues* (Washington, D.C.: Office of Technology Assessment, September 1986) suggested that service exports might be undercounted by as much as 50 percent. While the statistics have improved over the past decade, cross-border flows of services are inherently difficult to measure, and reported figures for both exports and imports no doubt continue to be substantially understated.

2. *Manufacturing Productivity* (Washington, D.C.: McKinsey Global Institute, October 1993).

3. This illustration is worked through in some detail in John A. Alic, Lewis M. Branscomb, Harvey Brooks, Ashton B. Carter, and Gerald L. Epstein, *Beyond Spinoff: Military and Commercial Technologies in a Changing World* (Boston: Harvard Business School Press, 1992), 38–40.

4. See, for example, John Jewkes, David Sawers, and Richard Stillerman, *The Sources of Invention*, second edition (London: Macmillan, 1969); Sumner Myers and Donald G. Marquis, *Successful Industrial Innovations: A Study of Factors Underlying Innovation in Selected Firms*, NSF 69–17 (Washington, D.C.: National Science Foundation, 1969); J. Langrish, M. Gibbons, W. G. Evans, and F. R. Jevons, *Wealth from Knowledge* (London: Macmillan, 1972).

5. "... in hindsight, it is clear that our present level of electronic sophistication is the result of luck, trial-and-error, and the beneficence of nature, aided in small part by a few critical theoretical insights." E. Yablonovitch, "The Chemistry of Solid-State Electronics," *Science* 246 (October 20, 1989): 347–351.

6. For a marvelously detailed illustration, see Walter G. Vincenti, "Technological Knowledge Without Science: The Innovation of Flush Riveting in American Airplanes, ca. 1930–ca. 1950," *Technology and Culture* 25 (1984): 540–576.

7. John A. Alic, "Computer-Assisted Everything? Tools and Techniques for Design and Production," *Technological Forecasting and Social Change* 44 (1993): 359–374.

8. Richard S. Rosenbloom and Michael A. Cusumano, "Technological Pioneering and Competitive Advantage: The Birth of the VCR Industry," *California Management Review* 29 (1987): 51–76.

9. On the science establishment in the former Soviet Union, see, for instance, Loren Graham, "Big Science in the Last Years of the Big Soviet Union," 2nd series, Vol. 7, 1992, pp. 49–71. Information on Soviet technical design practices is scattered and incomplete, but see, for example, Leon Trilling, "Styles of Military Technology Development: Soviet and U.S. Jet Fighters, 1945–1960," *Science, Technology, and the Military*, Vol. 1 (Dordrecht, the Netherlands: Kluwer, 1988), pp. 155–168.

10. Stewart Nozette and Robert Lawrence Kuhn, eds., *Commercializing SDI Technologies* (New York: Praeger, 1987).

11. David K. Henry and Richard P. Oliver, "The Defense Buildup, 1977–85: Effects on Production and Employment," *Monthly Labor Review* (August 1987). Table 7, p. 10, gives figures close to 15 percent.

12. When the National Research Council looked for signs that the Carter-Reagan defense buildup had diverted engineers from civilian industry, it found none. See *The Impact of Defense Spending on Nondefense Engineering Labor Markets* (Washington, D.C.: National Academy Press, 1986).

13. This discussion draws on *Beyond Spinoff*, pp. 257ff.

14. National Science Foundation, *National Patterns of R&D Resources: 1990*, Table B-13, p. 55; and R&D Spending of OECD Countries: Division of Scientific, Technological, and Industrial Indicators, Directorate for Science, Technology, and Industry, Organization for Economic Cooperation and Development.

15. The discussion that follows draws on *Beyond Spinoff*, and on John A. Alic, "Technical Knowledge and Technology Diffusion: New Issues for U.S. Government Policy," *Technology Analysis and Strategic Management*, Vol. 5, 1993, pp. 369–383.

16. *Worker Training: Competing in the New International Economy* (Washington, D.C.: Office of Technology Assessment, September 1990).

17. Linda R. Cohen and Roger G. Noll, *The Technology Pork Barrel* (Washington, D.C.: Brookings Institution, 1991).

2

Research and Development after the Cold War

Linda R. Cohen and Roger G. Noll

The most casual observer of government budgetary activity in either the Russian Federation or the United States knows that the dramatic easing of East-West tensions since the late 1980s has led to a substantial reduction in defense-related expenditures, including expenditures by government on weapons-related research and development (R&D). The data from the United States are instructive. Real (inflation-adjusted) federal expenditures on R&D peaked in fiscal 1988, and have fallen in all but one year since then. R&D expenditures as a fraction of gross domestic product—that is, the share of the domestic economy devoted to R&D—peaked in 1985, and likewise have declined more or less steadily ever since. Data from Russia are more fragmentary and less reliable, but anecdotal evidence about scientists and engineers in Russia working for starvation wages and trying to sell both their expertise and weapons components in their labs (including fissionable material) strongly suggests that R&D is declining in both countries.

The purpose of this chapter is to discuss in somewhat more detail the connection between international tensions, defense expenditures, and government R&D. Our investigation concerns both the historical pattern of R&D effort during the past century, and the detailed composition of government-supported R&D effort during the past two decades. Because our expertise lies in U.S. policy, we focus mostly on the American experience, but pay some attention to how the motives and patterns of R&D in the United States differ from those in other countries. Our principal conclusions are rather pessimistic: we see no natural break to the recent erosion of political support for R&D in the United States, and anticipate that within a few years the decline in U.S. R&D effort that is now under way not only will weaken American research institutions and the U.S. economy, but will reduce the rate of economic growth throughout the world.

Lest we be misunderstood, this somber assessment does not reflect nostalgia for the Cold War or any similar prospective future substitute. Instead, the message is simpler. The Cold War created a political consensus in support of generous government support for R&D, so that, until recently, the United States accounted for half or more of world R&D. As a result, the United States has been, for half a

The authors gratefully acknowledge financial support from the Markle Foundation.

century, the primary world source of technological innovations. But the end of the Cold War has caused consensus support for R&D to unravel, and no substitute bipartisan rationale for this generous support has emerged.

The History of Government R&D in the United States

Although both federal and state governments have supported R&D since the early nineteenth century, until quite recently this support was sporadic and targeted toward a handful of projects and technologies.[1] Whereas a few of the great technical innovations of the nineteenth century were subsidized by the federal government (for example, the telegraph, the railroad, and hybrid seeds), most emanated from work financed by industry (for example, electric generation, Bessemer converters, industrial machinery, and the telephone). To some degree, U.S. technology policy lacks a coherent focus, and still follows a pattern of pursuing the technology *du jour* for a few years, then moving on to the next fad. Examples are nuclear power in the 1950s, the supersonic transport and communications satellites in the 1960s, breeder reactors and synthetic fuels in the 1970s, and fusion reactors and "clean coal" electric generation facilities in the 1980s. But these projects, though highly visible, have long since ceased being the centerpiece of federal R&D effort.

American policy changed due to the experiences of World War II, when scientists and engineers developed new, higher-performance weapons and many other technologies that served military purposes but had much more widespread application. Examples of the latter are jet aircraft, microwave radio transmission, and electronic computers.

Unlike the consequences of the demise of the Cold War, the end of World War II did not bring a withdrawal of the government from supporting research, doubtlessly due in part to the tensions that developed between Eastern and Western Europe immediately after the war ended. Figure 2.1 shows U.S. federal and industrial R&D effort as a fraction of gross domestic product (GDP) since 1950, which is the beginning of reasonably reliable data. These categories do not exhaust all R&D; missing are individual lines showing trends in R&D that is paid for by state government or nonprofit institutions (including universities from their own funds), both of which have grown as a fraction of all R&D.[2]

Around 1960, U.S. R&D effort approached 2.5 percent of GDP, and ever since it has hovered between 2.2 and 2.8 percent. The growth in R&D effort that began during World War II came to an end with the escalation of the fighting in Vietnam in the mid-1960s. The federal government attempted to minimize the unpopularity of the war by financing it out of other defense-related expenditures, including R&D, rather than new taxes or cuts in domestic programs that were popular with the constituency of the Democratic Party, which controlled both the presidency and Congress during the mid-1960s.

Figure 2.1: R&D as a Share of GDP

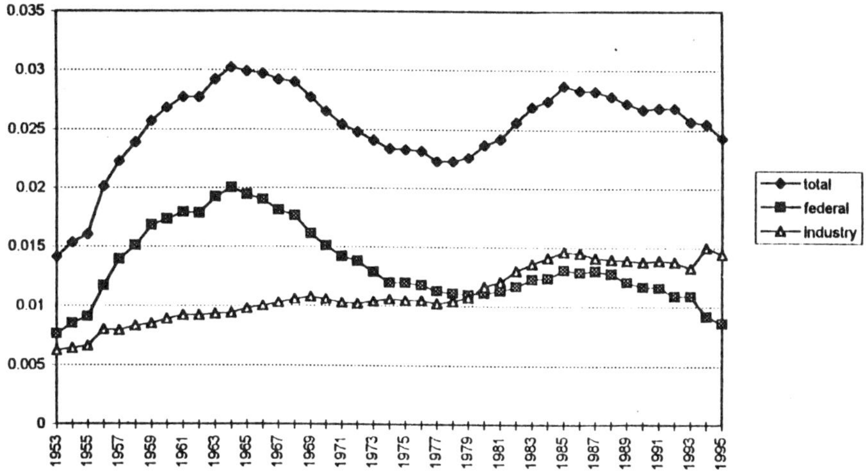

Sources: *National Patterns of R&D Resources*: 1992, NSF-92-330, Table B-2; and 1995, Table B-2. *Historical Tables, Budget of the United States Government*, FY1996, Table 9.7. *The American Almanac* 1994–1995 (The Reference Press: 1994), Table 965.

As the Vietnam War wound down in the mid-1970s, R&D effort was not restored. Note that in Figure 2.1, federal R&D spending as a fraction of GDP fell until fiscal 1979, which coincides with the beginning of the defense buildups initiated by the Carter and Reagan administrations. At this trough, federal R&D effort (measured by the fraction of GDP devoted to R&D) had fallen back to levels not experienced since the mid-1950s.

The data in Figure 2.1 show a steady decline in R&D as a fraction of GDP since the mid-1980s. A recent comprehensive survey by the Department of Commerce concludes that for several years its annual survey has underestimated private R&D expenditures, primarily because it underestimated R&D outside of manufacturing. The effect of the new estimation method is to cause the R&D/GDP ratio to be about .3 percentage points higher than was previously estimated for 1991–1993.[3] If the new survey method is more accurate, adjustments also would have to be made for data in earlier years, so that the general pattern of decline in national R&D effort since 1986 would remain. For example, the new method estimates that the R&D/GDP ratio has fallen about .1 percentage point per year since 1992, but from 2.8 percent (instead of 2.6 percent). After taking account of the new estimation methods, the most recent National Science Foundation report still concludes: "Not since the early seventies has there been a period of such protracted low growth in national R&D support."

In comparison with other countries, the United States has always been among the leaders in R&D effort. Table 2.1 shows the fraction of gross domestic product that is accounted for by R&D in the leading advanced industrialized countries. As measures of R&D that is relevant to economic growth, these data are not strictly comparable across nations. Although there are many differences, the two most important are the treatment of universities and the amount spent on defense R&D that is commercially irrelevant.[4]

Most other countries include subsidies from the national government to higher education for administration and teaching as part of R&D.[5] Subtracting these expenditures from total R&D, in 1992, the last year for which data are available, the estimated fraction of GDP devoted to R&D was 2.3 percent in France, 2.5 percent in West Germany (2.2 percent in all Germany), 2.4 percent in Japan, 1.2 percent in Italy, and 2.0 percent in the United Kingdom, compared to 2.6 percent (old method) or 2.8 percent (revised method) in the United States. But the United States spends a much larger share of its budget on aspects of defense that have little or no commercial application. In recent years, the United States has spent about .7 percent of GDP on defense. France spent about .5 percent, the United Kingdom .4 percent, and all other countries .1 percent or less.[6] If a quarter of defense R&D is commercially irrelevant, the United States probably lost its lead over West Germany and Japan in about 1993.

Until the mid-1980s, the data issues we have discussed would not have been sufficiently important to create doubt that the United States was the world leader in commercially relevant R&D. Prior to 1980, the United States was the clear leader, with only the United Kingdom a close competitor. R&D effort in the United States and United Kingdom reached 2 percent of GDP in the 1950s. This level was not attained by the Federal Republic of Germany until 1968, Japan until 1978, and France until 1981. Within the range of the data uncertainties discussed in the previous paragraphs, during most of the 1980s it appears that commercially relevant R&D effort was broadly similar in Germany, Japan, and the United States, slightly lower in France and the United Kingdom, and much lower in Canada and Italy.

Since 1990 the United States has taken a path that, if present trends continue for the rest of the decade, will cause U.S. R&D effort to drop to the level attained by Canada and Italy, at the bottom among the advanced industrialized economies. By 1993, U.S. total R&D effort had fallen to 2.6 percent of GDP (revised method). The data for 1994 and 1995 concerning federal R&D indicate that R&D effort fell by another .1 percent of GDP in each of those years, and the budget resolutions passed by Congress in the summer of 1995 presage another drop of .2 percent for 1996. These drops will place the United States below France and roughly equal to the United Kingdom.

The details of the federal R&D budget reveal that three categories account for nearly all of the growth in the 1980s: defense, medical research, and basic research in physical sciences and engineering conducted at universities and na-

Table 2.1: National R&D Expenditures as Percent of GDP for Large Advanced Industrialized Democracies*

Year	U.S.	Japan	West Germany	France	United Kingdom	Italy	Canada
1961	2.7	1.4	1.2	1.4	2.5	NA	NA
1966	2.8	1.5	1.8	2.1	2.3	NA	NA
1970	2.6	1.8	2.1	1.9	2.3	0.8	1.2
1975	2.2	2.1	2.2	1.8	2.0	0.8	1.1
1980	2.3	2.2	2.4	1.8	2.0	0.8	1.1
1985	2.8	2.7	2.7	2.2	2.3	1.1	1.5
1987	2.8	2.8	2.9	2.3	2.2	1.2	1.4
1989	2.7	3.0	2.9	2.3	2.2	1.2	1.4
1990	2.7	3.1	2.7	2.4	2.2	1.3	1.4
1991	2.6	3.0	2.8	2.4	2.1	1.4	1.4
1992	2.6	2.8	2.7	2.4	2.1	1.4	NA

* Except for France, 1960s data are the ratio of R&D to GNP. These ratios are usually the same, occasionally differ by 0.1, and never differ by more than 0.2 for these countries. West Germany entry under 1961 is for 1962. United Kingdom entry under 1970 is for 1969 and under 1980 is for 1981. Unified Germany R&D/GDP ratio was 2.6 in 1991 and 2.5 in 1992. U.S. data for 1991 and 1992 are based on the 1987 survey method of the Department of Commerce for industrial R&D to retain historical comparability. The 1992 method causes these estimates to increase to 2.8 percent. Because the new method differs from the old primarily because it surveys nonmanufacturing industries more completely, it would be likely to cause an upward revision in all prior years as well.

Sources: National Science Board, Science and Engineering Indicators 1988 (p. 287) and 1993 (p. 375), Washington: U.S. Government Printing Office; National Science Foundation, National Patterns of R&D Resources: 1994, NSF Document 95-304, Arlington, Virginia: National Science Foundation, 1995; and American Almanac Statistical Abstract 1994–95, 114th Edition, Reference Press, 1994, p. 607.

tional laboratories. Total nominal dollars spent by the federal government on R&D related to energy, resources, transportation, agriculture, social services, and international affairs actually fell between 1979 and 1989.[7]

The defense category of research carries a misleading name, because in reality these expenditures support a great deal of R&D that has military applications, but that has wider applicability in many high-technology industries, notably the industries related to microelectronics, computers, and communications. As defense weapons became more technically sophisticated, the industrial base that supported the defense industry broadened, so that a defense-oriented R&D effort, motivated solely by Cold War concerns, became a de facto industrial policy through the back door.[8] Specifically, defense has provided the rationale for an enormous U.S. R&D investment in aerospace and information technologies.

Since about 1980, the emphasis in federal R&D that is relevant for civilian purposes increasingly focused on three areas: basic research at national laboratories and universities, medical technologies, and federally subsidized industrial R&D joint ventures. At the peak, around 1990, the federal government was spending more than a third of its R&D budget in national labs and universities. Research in biological sciences, physical sciences, and engineering that was conducted at government labs and universities boomed throughout the decade of the 1980s, with the fastest growth in medical research.

In the mid-1980s, the federal government committed to a dramatic increase in its support for commercial research collaborations. Some of these collaborations took the form of subsidized industry R&D joint ventures, the prototype being SEMATECH—a consortium that initially included 14 manufacturers of semiconductors. Others were collaborations between companies and universities (the National Science Foundation's Science and Engineering Research Centers) or national labs (the Cooperative Research and Development Agreement, or CRADA, program).[9] In keeping with its traditional concern for small business, Congress created and generously supported the Small Business Innovation Research Program (SBIR) to assure that small business would eventually receive 2.5 percent of federal research expenditures. In a few years, these programs expanded from virtually nothing to more than 5 percent of federal R&D effort, with official plans being to increase them to as much as 20 percent.

The turnaround in defense R&D effort was almost simultaneous with the rise of Mikhail Gorbachev and the demise of the former Soviet Union; however, it did not spread immediately throughout the federal R&D budget. Gradually, almost every program area saw its real expenditure growth stall, and then turn negative. By fiscal 1995, almost every category of expenditures was heading south.

The Economics and Politics of Federal R&D

The most plausible conceptual model for understanding federal R&D policy involves three related but distinct factors that affect the attractiveness of R&D to government officials. The first is technical possibilities: technical opportunities that present themselves, in the form of either new insights about how to expand fundamental scientific knowledge or results from past research or production experience that suggest new avenues for developing applied technology. The second is economic incentives: market opportunities for new products and production processes, which depend on the costs of available unexploited technical opportunities and the potential demand for the results of developing them. The third is political feasibility: the willingness of citizens to pay tax dollars to take advantage of new technical and economic opportunities, and to give their political support to candidates for office who advocate active government involvement in R&D.

There is simply no good reason to believe that changes in either technical possibilities or economic incentives had anything substantial to do with the boom-and-bust pattern of federal R&D in the post-war era. Technical opportunities are not a plausible explanation because the decline in U.S. R&D effort in the 1970s and again in the 1990s is not paralleled by similar declines elsewhere among advanced industrialized countries. France and the United Kingdom, the other nations that spend a considerable amount of R&D on defense, did experience a similar pattern in the 1970s, but the decline in R&D effort was much smaller than in the United States, and neither of these countries has yet to exhibit a significant decline after the end of the Cold War. Presumably science and technology are not differentially revealing themselves according to the nationality of the researcher, so that the technical opportunities available to the United States in the 1990s are not wildly different from those available and pursued elsewhere.

The economy did not perform well in the 1970s, and it recovered in the 1980s, but the pattern of R&D expenditures is not explained by these general economic trends. The decline in R&D effort of the 1960s began long before the economy weakened, and the recovery in R&D effort began before the economy recovered. Moreover, the R&D boom in the 1950s occurred during a period when the economy was relatively weak, and the massive cuts in R&D since 1992 have taken place during a period of rapid economic growth and, by 1994, nearly full employment. As with technical possibilities, the economic explanation is belied by the continuation of past levels of R&D effort in other advanced industrialized nations, most of which have experienced softer economies during the late 1980s and 1990s than has the United States.

The most plausible candidate for explaining the turnaround in U.S. R&D effort is political. The root of the problem apparently is that the United States never was politically committed to a strong national R&D enterprise outside the context of national security concerns. A useful way to conceptualize the political basis of support for R&D in the post-war era is that a general political distaste for proactive government economic policy was overcome by, first, the demands of World War II, and then the perceived necessity to contain the Soviet Union militarily. Soon after World War II, the nation reached a political consensus that the Cold War must be fought, and that R&D had to play a central role in the fight.

As time progressed, and the political popularity of fighting the Cold War declined, elected political officials made accommodations in the budgetary trade-off between domestic and national security policies. For R&D, this accommodation meant that a consensus coalition of support for defense-related R&D required including those who sought federal support for other kinds of research. Supporters of military R&D, recognizing that a broad consensus in favor of Cold War policies was politically necessary to sustain the credibility of the threat of American military intervention, were willing to ensure the consensus for their policies by using support for other kinds of R&D as a means to solidify support for their programs. In this way, conservatives found reason to support R&D that did not serve defense purposes, and liberals found a way to expand the scope of the government's role in developing new technology and, thereby, shaping the pattern of national economic development.

The end of the Cold War weakened the consensus coalition concerning R&D policy. The perception that less defense effort was now needed to attain U.S. national security objectives did not lead to a reallocation of resources to other problems that remain as demanding as before (the classic basis for the belief in a "peace dividend" as the United States cuts its defense budget). Instead, it led to the budgetary outcomes of a coalitional agreement (the classic "share the pain" policy of reduced expenditures for all). Proponents of universities, national labs, commercial R&D subsidies, and other domestic R&D programs simply can now demand less in return for supporting a smaller defense R&D budget than was the case during the Cold War.

This interpretation of the pattern of federal R&D effort during the post-war era sheds new light on the innovations in R&D policy since the mid-1980s. The basic ideas behind these policies were, first, that commercial R&D was going to become an important part of the U.S. strategy for successful international economic competition (the "competitiveness" rationale), and, second, that because the new orientation was commercial rather than military, the emphasis should be placed on programs in which industry has more influence in selecting projects and coordinates business efforts to compete effectively with foreign firms (the "cooperative" rationale). As the Cold War wound down, government officials sought a replacement by likening international economic relations to international military threats as a means for mobilizing political support from across the ideological spectrum for continuing roughly the same level of R&D effort.

The past few years show that this strategy is simply not going to work politically. The first reason derives from the fact that this strategy is based on economic nonsense: the idea that international economic competition is conceptually equivalent to anything like the Cold War conflict. The Cold War was a negative sum game: if both sides spent hundreds of billions a year, they could retain the status quo of political and military stalemate. Economic competition is a positive sum game: international economic specialization and trade expands production possibilities and global wealth. Nations that engage in trade are not at war, but rather are conducting the joint enterprise of exchanging things that are valued more highly by the users of the product than their producers. To say that firms are rivals in a world market means that they compete for the opportunity to provide greater benefits to consumers, which is the mechanism by which competitive markets cause production efficiency. The social value of market competition in promoting efficiency is undermined if attractive rivals are excluded from the market by government actions (such as direct trade barriers or production subsidies), causing less efficient rivals to be advantaged.

The nonsensical core of the competitiveness rationale became apparent during the late 1980s and early 1990s, when the government simultaneously was justifying federal support for commercial R&D on the basis of the competitiveness rationale while simultaneously advocating major new initiatives in reducing trade barriers: the Canadian-American Free Trade Agreement, the North American Free Trade Agreement, and the new General Agreement on Trade and Tariffs. It simply made no sense economically to put forth a neo-mercantilist argument for commercial R&D while simultaneously espousing an anti-neomercantilist argument for free trade and, in particular, against Japan's international economic policies. Politically, citizens who believed the arguments for free trade and thought that the rationale for federal R&D policy was neo-mercantilist would support the former and oppose the latter, which in fact appears to be the pattern of 1990s political opinion in the United States.

Second, the cooperativist element of the strategy was also based on a fundamental misperception, although a far more subtle one than the error of the competitiveness rationale. The difficulty with the cooperativist idea arises from the implicit assumption that the government's objectives—more effective competition against foreign firms—constitute the most important objectives for the firms with which the government is cooperating. Although competition with foreign firms was not the most significant problem facing the American economy in the 1980s, even if it had been, the cooperativist idea would have faced major problems because the objectives of the private partners are not likely to parallel the objectives of government.

The purpose of R&D in for-profit firms is, by definition, profit. For the vast majority of American companies, their most important competitors are other American companies, not foreigners. Hence, the primary objective of each firm's R&D effort is going to be to eliminate effective domestic competition, not foreign competition. This can be achieved in two ways: by attaining a technological advantage over one's competitors, or by forming a domestic cartel. Unless international trade faces zero trade barriers and is costless (i.e., involves no greater transportation and transactions costs than domestic trade), a firm can obtain a permanent increase in its profits by using R&D to attain either of these effects. Of course, both domestic objectives create substantial political problems for the government.

If a company uses federal dollars to gain a competitive advantage over domestic competitors, the losers will, with some justification, object: why should the government be deciding who wins an R&D competition by favoring some competitors with subsidies or the gift of government technology? And, because the intellectual property rights in this form of federally supported R&D accrue to the private partner, why should the government make some companies fabulously wealthy? Precisely these issues arose very early in all of the new cooperative commercialization initiatives.[10]

Alternatively, the government can avoid this problem by encouraging not just public-private collaboration but also collaboration among competitors, as exemplified by the consortia that have been set up by the government in semiconductors, low-pollution automobiles, textile manufacturing, advanced aircraft, and several other areas. Here the risk is cartelization: that the effect of the collaboration will be a mutual agreement not to compete in R&D, thereby reducing the diversity of projects and causing federal funds to be offset by reductions in private R&D expenditures, and to use cross-licensing to enforce cartel prices. Again, exactly this claim has been made against the consortia that have been established. Excluded competitors, supplier industries, and industries that use the products all have complained that they are being harmed, not helped, by the new arrangements, and preliminary research indicates that in the case of at least two programs, SEMATECH and SBIR, federal funds seem simply to have displaced private spending, rather than increased overall R&D effort.[11]

The combination of the competitiveness and collaboration rationales created still a third irrationality. Federally subsidized cooperative R&D sometimes undercut the American firms that were the most successful internationally. The collaborative programs usually are limited to companies that satisfy narrow, nontrivial regulations defining domestic firms. Many large, internationally successful companies, including AT&T and IBM, have been rendered ineligible to participate in federally financed R&D collaborations because of their foreign subsidiaries and joint ventures. Likewise, some firms that undertake all R&D and production in the United States but that are subsidiaries of foreign firms have also been denied participation in these programs. Of course, a narrower definition of the firms that are eligible inevitably reduces the efficiency of these programs in generating net improvements in technology and economic welfare.

The political problems created by reorienting federal R&D policy toward commercial projects and programs are only loosely tied to whether these projects are actually worthwhile. In fact, if an R&D project is extremely successful, it is more likely to damage competitors, and more likely to enable the members of the consortium that developed it to build an effective cartel. The political problem stems primarily from the redistributive effect of these programs. Whereas a highly successful R&D project might increase productivity and profits in an industry by a few percent, it can also reallocate all or most of the sales and profits from previous industry leaders to successful innovators. Consequently, the redistributive effect can be very large compared with the benefits of even a highly successful program. Almost inevitably, the redistributive effect creates well-organized, politically influential losers.

Based on our conceptual account of the politics of R&D policy, and our observations of the problems that have arisen during the post-war era whenever the government has pursued commercial R&D projects, we conclude that this strategy does not constitute a politically feasible substitute for the Cold War as the galvanizing force behind a broad, generous federal R&D policy. The problem is partly the castle of sand—the competitiveness rationale—which forms the intellectual basis for these programs. But this problem is superficial. The existence of this rationale underscores the deeper problem of American politics, which is that one cannot develop an enduring political basis of support for a long-term investment in R&D. This problem is exacerbated by the high sensitivity of the American political system to government actions that cause someone economic harm, even if society as a whole derives substantial benefit from such actions.

Why Should We Care?

If we are right, the American research establishment is in for a very rough time. We anticipate continued erosion of expenditures on R&D as government struggles with the massive problem of balancing the budget without raising taxes or cutting the most rapidly growing part of the budget, expenditures on the elderly through social security and Medicare. Indeed, until something fundamental changes in domestic politics and policy, the United States is on a course to have nothing left of the federal government sometime in the next twenty years other than defense, programs for the elderly, and interest on the debt. Obviously, a broad-based federal R&D effort cannot survive this trend, nor can hardly anything else.

These trends raise an important issue: what precisely is the nature of the problem posed by the demise of federally sponsored R&D other than for items directly related to weapons development? The obvious effect is clear: the United States faces a vast shrinkage of its national R&D effort, including the closing of many if not most national laboratories and a vast reduction in research universities, which typically derive one quarter to half of their income from the federal government. While this will harm the people who work in those facilities and the communities in which they are located, the question remains: exactly to what extent is this a more severe problem than, say, the parallel process of closing large numbers of military bases?

The answer, of course, is that R&D makes a substantial contribution to improvements in economic welfare, and that R&D in the United States is especially important in this respect.[12] Here is what we know about the economic effects of R&D.[13]

First, most of the growth in per capita income in advanced industrialized nations arises from technological progress. Education and capital investment are important, not only in themselves but as factors that facilitate the development and adoption of new technology. But most growth can not be accounted for by increases in either factor.

Second, the total returns—counting the economic benefits to producers and consumers—to investments in technological innovation appear to be very high, and much higher than the returns to investments in physical capital. Moreover, a very large fraction of these benefits—the consensus estimate is around half—accrue not to the parties undertaking the innovation, but to others.[14] This finding constitutes the economic case for government support for R&D. Because private parties who undertake R&D derive, on average, only about half the economic benefits, the private sector is likely to underinvest in R&D, and because the total social returns to R&D are so high, federal R&D programs, in principle at least, have high potential benefit/cost ratios.

Third, the American system of research outperforms the systems of other countries, especially with respect to new product innovations and advancements in fundamental scientific and engineering knowledge. These categories refer to the products of basic research plus radical technical breakthroughs that revolutionize entire industries.[15] By any measure, the United States thoroughly dominates basic research, accounting for a much greater share of publications and prizes than its share of research effort. One plausible reason is the peculiar role of the American research university, which also dominates the world competition for the best students in science and engineering. In most nations, government support for scholarly research takes place almost exclusively in national laboratories. Outside of the United States, the research university is rare (although becoming less so). Other leading advanced, industrialized countries commonly have only a handful of universities in which research plays any significant role.

In the United States, not only is a substantial share of basic research undertaken by universities, but in addition, many national laboratories are also managed by universities, with considerable mixing of research employees of the labs and students and faculty from universities. The effect of the American system is to integrate more completely education and research. This arrangement allows students not only to take classes from accomplished researchers, but to participate in the research activities of their teachers. As a result, advances in research are more quickly and easily transmitted from the lab to the classroom, and, as students are employed, to practical applications in industry.

The intermeshing of research and education provides still another economic argument for government R&D. The public payoff to R&D is accomplished more rapidly, and so the returns to R&D are more valuable, because the American system of basic research is integrated with education. Even the private payoff to R&D hinges on this system, because private returns to R&D are greater if private projects have access to more recent technical information that is generated in public R&D projects. Moreover, the relative openness of the American system of basic research, including not just open publication of results but immediate dissemination in classes and seminars, permits a broader range of potential applications of new knowledge to be considered as students are employed through numerous firms and industries.

Finally, the integration of public research and education facilitates productive feedback from industry to researchers. The educational process not only provides industry with employees who are trained in the latest technical know-how, but creates permanent relationships between public researchers and their former students in industrial R&D. The two-way nature of research is extremely important: not only are advances in products and processes derived from more fundamental research in universities and national labs, but new problems and unexplained phenomena encountered in production feed back to generate interesting new puzzles for the basic researcher.[16]

The upshot of this discussion is that the consensus among economists who study technological change is that R&D strongly influences the growth of per capita income, and that in the United States the payoff to R&D has been high—partly because of the American system of higher education, in which scientific and engineering education is integrated with research. This system is threatened by the downturn in federal R&D effort.

The implications are grim for the United States. In a few years a substantial reduction in R&D effort is likely to cut severely the rate of economic growth. As a rough rule of thumb, a reduction of one percentage point in the fraction of GDP going to R&D should reduce the long-run rate of economic growth by about one-half of one percent.[17] In addition, the rest of the world will suffer as well. The contraction of the open system of basic research, with a technical education system that includes many foreigners, will undermine technological progress everywhere. The spillover benefits of R&D are not confined to the United States, but extend to the rest of the world. Indeed, one source of the declining political popularity of government-supported R&D is the fact that foreign firms and consumers derive benefit from it. Of course, why the United States would want to harm itself so that it could prevent others from benefiting from U.S. research is something of a puzzle!

Is There Hope?

Economics is not called the dismal science for no reason. Our assessment of the future for federal support for R&D in the United States is depressing. What, if anything, might reverse this decline?

At the core of the problem as we see it is the expression of public preferences through the American political system. Hence, to turn the system around requires a change in the policy preferences of public decision-makers, which can arise in two ways: from changes in the preferences of the public (voters, contributors, the press), or from changes in the way citizen preferences are given voice in the political process through the institutions of government.

To accomplish the first requires making more citizens perceive that they benefit from R&D. One way to do so is simply to spread the word: to disseminate more thoroughly and forcefully the information about the importance of R&D to the performance of the American economy that is summarized in this chapter. But we suspect that a more important reason for the fall from fashion of government programs in general, and R&D in particular, is that, for a generation, the middle of the American income distribution has not done very well economically.

A plausible hypothesis is that stagnant after-tax real wages for working-class Americans have caused them to be skeptical of the benefits of government programs to promote economic growth. R&D is embedded in a system that the American middle class perceives to be part of the problem. In the eyes of working-class Americans, R&D, in the first instance, goes primarily to pay the salaries of relatively well-off people, and secondarily, to benefit corporations, not workers and consumers. Thus, a necessary condition for a turnaround in political support for R&D—and for other forms of public investment, including education—probably is a broader dissemination of the benefits of economic growth than has been true since the 1970s.

The second way out is political reform: a means of aggregating citizen preferences in a different way, so that distributional concerns and interest-group politics play less of a role in making public policy. Again, this set of issues hardly pertains to R&D policy alone. It rears its head in procurement policy, agricultural price supports, and federal investment programs. If many government programs are perceived as benefiting only a small fraction of the population, most voters may simply withdraw support from public sector activity generally because they conclude that the part of government that benefits them (or, otherwise, that they like) can not be obtained without accepting with it an intolerably large set of programs that do not benefit them (or that they do not like).

Examples of broad institutional changes that would serve this function are campaign finance reform, reduction in the duration of campaigns, and even broad-based legislative representation (such as proportional representation). Whereas the national government has as yet demonstrated no great interest in adopting major changes such as these, narrower examples have been adopted. One is the base-closing commission, and the other is the "fast track" process for approving detailed agreements to reduce trade barriers. In both cases, Congress has committed in advance to a process that prevents it from amending the recommendations it receives, enabling the proposers to offer Congress an all-or-nothing bargain. This process, in turn, substantially reduces the ability of organized interests to protect themselves through legislative lobbying and contributing to campaigns. Thus, the prospects for institutional reforms that reorient government toward activities such as R&D, while far from certain, are not hopeless.

Before a reversal of policy can take place, however, further severe cutbacks in federal R&D are in store. A federal labs closing procedure, resembling the military base closing process, is imminent. Because many of these labs are almost exclusively oriented toward defense, shrinking this part of the national R&D effort need not be harmful to technological progress in nondefense areas. We suspect, however, that federal lab closings will be somewhat less extensive than is optimal, for lab closings will be especially disruptive events politically. Squeezing more from other institutions that will continue to exist, with fewer employees and lower salaries for those who remain, is more attractive politically than targeting a death blow to specific facilities. Some facilities will be closed, but probably not as many as economic efficiency would dictate.

The flip side of the previous argument is that bigger hits will be taken in other programs. For ineffective programs, this is, of course, a desirable result, but there is no reason to believe that cuts will be targeted at less effective programs. The same politics that supported bad programs in the heyday will make cutting them difficult in the contraction, so we doubt that the remaining cuts will do much to improve the overall efficiency of federal R&D programs. As an example, in 1995 probably the single least effective federal program—the Small Business Innovation Research Program—did the best in the budget process among all the new commercial R&D programs that were initiated in the 1980s.

Thus, we expect that extramural expenditures for research, including support for universities, will take a very large hit over the next few years. Especially vulnerable are highly leveraged universities that derive a large share of their research support from the federal government. Whereas the universities with large endowments and a large stream of gifts from corporations and wealthy individuals will experience some pain, the universities that derive nearly all of their income from tuition (including state support as well as private payments) and federal grants will be seriously hurt. We suspect that many will simply gradually withdraw from the ranks of research universities, leading to an increase in the number of institutions that place virtually all emphasis on teaching and a decrease in the total number of institutions of higher education. The most important effect of this transition will be to reduce the quality of education in more technical disciplines, and thereby reduce the supply of the best-trained technically educated workers.

Americans are unlikely to feel the impact of the cutback in federally supported R&D quickly, other than through the direct impacts that resemble the effects of base closings. Instead, the effects slowly will cumulate: slower economic growth, and declining quality of higher education. For these effects to become visible will take perhaps a decade. By then, the institutional damage to the U.S. R&D establishment will be fairly severe. Even if policy is reversed then, because of the cumulative effect on training in science and engineering it will take longer to reestablish these institutions than it took to destroy them.

The Cold War is over, and we do not want it, or anything like it, to return. But the demise of the former Soviet Union has had a largely unanticipated consequence. The R&D institutions that were put in place during and after World War II are being radically changed. Federal R&D will not disappear altogether, for the United States is unlikely to dismantle completely its military establishment, and a modern military needs advanced technology and enough well-educated technicians to develop it. Because R&D support is coalitionally based, most other areas of federal R&D will not die out, either. But future support is likely to be much lower. Stripped of the Cold War as motivator, technological progress has a fragile and weak constituency.

Notes

1. For a summary of the history of federal support for R&D, see Linda R. Cohen and Roger G. Noll, *The Technology Pork Barrel* (Washington, D.C.: Brookings Institution, 1991), Chapters 1 and 2. For a very interesting selection of case studies on the role of government in developing technologies for specific industries, see Richard R. Nelson, ed., *Government and Technical Progress: A Cross-Industry Analysis* (New York: Pergamon, 1982).

2. These data are not without serious measurement problems. Recently, the Department of Commerce has revised its estimates of industrial R&D, based on a new survey method that was first used in 1992, as discussed subsequently in the main text. To retain historical comparability, we show the earlier estimates here. In addition, data on R&D financed by nonprofit institutions other than universities are very poor. The last comprehensive survey of these institutions was conducted in 1973, and the last update of the earlier data was undertaken in 1983. National Science Foundation, *National Patterns of R&D Resources: 1994*, NSF Document 95-304, Arlington, Virginia: National Science Foundation, 1995.

3. National Science Foundation, *National Patterns of R&D Resources: 1994.*

4. Technically, improvements in the productivity of the defense sector is part of economic growth; however, because both hot and cold wars are negative sum games, arguably one should exclude defense expenditures from measures of national economic welfare. In any case, our purpose is to examine national efforts to advance technology outside of the defense sector.

5. In 1988, the fraction of national government expenditures on R&D that were accounted for by general support for higher education was 11.7 percent in France, 30.8 percent in Germany, 43.7 percent in Japan, 15.8 percent in the United Kingdom, 31.4 percent in Italy, and zero in the United States. As a fraction of total R&D effort, these numbers are about 6 percent in France, 11 percent in Germany, 13 percent in Japan, 6 percent in the United Kingdom, and 17 percent in Italy. National Science Foundation, *International Science and Technology Update: 1991.* Washington, D.C.: U.S. Government Printing Office, 1991, p. 11.

6. National Science Board, *Science and Engineering Indicators: 1993.* Washington, D.C.: U.S. Government Printing Office, 1993, pp. 375-6.

7. The best source for detailed data on the distribution of federal R&D expenditures is the biannual publication of the National Science Board, *Science and Engineering Indicators.*

8. For a history of the evolution of defense expenditures into an industrial policy, see Ann Markuson and Joel Yudken, *Dismantling the Cold War Economy* (New York: Basic Books, 1992).

9. For a detailed discussion of the CRADA program in the national laboratories, see Linda R. Cohen and Roger G. Noll, "Feasibility of Effective Public-Private R&D Collaboration: The Case of Cooperative R&D Agreements," *International Journal of the Economics of Business* 2 (1995): 223–240.

10. For a discussion of some examples of these controversies arising from the CRADA program, see Cohen and Noll, "Feasibility of Effective Public-Private R&D Collaboration." For a discussion of how the same problem led to the termination of the communications satellite R&D program in the 1970s, see Cohen and Noll, *The Technology Pork Barrel*, ch. 7.

11. For some examples, see Linda R. Cohen and Roger G. Noll, "Privatizing Public Research," *Scientific American* (September 1994): 72–77. For an account of the effect of SEMATECH on R&D effort, see Douglas Irwin and Peter Klenow, "High Tech R&D Subsidies: Estimating the Effects of SEMATECH," *NBER Working Paper Series*, Working Paper No. 4974, National Bureau of Economic Research, December 1994. For an assessment of SBIR, see Scott J. Wallsten, "The Small Business Innovation Research Program: Crowding Out or Stimulating Private Investment?" Department of Economics, Stanford University, 1995.

12. By economic welfare, we mean not just the standard measure of per capita GDP, but the overall ability of people, alone or collectively, to solve economic problems, including environmental pollution, resource scarcity, and disease.

13. For a more thorough summary of economics research on technological change and R&D, see Cohen and Noll, *The Technology Pork Barrel*, ch. 1 and 2.

14. For an excellent summary and analysis of economics research on the distribution of economic benefits from R&D, see Zvi Griliches, "The Search for R&D Spillovers," *Scandinavian Journal of Economics* 94 (1992): 29–47.

15. See, for example, Edwin Mansfield, "Industrial R&D in Japan and the United States," *American Economic Review* 78 (1988): 223–228, and "Basic Research and Productivity Increase in Manufacturing," *American Economic Review* 70 (1980): 863–73.

16. The concept of the innovative system, including the most fundamental and abstract research and the most pragmatic learning-by-doing on the assembly line, is discussed in a very illuminating fashion in Nathan Rosenberg, *Inside the Black Box: Technology and Economics* (Cambridge: Cambridge University Press, 1982), and in David C. Mowery and Nathan Rosenberg, *Technology and the Pursuit of Economic Growth* (Cambridge: Cambridge University Press, 1989).

17. Of course, to the extent that the least productive projects are cut first, and to the extent that government is less efficient than the private sector, the effect on economic growth will be less.

3

Defense Industry Conversion in the United States and Russia: Methods and Prospects

Jacques S. Gansler

The outlook for defense industries around the world is bleak. The end of the Cold War has meant drastically reduced defense budgets, the dramatic downsizing of military forces, and the flooding of the arms market with excess military equipment. With the Russian economy and political system in disarray, the picture is even bleaker for Russia's defense industry.

It is clear, however, that Russia is determined to maintain a strong defense posture. Some defense production will therefore continue to be required, albeit with significantly fewer and smaller industrial firms producing state-of-the-art military equipment. Several of the former defense enterprises may be able to downsize and remain viable. The rest—the overwhelming majority of enterprises—must significantly diversify, and/or fully convert, to non-defense production and research, or face extinction.

In the restructuring of any large industrial sector such as the defense industry, it is important to recognize that there is no single, "right" strategy that will work for every firm. There are two general approaches to restructuring, however. The first is the segregated defense-industry model, in which low-level production is maintained in a wide variety of defense unique plants—essentially those that have been downsized from the larger defense industry of the past, either in a single country or via multinational, defense firm consortia. The alternative is to totally restructure the defense industry into one that is effectively integrated with all aspects of commercial operations, including production lines, labor forces, engineering, and parts and materials. In this second model, the plants are viewed as "dual use": capable of military production when required, and, in order to maintain state-of-the-art defense capacity, continuing to engage in selected military R&D, including weapon prototypes. The key difference, however, is that the plants are largely operating in the commercial sector.

In determining which of the two broad approaches to take in restructuring a nation's defense industry, several key issues must be addressed. First, the military equipment of the future is likely to be significantly different from today's.[1] Thus, a strategy of simply downsizing current defense industry facilities would likely result in the maintenance of obsolete technologies and equipment. Conversely, for many defense sectors, the desirability of civil/military industrial integration is highlighted by the likely dominance of electronic equipment and new materials in future military systems, and the fact that commercial technology in these areas will represent the leading technical edge.

Second, any nation preparing for defense industrial base restructuring must determine the extent to which it intends to rely upon foreign military sales in order to maintain a domestic defense industry. The short-term economic and industrial rationale for such sales is quite clear. However, a focus on foreign military sales diverts critical management attention, and limited available resources, from the necessary longer-term effort toward diversification. In addition, today's foreign military sales market is not only shrinking, but is increasingly competitive.[2] Future defense strategies should seek to maintain a legitimately required, state-of-the-art industrial base that is not heavily dependent upon foreign military sales for its future survival. In light of shrinking domestic defense budgets, this objective also points to the desirability of a shift to civil/military integration.

Russia must now choose between an isolated defense industrial structure dependent on waning foreign military sales, and an integrated civil/military structure. The path Russia chooses will affect such overriding decisions as the commitment to and pace of defense plant privatization; which international firms are best as partners—defense or commercial; what new products to pursue; and what markets to focus on. Russia's course will determine the willingness of foreign investors to put resources into Russian defense firms. In seeking avenues out of its current defense industry crisis, Russia would do well to observe how the defense industry in the United States is coping with its similar, although nowhere near as dramatic, task of downsizing and restructuring.

U.S. Defense Conversion Efforts

The defense industry of the United States is an isolated, highly subsidized, high-cost sector of the U.S. economy. It is increasingly falling behind leading-edge commercial developments in areas such as advanced materials, electronics, software, and manufacturing technologies. New-product-realization cycles of weapon systems have increased to a prohibitively long average of 16.5 years from concept to first production, and are incompatible with commercial cycle times. At the same time, administrative barriers created by the Department of Defense's unique way of doing business make it extremely difficult for a defense-oriented plant to shift into the commercial economy.[3]

Faced with restructuring, U.S. defense industries have pursued one of three options: (1) getting out of defense; (2) weathering the storm; or (3) diversifying. In each of these categories, of course, there are wide variations. For example, General Dynamics, an essentially defense-only firm seeking to get out of defense, has taken the approach of simply selling off as many of its operations as it can. In liquidating the company, stockholders have benefited enormously, as has senior management; but tens of thousands of workers have been laid off. An alternate "get out of defense" model might be provided by General Electric, a highly diversified company that could simply sell off some of its defense operations to focus on its mainstream commercial business.

Choosing the second option are firms that plan to shrink and become the most competitive of the remaining defense firms. Many of these plants are shifting to new defense products for the next generation of weapon systems, and pursuing foreign military sales—which could not be sought without government support. Some firms, such as Loral and Lockheed Martin, are buying up the defense operations of companies that are getting out of defense, with the hope that they can achieve production efficiencies through consolidation. Since there will be a few sectors in which defense-unique capabilities will still be required, this strategy is viable but limited to a greatly reduced scale.

Soon, most U.S. defense firms will be forced out of business. In the meantime, the government has been buying billions of dollars worth of unneeded, older-production weapon systems in order to help failing defense firms and obtain the votes of their employees. In the long run, the Defense Department will be forced to decide whether to continue the inefficient subsidizing of a large number of firms or to buy advanced military equipment from only a few, efficient firms. There will not be enough money for both.

Finally, the third approach to defense industry diversification and/or conversion has proven to be extremely difficult. The failure of conversion efforts following World War II, the Korean War, and the Vietnam War indicates the difficulty of achieving even diversification, much less conversion, of defense facilities. In fact, the empirical evidence indicates that diversification, even to relatively similar products within commercial plants, has had only about a 50 percent success rate. And, due to the dramatic differences between defense and commercial businesses, defense diversification/conversion to commercial business has had only about a 20 percent success rate. Nonetheless, diversification/conversion is the only viable alternative for the overwhelming majority of defense plants. Moreover, there is considerable evidence that the chances of success can be significantly increased by learning from the legacy of prior conversion attempts. Encouragingly, approximately 35 percent of firms that have recently attempted conversion have been successful, and those in closely related fields have even exceeded a 60 percent success rate.[4]

In order to further expand the probability of success, the government must take an active role in defense industry restructuring. Unfortunately, during the critical post–Cold War period, the U.S. government chose a laissez-faire approach. The Bush administration essentially argued that free-market forces would rapidly achieve efficiency and effectiveness in the transformation of the U.S. defense industry. Because this industry is dominated by a single buyer, is severely regulated, and has very few suppliers in each sector, however, the natural forces of free-market operation were not present. Under such conditions, efficient downsizing was impossible. Both Congress and the administration provided and continue to provide funds for the extension of existing production lines, and actually encouraged increased foreign military sales. The refusal to establish a clear vision, objective, and strategy for the future structure of the U.S. defense industry meant that all of the government's budget actions were taken in an essentially random fashion.

Even with government assistance, any effective restructuring must allow market factors to be the driving force. Also, by assisting in the process of capacity reduction, government policy must resist the temptation to preserve inefficient and noncompetitive firms. Inevitably, a large number of jobs will be lost as a result of the required downsizing.

Today, the Clinton administration is attempting to face the realities associated with the need to radically restructure the U.S. defense industry. First and foremost, the administration recognizes that it must establish a clear vision, and has stated that it has chosen the path of civil/military integration.[5] Plants will primarily produce commercial items, along with a few advanced military R&D programs and a very limited amount of advanced defense equipment production. The defense industry will then simply be part of the overall U.S. industrial base, and the government simply another customer that orders different, though similar, goods and services. There will still be several defense-unique plants; but far fewer than before, and focused on unique production processes. For the majority, different products can be produced in the same plants, as long as the production processes are similar, especially as plants move toward flexible manufacturing. Integrated plants will produce a few defense electronics items that satisfy the military's unique needs, while manufacturing on the same production line large numbers of commercial electronics items, with both sets of equipment utilizing commercial components, commercial software, commercial materials, and commercial manufacturing equipment. Cannons will be produced on the same production line as railroad freight-car axles, for example, because they are both produced on the same rotary forge.

In order for this desired end state of integrated civil and military plant operations to be realized, dramatic changes must take place both in how the U.S. government conducts its business and in how America's defense industry operates. The government must eliminate all of its unique procurement practices, unique cost accounting practices, and unique military specifications and standards. There must no longer be laws or regulations applied solely to the purchase of defense goods and services.

This requires a very significant change in America's weapons acquisition practices, a process which the Clinton administration has begun under the heading of "defense acquisition reform." The plan to remove the current legislative and regulatory barriers to civil/military integration is of course being fiercely resisted by many entrenched institutions, such as government-owned defense facilities. However, it is an explicit policy priority of the Clinton defense team, and the plan includes support and incentives for the U.S. defense industry to move in this direction.

For example, the government increasingly funds "dual-use R&D," which allows firms doing government work to develop products for both the military and the commercial sector. Historically defense-oriented firms are encouraged to establish strategic alliances with commercial firms in order to develop cost-sensitive design practices and greater commercial market access and understanding. The ARPA Technology Reinvestment Project, funded at about $500 million in fiscal year 1995, is an example of such an effort. The administration is also moving ahead with a plan to fund more R&D in civilian sectors devoted to public works, such as infrastructure, environmental cleanup, transportation, energy, and education—areas in which the high-technology skills of defense firms are directly applicable. The new Republican-controlled Congress has recently begun to question some of these expenditures, however. And it is clear that, even if fully funded, these government actions are not sufficient. Industrial leaders must take the dominant role in the transformation of the U.S. defense industry.

Lessons Learned from Prior Industrial Transformations

The conversion of the defense industry is but one part of a much broader process of industrial transformation occurring worldwide as a result of growing international competitiveness. Large industrial bureaucracies like General Motors and IBM must totally transform their way of doing business if they are to survive. Even while attempting to capture larger market shares, competitive U.S. firms are dramatically downsizing their labor forces and taking advantage of modern information technology to "reengineer" their operations, eliminate large portions of middle management, and to empower their workers with greater decision-making responsibility. In each case, companies that have succeeded in these transformations have met two primary requirements. First, each recognized that it was in a crisis situation, and had no choice but to change; and, second, each had leaders who were committed to making the needed changes. Obviously, these are but the minimum prerequisites; yet without them, few firms have succeeded. Further, successful firms have used the lessons of the past as guidance for future steps. This is particularly critical for defense industry transformation, as the prior post-war cycles have indicated both clear barriers to conversion and the clear steps necessary for successful transformation.

A National Science Foundation study[6] of the conversion efforts of Western European defense firms following the Vietnam War indicated sixteen obstacles to defense industry diversification. This list includes economic as well as managerial factors:

1. Absence of an internal climate hospitable to change.
2. Priority of commitment to military production.
3. Shortage of capital.
4. Competition from existing commercial firms.
5. Obstacles from other (commercial) divisions of the firm.
6. Problems of commercialization (e.g., pricing, licensing).
7. Fragmented markets (vs. government markets).
8. Customer inability to specify needs (again, in contrast to the military).
9. Inadequate infrastructure (e.g., product support).
10. Incomplete knowledge and patents (in commercial area).
11. Labor force inflexibility.
12. Low profitability of initial commercial investments.
13. Scale of output (large, compared with defense).
14. Specialized institutional resources required (e.g., commercial contracting).
15. High cost of quality (particularly at the lower tiers).
16. Inadequate marketing skills.

These same barriers continue to exist in the current downsizing.

When one reviews the numerous case studies of prior conversion attempts, in a wide variety of countries,[7] a few conclusions stand out. First, there are no easy steps nor general rules to conversion; each case is different. Ultimately, each successful conversion requires a tailored approach, one that is primarily orchestrated by the firm involved. However, some lessons of prior conversion attempts appear to be universally true. Perhaps the most important of these is the fact that most successful transfers from military to civilian are at comparable, high-technology levels—they do not involve a step down in technology. Additionally, it is almost universally true that the primary ingredient of success is a "cultural reorientation," especially by senior management. In fact, it has been found that the primary barrier to successful transformation has been the reluctance of senior management to truly change their traditional way of thinking and their unique way of doing business—that is, to shift from a military to a commercial orientation. This includes a move away from the traditional organization segregation of military and commercial subdivisions. Since effective technology transfer from military to civilian sectors occurs primarily through people working together, integrating staff with commercial backgrounds and those with military backgrounds has been a very important strategy.

Finally, most conversion successes have been achieved by searching for growth markets and satisfying them, rather than by attempting to sell modified military products to commercial markets. Thus, "market pull," not "technology push," is key, and the longer the lead time, the better. Realistically, the process of commercializing technology must be viewed in terms of years, not months.

In making the transformation, firms must recognize that not all plants and facilities should be saved, and that the biggest manpower cuts required are likely to be in engineering and administration. Also, marketing and finance are dramatically different in the commercial world, and thus must largely be started anew. Engineering innovation, capital equipment, production labor forces, and skilled management are the principle existing assets that should be maximized.

Past conversion examples suggest several specific product-diversification strategies:

- Successful conversion is most frequently achieved when existing core competencies are utilized. These can be technologies and/or production processes, but they also can be management skills, and marketing distribution. For example, Kaman Corporation successfully applied its core competence in helicopter vibration-control techniques to the making of guitars. Large defense prime contractors are skilled in high-technology, large-systems integration; that is, they integrate and manage complex, state-of-the-art, advanced mission systems. If they view that capacity as their core competence, rather than defense systems, they have a greater advantage in entering the non-defense world. Thus, it is very important that a corporation interested in diversification/conversion recognize that it is often not similarities in end products, but similarities in core competence that can be critical to achieving success.
- Multiple product objectives should not be combined in a single product—for example, by adding some additional civilian capability to a military product. Thus, managers need to think in terms of a specific civilian product and a specific military product, even if the products are performing very similar functions. At the same time, these differentiated products should be built on the same production line to gain the advantages of economy of scale, far less labor training, and reduced capital equipment investments.
- In moving out of the military arena, it is essential to break the defense industry habit of depending completely on the buyer to define the characteristics of the product. Use of a variety of market research techniques, including focus groups, is essential in order to shape the new product to its potential market.
- Given the defense industry's traditional and almost exclusive focus on performance, another major problem faced by defense firms in conversion efforts is that their products are too expensive. Unfortunately, defense firms have had little engineering experience with the commercial practice of design-to-cost, and tend to overspecify their systems. Many defense conver-

sions failed because more attention was given to retooling the factory than to retraining the company's design engineers to focus on cost.

- Finally, studies have shown that two years is a reasonable amount of planning time for all the work that must be done to blueprint a product changeover. A reliance on crash operations rather than careful planning usually leads to failure.

The Prospects for Russian Conversion

The kinds of problems encountered in U.S. defense industry conversion are greatly compounded in the Russian case. Unlike in the United States, defense was the dominant sector of the Russian economy. At the time of the revolution of 1989, up to 40 percent of all machine-building output and 75 percent of all R&D were defense related. In many provinces, defense production represented more than 50 percent of the total industrial output,[8] and in numerous cities the sole employer of the tens of thousands of residents was a large defense production plant. With Russian defense budgets plummeting by 80 percent (beginning in 1989, but greatly accelerating in 1992),[9] the magnitude of the conversion problem is of a scale never before experienced. Even the U.S. conversion effort after World War II was totally different, in that the national economy could support the pent-up civilian demand, and the economic infrastructure (from commercial banking to contract law) was in place. Russia is on the horns of a dilemma. Downsizing, or even closing down, defense plants means huge unemployment and creates economic and political unrest. But continuing current defense production will waste the very limited, and extremely valuable, resources available. Although it will be very difficult and time-consuming, conversion is the only answer.

Prior to the breakup of the Soviet Union, then-President Gorbachev had begun a conversion effort in an attempt to lighten the military burden on the economy and to stimulate civilian growth. By 1992, many of the former missile factories had shifted part of their capacity to civilian production, including metal cutting machines, drilling rigs and other oil industry equipment, washing machines, bicycles, and even baby carriages. The intent was clearly for these plants to remain as "dual-use" factories. An increasing percentage of production was to be for the civilian market, with a mandated target of more than 60 percent by 1995. Unfortunately, simply increasing the civilian share of the output of military industries without any significant reorientation is likely to result in noncompetitive products—with regard to both price and quality. Earlier, similar attempts at Russian industrial conversion produced washing machines that cost twice as much as those of the same design produced under a civilian ministry. Or, to cite another well-known example, television sets were produced that exploded and caught fire during use. Additionally, many of the civilian products selected have not been a good match for the resources of the factories that were targeted for conversion. Of the

585 consumer goods to be manufactured by the military factories, only 23 were actually being made by the end of 1989. Only 15 percent of these new products met international quality standards.[10]

Russia's conversion problems stem largely from having the central bureaucracy dictate what individual items were produced in each of the military enterprises, rather than employing "market pull" as the guiding principle. In some cases, there was a gross mismatch of military-related inputs and civilian requirements. Again, a frequently cited example is the production of garden spades from titanium, a very high-cost, exotic material used in aircraft to achieve light weight and heat resistance. It is interesting to note that the Russian leadership apparently chose to dissolve its commercial ministries and use the defense industry—with its skilled labor, materials, equipment, and technologies—as the leading sector for the transformation of the civilian economy. They recognized that military plants were well ahead technologically and had the best of the labor force. Unfortunately, it is not recognized that the defense industry's lack of familiarity with commercial needs, as well as cost and quality sensitivity, will be great hindrances to successful conversion. This lack of customer sensitivity and customer service is one of the most critical cultural transformations required if the Russian plants are to convert successfully. Finally, world-class, internationally competitive firms are today moving toward tight integration between their R&D and production activities. The historic separation between R&D and production in many Soviet military activities works counter to this growing competitive practice, known as "concurrent engineering." So here, too, change is clearly required.

If industrial conversion is to be achieved in Russia, it must be predominantly driven by local decision-making at the plant level, rather than through centralized conversion planning. While the government can provide incentives and support, market sensitivity and detailed conversion planning must be accomplished by local plant management.

Industry Strategies and the Role of Government

There are three broad models for conversion of the Russian defense industry. While interrelated, there are clear differences among them; the specific approach selected must be tailored to the individual plant circumstances. The first, and perhaps the easiest and fastest, option is setting up small, private, entrepreneurial "spin-offs," often physically located directly inside of a large defense plant and utilizing the same labor force and equipment. While this approach has been the most successful to date, its scale has been quite small, and it has focused heavily on employment of engineers, rather than on production. The benefits of this model are its ease and speed; however, it is not likely to rapidly solve the employment needs of the tens of thousands of workers at any given defense operation.

The second model involves privatization of the complete defense plant and a widespread diversification effort in related civilian business. Since one key to successful conversion is the leadership of individual plant managers, they must be relieved of government control and set free to pursue diversification. Naturally, the government will still be a customer for the plant's military products, and it will still exercise control over its foreign military sales, but all other activities will be decided by plant managers. Generally, this option appears to be the choice of the defense industry managers, but there is widespread government concern about losing control over critical defense plants.

The third model has been advocated by many in the Russian government.[11] It is modeled after the Korean and, in some cases, Japanese approach. It advocates forming large industrial groups made up of a major defense firm or firms, one or more large commercial firms, a major financial institution, and a large trading company. This industrial group, aided and initially supported by the government, would have all of the required elements for successful diversification and should be competitive with large, Western multinational firms.

All three approaches in Russia should incorporate these critical lessons:

1. Focus on development of the required business skills and understanding of the growth markets, not on selling of existing products.

2. Stick to selling comparably complex, high-technology products (not sleds, pressure cookers, and children's bicycles, for example) in order to take advantage of existing engineering and management skills. Significant growth markets exist in such areas as oil and gas extraction and distribution, nuclear waste reprocessing, prefabricated housing, food processing and distribution, and, particularly, infrastructure needs such as air traffic control, telecommunications, environmental cleanup, and modern transportation systems.

3. Make maximum use of strategic alliances, especially with foreign commercial firms. The latter bring market knowledge and entry as well as price sensitivity, and knowledge and credibility for product service and support.

4. Finally, start with familiar markets within which credibility is already established. This would tend to favor the domestic markets. Even though foreign exchange from exports is desirable, those markets will be much harder to penetrate than the Russian infrastructure market or the large, domestic commercial market.

There have already been a number of successful Russian defense industry diversifications, and these must be built upon. For instance:

- The Saratov Aviation Plant was privatized in 1991 and now more than 90 percent of its production is for the civilian market,[12] including passenger airplanes and consumer goods.
- The Saratov Electro-Mechanical Production Organization was also privatized in 1991, and more than 95 percent of its production is consumer goods.[13]

- The Impuls State Research and Production Enterprise now receives about 80 percent of its revenue in its areas of advanced optics and electronics from commercial orders. [14]
- The Ilyushin Design Bureau and Pratt & Whitney have formed a joint venture to develop and produce commercial airliners using Russian airframe technology and manufacturing facilities and U.S. jet engines and avionics.[15]
- The Moscow Center for SPARC Technology was established as a new private company, with a contract from Sun Microcomputers for employment of more than 80 engineers. The new company is located within the Institute of Precision Mechanics and Computers.[16]

Although the greatest need is for the Russian government to free industrial managers to make and fully implement individual facility plans, there are some significant steps that the Russian government can take to facilitate this transformation. Perhaps the most critical, near-term decision facing the Russian government is the initial question raised in this chapter: will Russia's defense industrial base remain isolated and defense-unique, fully controlled and subsidized by the government, or will the government obtain its defense goods and services from a private-sector, dual-use industrial base that treats the government as but one of its many customers? There may be a few plants that even in the dual-use model still must remain as defense-unique and government controlled. But in this model, the overwhelming share of plants will be privatized and free to pursue diversification, with government-funded defense sales a critical portion of their overall business.

Once this decision is made, it is essential that the Russian government be willing to assist the firms that want to convert. It would be appropriate for the government to provide investment incentives to companies that do not choose to rely on the export of arms but, rather, want to expeditiously move to a dual-use capability. Here, the government can provide funds for plans and actions associated with conversion. It can also fund, and provide assistance for, the needed market research, R&D, and capital investment required for the transformation. For conversion to be of greatest benefit to the Russian nation as a whole, funding should be primarily directed to high-technology infrastructure needs, such as telecommunications, advanced railroads, air traffic control, airline modernization, and environmental cleanup. This would allow defense plants to move into civilian sectors with comparable high technology, while contributing to the infrastructure modernization so critical to Russia's economic growth.

Conclusion

The need for dramatic downsizing and indeed a total transformation of the defense industries of the United States and Russia is indisputable. And there is clearly a time urgency, especially in Russia. The question is simply whether the required transformation will be done efficiently and effectively, or whether short-term politics will dominate these processes. If the latter continues to be the case, then instead of achieving the desired long-term economic and security objectives, there will be an enormous waste of both dollars and time, with a corresponding significant economic and security loss to both nations. The move to an integrated civil and military industrial base would allow the resources of the defense sector in each nation to stimulate economic growth and still maintain needed defense capabilities. The choice is clear, but the path will be difficult; and the more that both governments can assist and cooperate in this transformation, the more successful both nations are likely to be.

Notes

1. A. Carter, W. Perry, and J. Steinbruner, *A New Concept of Cooperative Security* (Washington, D.C.: Brookings Institution, 1992): 29–30.

2. Randall Fosberg and Jonathan Koon, "The Global Arms Market: Prospects for the Coming Decade" (Cambridge, MA: Institute for Defense and Disarmament Studies, August 1993). (Based on SIPRI Yearbook 1993 and SIPRI Worksheets.) Also see W. J. Durch, S. M. Irwin, and J. Henick, "Overview of the International Arms Market, with Emphasis on Variables Affecting Russian Sales Prospects" (Washington, D.C.: Brookings Institution, September 1993).

3. "Integrating Commercial and Military Technologies for National Strength, an Agenda for Change," Report of the CSIS Steering Committee on Security and Technology, Washington, D.C., March 1991. Cochairs Jeff Bingaman, Jacques Gansler, and Robert Kupperman.

4. Paul Blumhardt of Martin Marietta reported a 32 percent success rate from data over the 1973–1993 time period. (Naval Research Advisory Committee Report, "Defense Conversion." Washington, D.C., November 1993, 55.) Also, a survey of 148 firms in 1991 reported a success rate of 36 percent. ("The Commercialization of Defense Technology: A Survey of Industry Experience." Lexington, MA: DRI/McGraw-Hill, November 1991, 2.) Blumhardt's data showed that firms that converted into closely related areas had much higher success rates; e.g., commercial aerostructure subcontracting enjoyed a success rate of 71 percent, and civil government information systems 67 percent.

5. National Economic Council, National Security Council, White House Office of Science and Technology Policy, "Second to None: Preserving America's Military Advantage Through Dual-Use Technology" (Washington, D.C., February 1995).

6. National Foundation Study, "The Western European Experience" (Washington, D.C., 1972).

7. These are covered in detail in Jacques S. Gansler, *Defense Conversion: Transforming the Arsenal of Democracy* (Cambridge, MA: MIT Press, May 1995).

8. Daniel Yergin and Thane Gustafson, *Russia 2010—And What It Means for the World* (New York: Random House, 1993).

9. Julian Cooper, "Defense Industry Conversion in the East: The Relevance of Western Experience." In *External Economic Relations in the Central and East European Countries*, NATO Economics Directorate Colloquium, April 1992.

10. Greg Bischak, "International Roundtable Meeting: Survival of Civilization; Disarmament for Development and Conversion" (report of the meeting sponsored by the All Union Central Council of Trade Unions, Moscow, November 23, 1989), 25.

11. Julian Cooper, "Transforming Russia's Defence Industrial Base," *Survival* 35, no. 4 (Winter 1993): 147–162.

12. John Battilega, "The Saratov Aviation Plant." In *Can the Russian Military-Industrial Complex Be Privatized? Evaluating the Experiment in Employee Ownership at the Saratov Aviation Plant*, edited by Michael McFaul (Stanford, CA: Center for International Security and Arms Control, 1993): 41ff.

13. "Controls on Arms Exports: How Can Privatization and Conversion Help?" In *Conversion*, no. 1 (Stanford, CA: Center for International Security and Arms Control, March 29, 1993): 8.

14. Ibid, pg. 9.

15. Michael McFaul and David Bernstein, *Industrial Demilitarization, Privatization, Economic Reform, and Investment in Russia: Analysis and Recommendations* (Stanford, CA: Center for International Security and Arms Control, 1993), 11.

16. Ibid., pg. 12.

4

Dismantling Russia's Technotopia: Six Ministries in Search of an Industrial Policy

Harley Balzer

It would be surprising if a nation as addicted to technological mythology as Russia did not seek to alleviate its economic crisis by turning to some form of industrial policy.[1] Not only was the USSR highly, albeit unevenly, industrialized, it was ruled by a leadership committed to a "scientific" ideology that elevated technology to the status of a cult.[2] According to Stalin, "technology decides everything." In the Brezhnev era, entire research institutes discussed the "scientific-technical revolution" and its meaning for Soviet development. Political rectitude could be calculated by the number of times the mantra "science-intensive industry" appeared in an economist's writings. This heritage of what I have called a "technotopian" orientation has had a strong and for the most part pernicious influence. Even many of the leaders in Russia's younger generation who know something about real (as opposed to command) economics are captivated by the myth of high-tech capabilities.[3]

The myth of Russian high technology prowess has had a detrimental effect on discussions of science, technology and industrial policy. It has proven extremely difficult for Russian policymakers to formulate an approach based on rational assessments of the nation's industrial and technical capabilities. In a context of organizational chaos and institutional rivalries, the high technology myth inhibits elaboration of a coherent and effective policy. It has also left Russian policymakers susceptible to well-meaning but seriously misguided attempts to impose developmental models based on the experience of nations with vastly different cultural and institutional conditions.

This essay will sketch the context for Russia's industrial policy debate with a brief review of the situation in applied R&D and conversion—the two realms that are the key to any successful Russian industrial policy. We will then turn to an

This essay draws on material collected in the preparation of an OECD assessment of Science and Technology Programs of the Russian Federation, and also uses material from a research project on "The Changing Russian Technostructure" supported by Carnegie Corporation of New York. The author is grateful for research assistance provided by Christine Zapotocky

analysis of the debate over industrial policy within the Russian government. To help clarify the issues of the Russian debate, we will also examine a parallel discussion taking place among Western specialists. Examining these competing proposals against the background of Russia's technotopian orientation should help to clarify the prospects for different lines of development.

Applied R&D

Technological hubris resulted in a serious overestimation of the capabilities of Russian applied technology. In Gorbachev's reform plans, and in the various programs for transition to the market, everyone assumed industry and innovation would play a key role (Balzer, 1989; *O korennoi perestroike*, 1987; *Perekhod k rynku*, 1990). When the extent of Russia's economic crisis became apparent in 1991–92, competing claims on scarce resources helped encourage the prevailing technological myths. The operative assumption was that applied R&D would "take care of itself." That is, industry would finance the R&D it needed. State science policy was devoted to financing fundamental science, which was perceived as defenseless and seriously threatened.[4] Basic science was also generally regarded as the flagship of Soviet S&T, the area of highest-level performance that achieved the greatest international recognition. Given how naive the belief in the self-sustaining capabilities of applied R&D has turned out to be, it is remarkable that the view was so widely—indeed, almost universally—held.

By early 1995 it became clear that few enterprises were in a position to devote resources to R&D and that most Russian technology is not internationally competitive. Prototypes of many Russian/Soviet inventions exist in single copies that were developed with little concern for cost, marketability, or the requirements of serial production. Most of the small number of successful technology-based firms have worked with foreign partners who can assist with marketing and other business skills, as well as provide experience in cost-effective production.

The Russian government has made some attempts to provide assistance to applied R&D, but these steps have been limited and halting. The government's major initiative, the Foundation for Technical Development, has foundered. Based on a plan requiring enterprises to contribute 1.5 percent of their gross turnover to a technical development fund, it has failed due to the nearly universal absence of profits (or at least profits to which any firm will admit), and firms' inability or unwillingness to pay. It illustrates the difficulties in administrative conceptions and enterprise behavior. Russian bureaucrats still look for ways for the state to determine how firms will utilize their resources.[5] There are no effective enforcement mechanisms, and most enterprise directors tend to behave in ways that maximize immediate returns in Russia's chaotic economic environment. Long-term investment in R&D-based technologies is not an attractive option. This may change over time, but even if it does the results will not be visible for many years. A second state institution, the Russian Bank for Reconstruction and Development, was created to foster investment in promising enterprises. It has helped finance some 100 projects, virtually all involving natural resources (*Poisk*, 1995).

The Russian government has also selected a number of key R&D facilities to become Federal Science Centers. Thus far about 60 institutions have received this designation, which carries the promise of priority financing. In most cases, being designated as a Federal Center has encouraged administrators to maintain their staffs rather than making targeted reductions (Schweitzer, 1995). The Ministry of Science has plans to raise the number of Centers to 300, provided adequate resources are available.

Russian fundamental science turned out to have more supporters: it was internationally recognized as a realm of high quality; many individual scientists had good contacts with their colleagues abroad; and it was possible to argue that basic science makes an international contribution. This matrix of factors induced both individuals (most prominently George Soros) and governments (particularly the European Community through INTAS) to provide substantial support for Russian basic science. Virtually nothing comparable has been done in the area of applied R&D. This is to some extent a judgment of the quality of Russian technology, especially in the nondefense sector. But it also reflects concerns about intellectual property and national interest.[6]

The recent history of efforts to carry out some sort of critical review or evaluation of Russian S&T facilities provides instructive lessons for those who advocate an activist industrial policy. In 1993 an OECD evaluation of Russian science proposed creating mechanisms to fund science on a competitive basis (OECD, 1994). Members of the OECD team who recommended cuts to a level the state budget could support were told by the Minister of Science that they "failed to appreciate the political situation in the country." Russian science officials believed that it would be impossible to significantly reduce the number of science workers. Now, even following a substantial spontaneous movement out of S&T, one of the major items on the policy agenda for autumn 1995 is a critical review of scientific institutions and researchers.[7]

The Russian discussion about spin-offs is equally disheartening to those who propose "picking winners." The still-powerful ministries and many prominent political voices argue that it is irresponsible to break up the vertical chains that exist in the Russian economy. But as long as potentially viable firms remain embedded in large vertical structures they will have little chance to compete effectively in international markets. Recent efforts in some areas to replace the vertical structures with regional structures focused on preventing unemployment and guaranteeing economic survival for all enterprises will not provide a better solution (Bernstein, 1995; Prokop, 1995).

Conversion

One of the articles of faith among Russian commentators on economic policy has been an expectation of contributions from the defense sector. Virtually everyone, both in Russia and the West, agreed that Soviet military technology was superior to civilian technology, but there was a debate about the sources of military superiority. Some believed that the MIC was an island of competence and advanced technology in the Soviet system; others suggested that the Soviet MIC achieved its relatively better performance by "milking" the entire economy. We now know that both explanations were partially correct, but that only a few military facilities approached top levels. The Soviet MIC, like the entire economy, was uneven (Balzer, 1985). Perhaps more important, we also now have evidence that the hidden and opportunity costs imposed on the economy by the defense sector—the externalities—were even greater than some of us previously suggested (Kuznetsov, 1994).

This digression into the character of the Soviet defense sector is important if we are to understand the potential for defense conversion in Russia. Russia inherited the lion's share of Soviet defense industry, along with all of its problems. And these problems exacerbate the difficulties in realizing a conversion policy that is everywhere a difficult challenge.[8] In a recent book, Jacques Gansler (1995) provides a good summary of the current state of our knowledge about conversion in the United States and Europe. The picture he presents is sobering for American defense firms, and by implication much more daunting for Russian enterprises. Three out of four conversion initiatives fail in their first three years. According to Gansler, the success rate "used to be around 20 percent, increasing to 35 percent in recent years." The most serious obstacle to conversion is "the significant cultural differences between the defense world and the commercial world" (Gansler 1995, 71).

And it is not just that half or more initiatives fail. Surviving past the three-year marker is no guarantee of eventual profitability. Long time periods are required before most firms produce genuine returns: eight to ten years is a reasonable estimate for firms that succeed. Gansler (1995, 75) cites the experience of the French missile manufacturer Matra, which failed in its initial effort to move into watchmaking in 1970. A decade later it was producing 50 percent civilian products for cars, subway cars, telecommunications, and space. After another decade, its product mix was three-fourths civilian. In other words, conversion is a lengthy process with no guarantee of positive results, and it requires substantial investment.

Particularly noteworthy for the Russian context are the difficulties American technical personnel, especially higher-level designers and engineers, have experienced in making the transition from military to civilian industry. Gansler finds the main obstacles to this transition to be (1) the lack of cost consciousness in defense; (2) the view of business managers that defense engineers are not well suited to commercial work; and (3) the belief among engineers that the defense environment requires more specialists, and the commercial environment more generalists.

Organizational inflexibility constitutes a major obstacle. The experience of Spectra Enterprise Associates included several failures stemming from a lack of strategic thinking: "They didn't think products. They think solutions [like engineers], but they're not trained to finish a task" (Gansler 1995, 78).

Programs of defense conversion for Russia were discussed as part of perestroika in the Gorbachev era. The inherent difficulties were apparent, and were analyzed, at the time:

> Again, the high-quality batching items [for the Soviet space industry] . . . are achieved at the price of immense expenditures and strict military acceptance. The rank-and-file consumers would simply not be in a position to buy them, for they would transfer the cost of these items to the final product. The excessive increase in the cost of technical equipment for light and food industry products produced in the defense industries has already been reported in the press.[9]

Depending on which commentators one relies upon, Russia has had either too many conversion policies or no policy. The one thing most agree upon is that thus far there has been no effective policy.[10] While much has changed since the years of perestroika, the structural problems and difficulty in addressing them have, if anything, become more daunting. Yet few Russian analysts have been able to confront the serious nature of the challenge or the real economic implications. Many of the analyses of conversion written in Russia have a prescriptive character. Vladimir Fal'tsman (1994), in an approach typical of many Russian experts, advocates preserving the scientific and technical capacity developed in Russia's military-industrial complex, but gives few clues about how this is to be accomplished. Yuri Yaremenko and Victor Rassadin (1994) plead for a policy to preserve the high tech defense machinebuilding industry by integrating its capabilities with the civilian sector, but the mechanisms for financing and facilitating this process are not elaborated. Aleksandr Ozhegov notes the important potential in the defense sector, but he too avoids analysis of the trade-offs involved in a program of genuine conversion during a period of economic stringency.

Some of the clearest statements have come from Yevgeny Yasin (1994, 131). He discusses the need to set priorities, notes the role of an industrial development fund, and mentions creation of a state investment agency and the Russian Bank for Reconstruction and Development. He is aware of the choices and contradictions in any policy mix, and comes down firmly on the side of the market as the primary decision-maker. But while more clearheaded in his analysis of the issues involved, he too lacks a solution to the resource and policy equation.

One of the serious difficulties in mobilizing the productive resources of the MIC for commercial production is that defense plants tended to be especially large and heavily staffed. Ozhegov (1993, 5) notes that there were virtually no small enterprises. And the large enterprises were embedded in gargantuan vertical structures which seek to perpetuate themselves.

Most Russian commentators assumed that if the defense sector was enormous and relatively advanced, it would automatically be the leader in Russia's shift to the market. Yet Kuznetsov's data suggest that correcting the damage stemming from externalities imposed by the MIC—opportunity costs, ecological destruction, etc.—would have a far greater impact on the Russian economy than any industrial or sectoral policy could achieve.

As the difficulties in converting to civilian production and commercializing products never intended for the civilian market have become apparent, many Russian commentators have become shrill in their criticisms of Western barriers to their technology. Cronberg (1994, 197) cites the example of one Russian manager from an aviation enterprise who attended a Birmingham exhibition in May 1993 and spoke angrily in response to Western suggestions that Russia start with "small" products: "We will enter the world market as a high-tech nation, no matter what Western partners think." Cronberg notes that "Russian enterprises' attempts to cooperate with the West are becoming more frustrated, pleading and aggressive at the same time."

The shrillness is understandable. After decades of being told (and telling themselves) that they had superb science and were leaders in high technology, Russians find it difficult to accept the painful lessons learned in other nations about the difficulty of shifting military industry to commercial competition. Rather than creating conditions that might permit some firms to operate in the international arena, many Russian officials have found it tempting to blame the West for blocking their access to world markets.

We see here a classic example of the conceptual problems in Russia's transition. Many Russian officials see the solution to their difficulties in promulgating a law on conversion. This is a vivid contrast to Gansler's (1995, 82) conclusion based on the Western experience: "There is no evidence at all that conversion can be legislated." The Russian proclivities are a legacy of the planning mania that accompanies Russian technotopian thinking: the solution to any economic problem was sought in more and better planning. It has proven extremely difficult for government officials to visualize their role in terms of facilitating rather than controlling and administering. This approach clearly underlies much of the Russian discussion of industrial policy.

The Search for an Industrial Policy

Discussions of industrial policy within the Russian government have mostly been a not very well-defined subset of general economic policy. The major participants have been the Ministry of Science and Technology Policy (*MinNauka*), the State Committee for Industrial Policy (*Gosudarstvennyi komitet po promyshlennoi politike*, or GKPP), the Ministry of Foreign Economic Relations, and the Ministry of Economics. A generally separate but parallel discussion has been underway in the Ministry of Defense (MOD) and State Committee for the Defense Industry (*Gosudarstvennyi komitet po oboronnoi promyshlennosti*, or *Goskomoboronprom*). These are the "six ministries in search of an industrial policy."[11] Thus far, it has proved elusive.

The closest Russia's new administration came to formulating a coherent industrial policy was during 1993. The driving force in this effort was Sergei Glaz'ev, then Minister of Foreign Economic Relations.[12] Drawing on his political and personal relationships with Dmitry Piskunov (then Deputy Chairman of the State Committee for Industrial Policy) and a number of senior staff of the Ministry for Science and Technology Policy, Glaz'ev and his colleagues formulated a design for an agency consciously modeled on Japan's Ministry of Trade and Industry (MITI). There is little doubt that Glaz'ev visualized himself as the future minister in this new institution. Despite the generally accepted need to establish a coherent policy combining military and civilian industry, the military remains separate and neither the MOD or *Goskomoboronprom* ever took an active part in these efforts.[13]

The early effort to formulate an industrial policy included making use of European consultants. The "European Expertise Service" on behalf of the Commission of the European Communities contracted with Coopers and Lybrand, Europe to produce a study which was completed in April 1993. The consultants' report is somewhat divorced from the on-the-ground realities of Russia's economic plight, political situation, and enterprise structure. They cite the vertical organization of much of Russia's industrial structure, and the negative consequences that would result from removing links from these complex chains. This issue represents a serious problem in current economic debates in Russia, and the value of the vertical (ministerial) chains is being intensely debated. Preserving the chains precludes spinning off economically viable firms, and may doom the entire structure.

The European consultants made errors common to most foreign and a great many Russian experts who assumed privatization would automatically mean changes in the ownership and behavior of enterprises. Privatization should eventually result in the restructuring of many enterprises, but the impact has not been rapid, and the losses in the process have been quite severe. The consultants also assumed that restructuring would create "financially independent organizational structures based on market economy principles and 'the Western model,'" and that these new entities would "ensure the focused concentration of R&D capacities." The implicit assumptions of market-oriented management, financial viability, and availability of investment funds to support innovation all proved to be wildly optimistic.

In preparing policy prescriptions the GKPP began with a more realistic assessment of Russia's situation.[14] They consider it imperative to retain a basic core of national industry in the extractive industries, the manufacturing sector, and high technology branches.[15] They note that these changes will be complicated by the persistence of high levels of inflation, which will deter investment and innovation, and by the need for strict deflationary policies that will make it impossible to support enterprises and R&D organizations in the industrial sector. In light of these daunting difficulties, the GKPP suggested that basic directions for 1994 and the middle-range time period should be

- Creating a mechanism for establishing state priorities in industry and scientific-technical development.
- Analyzing the crisis in Russian industry to formulate measures to arrest the fall in production.
- Elaborating basic directions for restructuring and increasing the effectiveness of industrial production, including developing procedures for reprofiling and liquidating unprofitable enterprises, improving the tax system, and formalizing credit policy and antimonopoly and currency regulations.
- Reforming the structure of industry and industrial science, in particular creating Financial-Industrial Groups (FIGs), federal scientific-technical centers, technopoleis, and a program to support small and medium R&D enterprises.
- Establishing a new relationship between the state and industrial enterprises, including both those that are state property and those with mixed forms of property, by means of a federal contract system for purchase of goods and services for state needs.
- Regulating the external economic activity of industrial enterprises through an improved mechanism of currency controls, stimulating export and import, creating a system of state support for exports, and facilitating entry of Russian firms into foreign markets.
- Coordinating the activity of enterprises, sectoral administrations, and regions to introduce effective conversion of military enterprises, including measures to deal with growth in unemployment by creating additional job opportunities and retraining in regions facing critical levels of unemployment.

This outline makes it clear that at least some specialists in the Russian government were thinking about most of the elements of an industrial policy, and sought to place it in the context of the prevailing economic crisis and the status and potential of Russia's economy. But it is equally clear that many of the specific elements of the policy remained to be worked out, and that the political and administrative barriers to effective action were massive.

A number of influential members of Yeltsin's administration proposed creating a single ministry or state committee responsible for all science, technology, and industrial policy, but this proved impossible to realize. A "genuine" industrial policy would require a centralized system, but there was neither the willingness to establish such a system on the part of the participants nor the capacity to impose one by the government.

It became clear that the GKPP had high hopes but minimal clout. Overall, the committee had some 340 staff in 1993, and planned to place 200 more in eight regional offices, at locations to be determined.[16] They believed that the most basic problem preventing the GKPP from playing an active role was lack of financing. Other than a budget of some one billion rubles to finance work on their "general conception," the committee was completely dependent on the percentage of enterprise funds that was supposed to be contributed to the Russian Fund for Technical Development. Actual receipts have run at about 10 percent of the expected amount. The GKPP has no "stick" to use to exact compliance with its policy guidelines, and failure to collect the 1.5 percent means it has had no "carrots" to offer.

In 1993 the committee did begin to identify "losers" in the form of mines and enterprises that would have to be closed down. They concluded that about one-fourth of Russia's 200 coal mines would have to be shut immediately, and Vice Premier Shumeiko requested a list of the factories that should quickly declare bankruptcy. None of these policies proved easy to realize, as the government lacked the will, the power, and the legal apparatus to follow through with closings.

In 1994 a certain amount of work was carried out in formulating what an industrial policy should be, but the goals were still unclear and little was being implemented.[17] The Russian government arranged for a translation of the U.S. Council on Competitiveness volume *Gaining New Ground*, and this was circulated as an example of the type of analysis they wished to emulate. The most complete formulation of thinking about Russia's industrial policy was embodied in a series of reports commissioned by the committee. Judging from items in the Russian press in early 1995, these documents appear to reflect the main lines of thinking by Sergei Glaz'ev and other proponents of a greater state role—an "industrial policy approach" to Russia's economic crisis.[18]

The Glaz'ev/GKPP Program

The Glaz'ev program, formulated by scholars working at the Analytical Center for Science and Technology Policy and the Russian Academy of Sciences, is characterized by a belief in the potential of Russian science and industry, heavy reliance on state spending, and a strong dose of Russian patriotism. The document makes a case for heroic Russian managers struggling to preserve their enterprises in the face of a disastrous economic collapse, unremitting pressure from their workers to provide employment for all at higher wages, and a foreign-advised reformist government that was at best incompetent and more often predatory.

In assessing Soviet policies in R&D the report is contradictory. The authors note the serious defects in Soviet organization and performance but also stress the enormous S&T potential that existed. Policy recommendations based on this assessment include a rigorously selective approach to applied science combined with state measures to guarantee support from the "private" sector.

The report concludes with a review of macroeconomic factors related to innovation and technology. In the authors' assessment:

> At the end of the second year of reform the Russian economy was in a relatively stubborn condition of stagflation, the characteristics of which included an overwhelming decline in demand and production along with fairly intensive inflation. From the standpoint of institutional change the results of the previous two years saw the beginning of development of market relations, to which enterprises in large part had already managed to adapt.

These changes in the economy resulted in formation of a new type of production, capable of sustaining the economy for a limited period of time, but lacking perspective for long-term development.

The authors sketch three scenarios for Russia's economic future. These same scenarios reappeared in 1995 in an analysis by the Russian Academy of Sciences.[19] The first scenario would be a continuation of the government's strict anti-inflation policy. The authors believe this approach ignores Russia's unique conditions. It is producing a diminution of production with no decline in prices. If continued, the production drop will become so severe that it will be impossible to conduct a coherent economic policy, and the result will be massive social instability. Industrial restructuring is taking place in conditions that foster competition for redistribution of state property rather than efforts to promote investment or increase production. Given sufficient time, the anti-inflation program of strict financial and credit policy might lead to a normal economy, but it will destroy the S&T base of the nation and would almost certainly provoke social tensions before a real market could emerge.

The second scenario is an "adaptive" or compromise policy that balances between the drop in production and hyperinflation. The authors contend that this is what was done during 1991–94, with intermittent retreats from the worst consequences of fiscal restraint. While it may facilitate stability, the compromise approach is incapable of generating growth. Over time the production capacity of the nation degrades, investment and innovation cease, social differentiation grows, and more tensions are produced.

The Glaz'ev/GKPP group sees as the sole significant difference between these first two variants the length of time it will take to reach disaster. Their alternative is a third scenario, an "active coordination policy," which they claim takes into account the "specific features of the Russian economy," and allows them to undertake active structural reform at "branch, technological and institutional levels" while maintaining social stability.

Realization of the "active coordination policy" requires sufficient resources, both budget and credit, and this cannot be achieved through tax revenues. Therefore they proposed continuing inflationary sources of financing state needs through

1994, while shifting the inflation to a "regulated and predetermined character." This would slow the rise in prices, and then inflation could be brought "down to the level of developed nations." They also see a need to limit relative prices, since their rapid approach to world levels would destroy most of Russia's processing industries.

Under the Glaz'ev/GKPP program the state would finance four main directions of industrial activity: (1) Selective development of existing focal points of growth and creation of new ones able to stimulate demand;[20] (2) An active policy on the federal and local level regarding bankruptcy and critical review (*sanatsii*) of individual enterprises evaluating their form of property, individual conditions, and relationship to other enterprises; (3) Social support equal to the rate of inflation, while shifting the population to more productive activity through requalification, relocation, and similar assistance; and (4) Measures to critically evaluate production and to carry out restructuring and the retraining of personnel, including development of the system of education.

The GKPP program proposes a financial policy of continuing budget deficits at the level necessary to finance required tasks. Credit policy should allow for modernization of enterprises on the basis of privileged credit.

The difference between the second and third scenarios, according to the authors, is that under the second (government) variant, inflation is "destructive," while under the third variant there is a place for "constructive inflation, which becomes a source of financing for measures to change the structure of the economy, so that in the end inflation is suppressed." While Glaz'ev and his colleagues clearly believe that the third variant is preferable, they do note some of the obstacles to its implementation: "One of the basic limitations on realizing the third variant could be the inadequate personnel available and the limited political will for adopting an active and goal-directed policy." The second variant requires less effort.

Thus, the Glaz'ev/GKPP group concedes that the most likely scenario is fluctuation between a strict anti-inflation policy and an adaptive policy. But at a minimum they would reorient state investment priorities. In 1993 state investment policy was not used to support what they consider the productive sector: more than 70 percent of investment went to the fuel-energy and agro-industrial complex. Not more than 10 percent went to machinebuilding, and less than 2 percent to light industry.

If the Glaz'ev/GKPP program does a good job identifying the contradictions in Russia's economic condition and the weaknesses in the government's program, it also has its share of both contradictions and weaknesses. It mixes what might be termed "mainstream" industrial policy proposals with dubious economic formulations.[21] And it demonstrates no awareness of recent cautionary notes about industrial policy.

The Gaidar Program

Architects of the Russian government's anti-crisis economic program did not devote much attention to industrial policy. Yegor Gaidar and Boris Federov were solidly in the camp of those believing that the state should get out of the way. But despite criticisms claiming that it is fostering the deindustrialization of Russia, the Gaidar program did emphasize creating conditions to stabilize production.[22] The Gaidar team believes this can be done through tax, investment, and anti-monopoly policies, institutional and organization reforms dealing with the non-payment crisis, some protectionist measures, and policies focused on ecological issues. In sharp contrast to Glaz'ev, Gaidar advocates a drastic withdrawal of state funds from the economy. Gaidar's group proposes to relieve the tax pressure on the economy by reducing spending on state administration by 20 percent, and also by cutting expenditures for defense, subsidies, and capital investment. The one place Gaidar's program proposes an increase is in spending for social welfare programs and science, which would be raised to 2–3 percent of GDP. Otherwise, the basic mechanism for expending state resources should be issuing securities, with special credits made available only in extreme circumstances.

Gaidar's team argues that reducing taxes in the hope this will spur the economy will not be successful: it would mean that the government would have to finance a larger deficit, either through inflationary emissions or by foreign borrowing. Russia's foreign debt is already enormous, and raising money on the internal market is increasingly expensive. The steps they recommend to rationalize the tax system include regularizing regional and local taxes and prohibiting local governments from imposing any additional taxes on profits. At the same time, real estate taxes should be put under local jurisdiction and used to finance local needs.

The number one priority of the Gaidar group regarding investment is to increase it, particularly in the competitive industrial sectors. They argue that investors must hold a share of the property involved, and state investments must be made solely on a competitive basis. Tax incentives should be offered for reinvestment of profits.

Gaidar's program concedes that even with a complete rationalization of the system, it is difficult to imagine a solution based solely on Russian resources. Short of instituting a "national regime," which would probably not provide adequate investment to alleviate the crisis or restructure industry, it will be imperative to offer privileges for foreign investment. Russia will need some $20–30 billion of investment per year for the next several years: just to stabilize the oil industry will require $6–10 billion per year during 1995–97. Since the Duma has capped foreign debt at $4.7 billion, Gaidar and his colleagues see the sole recourse in foreign direct investment.

Gaidar's program also sees a special threat in the FIGs.[23] Most of the FIGs are based on former ministries or on regional associations supported by local governments, and focus on state support, continued state programs, and various types of state privileges, particularly tax breaks.[24] The greatest danger of the FIGs is that they will "lock in" current structures, which are either completely unreformed or at best transitional. If permitted to develop they would solidify existing property relations and cooperative arrangements, with no significant restructuring or optimization. While branch concentration in Russia is not greater than in the "developed" nations and large firms can facilitate technological development, the government should be strict with anti-competitive activity by cartels and other post-privatization combinations. There will be a certain number of "natural" monopolies in areas like transportation, utilities, and communication, and these should be separated from large financial-industrial structures and regulated by independent agencies free from ministerial tutelage.

Finally, Gaidar's program calls for addressing Russia's massive environmental problems by making pollution charges a part of enterprises' "natural" costs, along with electricity, natural resources, etc. Without such a policy, there seems to be little way to reverse industry's long-standing disregard for environmental consequences.

The Western Debate

If the debate within Russia is at times opaque, confusing, and conducted in a dialect that includes code words not always fully understandable by outsiders, the "proxy" debate that has developed in Western academic and policy circles is more transparent. Much of the criticism of shock therapy has been articulated by those who think the costs have been excessive, and at times needlessly so. I will focus on two strains of this criticism, which I have labeled the social democratic and the neo-orientalist.[25]

The social democratic argument is both economic and political. The economic critique begins from the stance that markets are imperfect, and therefore wise state intervention is necessary to remedy the defects. The political argument is that neo-liberal structural reforms have often been rammed through in ways seriously damaging to democracy.

The social democrats attack the Washington consensus for overemphasizing markets and ignoring the role of the state: "the state played a major role in all cases of successful development, including Western Europe and the Far East, by mobilizing savings, providing infrastructure, shaping sectoral priorities, and in many cases forcing individual agents to engage in market-oriented activities through taxation" (Bresser Pereira et al., 6). In their view, stabilization and increased competition by themselves do not guarantee growth: it is necessary "to coordinate individual actions toward intertemporal efficiency and other normatively desirable and politically desired goals." Or, as they summarize the case, "market orientation is not sufficient to generate market coordination toward collective prosperity" (Bresser Pereira et al., ibid.).

On the political side, the social democrats see unpopular and destructive economic policies as a serious threat to democracy:

> . . . since the neo-liberal strategy entails significant social costs and hence political opposition, reforms tend to be initiated from above and launched by surprise, independent of public opinion and without the participation of such organized political forces as there may be. They tend to be adopted by decree or rammed through legislatures without modifications to reflect the divergence of interests and opinions. The political style of implementation tends to be autocratic; a government seeks to demobilize its supporters rather than compromise its program by public consultation. In the end, the society is taught that it can vote but not choose; legislatures are trained to think that they have no role to play in policy elaboration; the nascent political parties, trade unions, and other organizations are taught that their voices do not count. (Bresser Pereira et al., 9)

The social democrats' argument that markets do not clear and require a guiding hand to encourage savings understates the difficulty of implementing the policies they propose. Bresser Pereira and his co-authors do note that an industrial policy must be "judicious and carefully targeted," but they do not explore how realistic such a prescription might be. Aslund (1994) has discussed the difficulties of implementing an industrial policy in the conditions prevailing in Russia:

> The necessary information for any structural policy is not at hand. Nor do state officials possess the necessary skills, and they are neither motivated, responsible nor honest. Any structural policy would be costly, but the scarce funds available should either be invested in viable private enterprises or infrastructure or given to suffering citizens. Under such conditions, a great belief in the competence of the state is necessary for any advocacy of industrial policy.

Aslund's concerns would appear amply justified by evaluations of the "Asian miracle" policies:

> The use of contests in Japan and Korea required competent and insulated civil servants. In parts of Sub-Saharan Africa and Latin America, and elsewhere in Asia where such institutional conditions are lacking, activist government involvement in the economy has usually gone awry. (World Bank 1995, 26)

Proponents of shock therapy would point to how badly awry activist government policies went in the USSR and caution the social democrats to exercise extreme caution in mandating a state role in economic policy.

The neo-orientalist critique sees the problems in the post-Soviet space as stemming from these nations' "copying the wrong capitalist model" (Amsden et al., 1995). In place of laissez-faire liberalism, the model should be the Asian experience, which "represented an attempt to customize a capitalism to suit the realities and exigencies of having to catch up with technologically advanced countries in a world economy that espoused free competition but exhibited pervasive oligopolistic business tendencies" (Amsden et al., ibid.).

To the "market fundamentalism" chosen by (or imposed on) Eastern Europe, the neo-orientalists oppose a new institutionalism:

> The market mechanism in the East Asian case, by contrast, has been relied on instrumentally rather than as a matter of ideological conviction; institutions—including a bureaucratic, 'developmental' state—assume a major role in allocating investment resources. Institution building in this context refers not just to the establishment of clear property rights and contract law but also to the creation of private and public organizations capable of carrying out expansionary macroeconomic policies as well as investment, trade, competition, and technology policies, all operating under the umbrella of what has loosely come to be called industrial policy. (Amsden et al., ibid.)

Like the social democrats, the neo-orientalists avoid discussion of what post-Soviet governments might be capable of administering in a reasonable way. They vastly overestimate both the potential of state enterprises and the quality of the human capital accumulated under the Soviet education system. They make the same error as holdover Soviet administrators who talk about "science-intensive production" and yearn for investment in obsolete machine tool plants. Most of Soviet industry was badly conceived, poorly managed, and economically inviable in real prices. The planners and managers were economically illiterate and abysmally unaware of industrial sociology. Whereas learning physics and chemistry might have helped one attain a sinecure in an oversized Academy institute doing fundamental research, it did not help managers design or market products that could compete with the very Asian tigers they are now being told to emulate.

The neo-orientalists share with the social democrats a moderately positive assessment of the achievements of socialism. This would appear to be in part ideological, and in part a function of their emphasis on Eastern Europe. They conclude that "on its own terms socialism *did* function" (Amsden et al. 1995, 23, emphasis in original). Yet the Soviet planning system was not merely a sub-optimal economic system, but downright life threatening (Graham, 1993; Feshbach, 1992). At bottom, the argument is more about what a given state is capable of doing rather than what it should be doing. Aslund (1994) notes that the state lacks the capacity to do everything that it is called upon to do in the transition, even without trying to create a new, "leaner and meaner" version of S&T coordination. The USSR sought for more than three decades to establish a coordinated S&T policy, and never succeeded (Balzer, 1989; Nolting, 1979).

Beyond ideological inclinations and different evaluations of the capabilities of administrators, the debate between proponents of shock therapy and social democracy rests on sharply differing views of the state-owned enterprises and their managers. Like Glaz'ev, the social democratic authors portray managers in almost heroic terms, striving to save their firms from the depredations of incompetent and venal government officials.[26] Gaidar's supporters see a managerial stratum addicted to state credits and resistant to restructuring until they are driven to it by stark necessity, or driven out of their positions by more market-oriented owners. Clearly, not all managers belong to either category, but Gaidar's view strikes me as an accurate depiction of the majority. And the Gaidar view is certainly more accurate in assessing the prevailing ethics.

Pervasive corruption presents an additional obstacle to implementing an effective industrial policy. The USSR was the world's first kleptocracy.[27] During the decade of perestroika, "official" corruption and the second economy began to merge with new forms of criminal behavior, a phenomenon that gathered momentum as new varieties of economic activity developed and authority evaporated. Organized crime in Russia has already sunk deep roots, in many cases beginning with Soviet-era economic controls and continuing in myriad varieties of licensing, banking, export, and other realms where the government plays a role in the economy (Handelman, 1995). As Anderson (1995, 356–360) convincingly argues, this is not a transitional phase of robber barony on the way to capitalism: it is a full-fledged mafia phenomenon, and "mafias tend to be extraordinarily persistent." Inventing additional direct roles for the government in this context is much more likely to foster rent-seeking behavior than "judicious and carefully targeted" industrial policy.

Conclusion

It has proven impossible to organize a coherent administrative operation capable of formulating an industrial policy for Russia. This essay takes its title from the efforts of at least six ministry-level administrative departments to articulate an industrial policy. Absent the political will to coordinate action in the national interest, the result is an inability to plan at a level above that of individual firms and, perhaps, FIGs. Even if a coherent administrative system could be established, the information available to economic policymakers is hardly adequate to formulate an intelligent structural policy.

Because the system is utterly riddled with corruption, there has been resistance by reformers to anything that increases the role of the state. No one should have expected liberalization of economic relations in conditions of political chaos to result in a lessening of the role of illegal activity. In 1992 Boris Federov's advisors opted to eliminate state involvement in as many areas of economic activity as possible as a way to reduce opportunities for corruption.

The post-Soviet environment is beset with an overwhelming scientific-technical hubris. Russians insist that the USSR built a world-class R&D establishment, and that Russia inherited almost all the best portions of that establishment. Yet most studies suggest that far from achieving revolutionary proportions, scientific and technical progress in the USSR was slowing down (Kontorovich, 1992;

Bergson, 1978, 1983). Although Russia is heir to a gigantic R&D system, the very size of that system is part of the problem in the transition. It was of a magnitude that could not possibly be supported by a "normal" country paying real-world prices for the personnel and resources involved (Kontorovich, 1991).

In the euphoria following the collapse of the USSR, many political and economic problems that should have been addressed were ignored—it seemed that everyone was a democrat and the market would take care of itself. To the extent that a coherent economic policy was needed, it was considered to require rapid marketization, anti-crisis measures, and a concerted effort to eradicate the massive system of state involvement in economic activity.

Yet this policy was founded on a political and economic optimism that proved premature. Failure to hold elections and consolidate political parties opened the door to creeping authoritarianism, personalism, and clientelism in the political system. In the economy, rapid inflation, declining purchasing power, and absence of investment resources fostered a massive and continuing decline in production. In a climate where little is being produced and most firms are focused on basic survival, few see the need or capacity for fostering technology.

Concern about deindustrialization and belief in advanced technology are creating enormous pressure to adopt an industrial policy (Lomanov, 1994; Sapir, 1995). Yet the training and previous experience of most industrial administrators raise concerns about their economic knowledge and suggest that it will be extremely difficult for them to resist using industrial policy to bail out failing enterprises. Some commentators have suggested a focus on regional industrial policies, allowing local authorities who are closer to the situation to foster regional economic networks.[28] But regional authorities are also more directly affected by the problems of unemployment, and some regions have sought to organize their local economies in what amount to mutual assistance societies to avoid restructuring and its attendant pain (Starodubrovskaya, 1995; Prokop, 1995).

A good example of the powerful if muddled appeal of technotopian logic can be found in an article by one of Yeltsin's advisors:

> . . . it is necessary to reconsider the economic conceptions underlying our state economic and scientific-technical policies. I do not deny the importance of such ideas as "the market," "inflation," "securities," "capital," "macroeconomic regulation," and so forth. But to make them priorities, the basic building blocks of economic theory, is the result of a specific historical experience and economic models that have proved their effectiveness in far from all the nations of the world. I think that taking into account Russia's specific conditions, creating a multi-structured market economy in the extremely unique situation of a transition from complete centralization to a decentralized economic system in a situation of general crisis and the need for reindustrialization, the basic understanding necessary for our conditions and for a nationally oriented economic theory must be "knowledge," "high technology," "rational expediency," "innovation," "modernization," "stable development," and "social welfare." Other economic categories that form the basis of currently fashionable

> theories must be viewed as following from these fundamentals. In this case "inflation," "goods deficit," or "market saturation," "tax," etc. take on their genuine character as reflections of the processes and mechanisms impeding or aiding Russia's transmutation into an economic superpower on the basis of the creation and introduction of the most advanced science-intensive technology. (Rakitov, 1995)

Ultimately, the social democratic variant articulated by a growing number of Western economists and by the Russian proponents of industrial policy may prove too attractive to resist. Some variant, involving a prominent role for corporatist structures, such as the FIGs, is emerging as a likely stage in Russia's near-term evolution. It could come about either as a result of the next parliamentary elections, or as a "compromise" between Yeltsin's administration and some portion of the different industrial lobbies. If such a program is implemented, the crucial indicators to watch will be the mix of "science intensive" as opposed to "appropriate" technology and the extent to which the bailouts are tied to a policy of restructuring that supports a reasonable number of deserving, potentially successful innovators.

With the plethora of important economic tasks facing Russia's economists, "getting the fundamentals right" appears far more important than trying to pick winners or foster key industries. Stabilization, privatization, and taxation policy constitute a daunting agenda. Policies that encourage savings and long-term investment would seem to hold more promise than attempting to deal with specific industries or enterprises. Given the Soviet technotopian legacy and Russia's current conditions, the only thing worse than not adopting an industrial policy could be introducing this type of policy.

Notes

1. The best definition of industrial policy is provided by Johnson, 1982. For the American variant see Branscomb, 1993.

2. The literature on this subject is large and growing. For examples see Balzer (1989); Graham, ed. (1990); Stites (1989), Ch. 7.

3. The mythology is of two types. One aspect concerns the technology itself: a belief that Russia possesses all sorts of advanced products and processes that will be easily marketable. When it has turned out that most of these examples of advanced technology were either not so advanced or not commercially competitive, some Russians have blamed the "outsiders" for denigrating their superior technical capabilities.

The second myth involves the millions of technical specialists in Russia. Many of these individuals are highly skilled, but few of them know much about commercial development, marketing, advertising or the other business skills required to successfully enter competitive markets. Many Russian technical specialists have found that their technical capabilities made them employable in Western companies where all the business acumen already existed, and they could deal with technical tasks only. But no Western companies have recruited management personnel in Russia.

4. See the remarks by Boris Saltykov and Dmitry Piskunov in *Science, Technology and Industrial Policy in the Former Soviet Union: A Conference Report*. Washington, D.C.:

Occasional Papers of the Georgetown University Russian Area Studies Program, No. 1, September 1992. For a recent discussion of the issue, see Anatolii Sautin, "Bez protektsii gosudarstva u otechestvennoi nauki net budushchego." *Finansovye izvestiia*, January 12, 1995, p. 8.

5. "Kto rulevoi v otraslevoi?" *Poisk* 293, no. 51 (24–30 December, 1994): 1.

6. When we were organizing the International Science Foundation, it was necessary to emphasize repeatedly that foreign funding of basic research would entail no claim on patentable results, or any other proprietary claims. Despite repeated assurance on this count, criticisms persist in the Russian press suggesting that Soros, the Americans, or unidentified "others" have some mercenary purpose in providing such aid.

7. Statistical data on employment in science are difficult to obtain and of questionable accuracy. Many individuals remain officially on the rolls of research institutions even though they perform no work. Not only does this permit the employees to retain rights to social benefits and perhaps even salaries if the government restores elements of the old system, but it also permits the institutions to avoid paying the excess wage tax if those who continue to work receive above-average salaries.

8. One of the few works to approach questions of conversion in a comparative framework is Alexander, 1990.

9. S. Elekoe, "Poslednyi kozyr'," *Sotsialisticheskaia industriia*, 28 January, 1989, p. 2. For analysis see Balzer, 1989; Cooper, 1991; and Malleret, 1992.

10. For a good summary of policies before 1993, see Filatochev et al., 1993. The discussion here also draws on Balzer, 1989: 198–202; Cooper, 1991; Gates, 1995; Kuznetsov, 1994; and Sapir, 1995.

11. Formally, while only four of these bodies are called ministries, all but one have the status of ministries in the Russian government. The GKPP has the rights of a ministry, as does the State Committee for Higher Education. *Goskomoboronprom*, while called a "state committee," lacks the ministerial prerogative to send legislation to the legislative branch.

12. Sergei Glaz'ev was an economist who wrote about R&D prior to the collapse of the USSR. He served as Deputy Minister of Foreign Economic Relations, and subsequently as minister. He was the only member of the government to resign following the October 1993 crisis. This did much to enhance his political standing, and he was elected to the Duma in December 1993. He currently chairs the Duma's Committee on Economic Affairs, and has used this post to criticize the government's economic proposals.

13. During visits to the GKPP, I was assured that the military agencies "participate when it is important for them to do so." It appears that a significant amount of informal coordination has been provided by discussions between Glaz'ev and Deputy Minister of Defense Kokoshin, who is also an advocate of industrial policy.

14. This section is based on GKPP, 1993a, and 1993b.

15. Processing industries (*obrabativaiushchie otrasly*) are discussed separately from high technology, reinforcing the impression that the latter is frequently a euphemism for the defense industry.

16. For perspective, at that time *Goskomoboronprom* had about 1,000 personnel; the Ministry of Machine Building 850; and the Ministry of Metallurgy 250. If the investment departments of the Ministry of Economy were transferred to GKPP, it would have a total staff of about 1,000.

17. Interview with Dmitri I. Piskunov, deputy chairman of the State Committee for Industrial Policy, Moscow, March 1994.

18. See the two reports (GKPP, 1993a and 1993b) by the Analytical Center for Scientific and Industrial Policy prepared for MinNauka, GKPP and the Russian Academy of Sciences. A recent article in *Ekonomicheskaia gazeta* (No. 6, 1995) appears to be a summary of the views in the report *"Tekhnologicheskie i institutsional'nye izmeneniia v ekonomike."*

19. "Tri stsenariia vykhoda iz krizisa." *Ekonomicheskaia gazeta* no. 6, 1995, p. 9.

20. This is a typical example of the persistence of "old thinking." In the view of these policy specialists, demand is something the state and the enterprise must stimulate rather than something to which the enterprise responds.

21. The Glaz'ev program covers the essentials of what would be a genuine industrial policy. There are two questions: first, whether the Russian government is capable of implementing such a policy, with the available personnel and under current conditions; and second, whether it could be done without undermining the modest successes achieved thus far in macroeconomic stabilization. The government's record in dealing with S&T is thus far not encouraging. And the proposals are couched in dubious reasoning about "creative" inflation and the need to have budget deficits dictated by spending needs rather than available resources.

22. On the government's economic program and role of Gaidar and Federov see Aslund, 1995; Mau, 1995. The recent Gaidar industrial policy program is outlined in "Komissiia po promyshlennoi politike pri politsovete DVR. Institut ekonomicheskikh problem perekhodnogo perioda." *Programma promyshlennoi politiki DVR* (Moscow, 1995). The principal author was Yu. N. Bobylev.

23. The number of officially registered FIGs is small, due to the stringent requirements and limitations imposed by the government. There are probably more de facto FIGs operating not in accord with the existing legislation. (See Starodubrovskaya, 1995; Freinkman, 1995; Prokop, 1995.)

24. It is difficult to resist forming the impression that these new structures bear an uncanny resemblance to the Interbranch Scientific-Technical Complexes (MNTKs), Associations, and some of the other institutional innovations of the 1970s and 1980s. Particularly when coupled with suggestions to restore a fixed exchange rate and regulate some industrial prices (Glaz'ev; Sapir, 1995: 160–161), it is not difficult to imagine these large entities becoming new sources of informal mechanisms to avoid market influences. At a minimum, they are invitations to anti-competitive and even predatory behavior.

25. Leading examples of these two groups of critics are Bresser Pereira et al., 1993; and Amsden et al., 1995. Bresser Pereira et al. call their approach "pragmatic" or "social democratic." The pragmatism claim is reminiscent of the attempt by Cold War hawks to call their policies "realistic," which suggested that less hard-line approaches were unrealistic. The "social democratic" label comes closer to truth in packaging.

Both the social democratic and neo-orientalist critiques gloss over the difference between initial industrialization of agricultural and craft-based economies compared with restructuring an enormous, antiquated, and inefficient industrial base.

26. An argument more sympathetic to the managers' point of view is presented in Sapir's contribution in this volume.

27. I first heard the phrase from the late Hugh Seaton-Watson. It has been used in print by Grossman and Simis.

28. This approach is implicit in Anderson (1995), and is reflected in the reorganization of MinNats as Ministry of Regional and Nationality Policy in autumn 1993.

Bibliography

Alexander, Arthur. "Defense Industry Conversion in China, the Soviet Union and the United States." Ms., 1990. Subsequently published as "National Experiences: A Comparative Analysis." In *Disarmament. Conversion: Economic Adjustment in an Era of Arms Reduction.* New York: United Nations Department for Disarmament Affairs, 1991.

Alimova, Tat'yana, Vladimir Buev, and Victoriya Golikova. "Problems of Small Business Formation in the Transition Period." *Communist Economies & Economic Transformation* 7, no. 1 (March 1995): 67–82.

Amsden, Alice H., Jacek Kochanowicz, and Lance Taylor. *The Market Meets Its Match: Restructuring the Economies of Eastern Europe.* Cambridge: Harvard University Press, 1994.

Anderson, Annelise. "The Red Mafia: A Legacy of Communism." In *Economic Transition in Eastern Europe and Russia: Realities of Reform,* edited by Edward P. Lazear, 340–366. Stanford, CA: Hoover Institution Press, 1995.

Aslund, Anders. *How Russia Became a Market Economy.* Washington, D.C.: The Brookings Institution, 1995.

———"The Role of the State in the Transition to Capitalism." In *Legacies of the Collapse of Marxism,* edited by John H. Moore, 181–197. Fairfax, VA: George Mason University Press, 1994.

Balzer, Harley. *Soviet Science on the Edge of Reform.* Boulder, CO: Westview Press, 1989.

———"Is Less More? Soviet Science in the Gorbachev Era." *Issues in Science and Technology* 1, no. 4 (1985): 29–46.

Battilega, John A. "A Case Study of Russian Defense Conversion and Employee Ownership: The Saratov Aviation Plant." In *Privatization, Conversion, and Enterprise Reform in Russia,* edited by Michael McFaul and Tova Perlmutter, 169–187. Boulder, CO: Westview Press, 1995.

Bergson, Abram. *Productivity and the Social System—the USSR and the West.* Cambridge: Harvard University Press, 1978.

———"Technological Progress." In *The Soviet Economy: Toward the Year 2000,* edited by Abram Bergson and Herbert S. Levine, 34–78. London: George Allen and Unwin, 1983.

Bernstein, David. "Spin-offs and Start-ups in Russia: A Key Element of Industrial Restructuring." In McFaul and Perlmutter, 201–215.

Branscomb, Lewis M., ed. *Empowering Technology: Implementing a U.S. Strategy*. Cambridge: MIT Press, 1993.

Bresser Pereira, Luiz Carlos, Jose Maria Maravall, and Adam Przeworski. *Economic Reforms in New Democracies: A Social-Democratic Approach.* Cambridge and New York: Cambridge University Press, 1993.

Cooper, Julian. *The Soviet Defense Industry: Conversion and Economic Reform.* New York: Council on Foreign Relations Press, for the Royal Institute of International Affairs, 1991.

Cronberg, Tarja. "Enterprise Strategies to Cope with Reduced Defense Spending: The Experience of the Perm Region." In McFaul and Perlmutter, 189–200.

Fal'tsman, Vladimir. "The New Role of Defense R&D in the Russian Economy." In Rowen, et al., 83–98.

Feshbach, Murray, and Alfred Friendly, Jr. *Ecocide in the USSR: Health and Nature Under Siege.* New York: Basic Books, 1992.

Filatochev, Igor, Trevor Buck, and Mike Wright. "The Military-Industrial Complex of the Former USSR: Asset or Liability? *Communist Economies & Economic Transformation* 5, no. 2 (1993): 187–204.

Freinkman, Lev. "Financial-Industrial Groups in Russia: Emergence of Large Diversified Private Companies." *Communist Economies & Economic Transformation* 7, no. 1 (March 1995): 51–66.

Gansler, Jacques S. *Defense Conversion: Transforming the Arsenal of Democracy.* Cambridge: MIT Press, A Twentieth Century Fund Book, 1995.

Gates, Susan. "Defense Conversion and Economic Reform in Russia." In *Economic Transition in Eastern Europe and Russia: Realities of Reform,* edited by Edward P. Lazear, 367–388. Stanford, CA: Hoover Institution Press, 1995.

GKPP, MinNauka, RAN. *Sotsial'nye osnovy novoi promyshlennoi politiki Rossii: Analiz, prognozy, predlozheniia Analiticheskii doklad (Sokrashchenyi variant)* Moscow: Analiticheskii tsentr po nauchnoi i promyshlennoi politike, (December) 1993a.

GKPP, MinNauka, RAN. *Tekhnologicheskie i institutsional'nye izmeneniia v ekonomike. Analiticheskii doklad.* Moscow: Analiticheskii tsentr po nauchnoi i promyshlennoi politike, (December) 1993b.

Gomulka, Stanislaw. "Economic and Political Constraints During Transition." *Europe-Asia Studies* 46, no. 1 (1994): 89–106.

Gorbatova, Larisa. "Formation of Connections between Finance and Industry in Russia: Basic Stages and Forms." *Communist Economies & Economic Transformation* 7, no. 1 (March 1995): 21–34.

Graham, Loren R. *The Ghost of the Executed Engineer: Technology and the Fall of the Soviet Union.* Cambridge: Harvard University Press, 1993.

———, ed. *Science and the Soviet Social Order.* Cambridge: Harvard University Press, 1990.

Handelman, Stephen. *Comrade Criminal: Russia's New Mafiya.* New Haven: Yale University Press, 1995.

Islam, Shafiqul. "Conclusion: Problems of Planning a Market Economy." In *Making Markets: Economic Transformation in Eastern Europe and the Post-Soviet States*, edited by Shafiqul Islam and Michael Mandelbaum, 182–215. New York: Council on Foreign Relations, 1993.

Johnson, Chalmers. *MITI and the Japanese Miracle: The Growth of Industrial Policy.* Stanford, CA: Stanford University Press, 1982.

Kontorovich, Vladimir. "Technological Progress and Research and Development." In *The Disintegration of the Soviet Economic System,* edited by Michael Ellman and Vladimir Kontorovich, 217–238. New York and London: Routledge, 1992.

———"The Future of Soviet Science in a Normal Civilized Society." Ms. presented at a conference at Birmingham University, 1991.

Kosals, L. Ia. *Oboronnye instituty v usloviiakh peremen. Doklad po materialam vyborochnykh sotsiologicheskikh oprosov rukovoditelei.* Moscow: mimeo, 1993.

———*Why Doesn't Russian Industry Work?* Translated by John Crowfoot, Igor Poluyan, and Veleriya Sedova. London & New York: I. B. Tauris Publishers, 1994.

Kozminski, Andrzej. "The Main Issues of Industrial Policy for Poland." *Communist Economies & Economic Transformation* 4, no. 2 (1992): 237–248.

Kuznetsov, Yevgeny. "Adjustment of Russian Defence-related Enterprises in 1992–94: Macroeconomic Implications." *Communist Economies & Economic Transformation* 6, no. 4 (1994): 473–513.

Lazear, Edward P. "Economic Reform: Appropriate Steps and Actual Policies." In *Economic Transition in Eastern Europe and Russia: Realities of Reform,* edited by Edward P. Lazear, 3–34. Stanford, CA: Hoover Institution Press, 1995.

Lomanov, G. "Deindustrializatsiia rossii nabiraet tempy." *Inzhener*, no. 9 (1994): 1–3, 12–13.

Malleret, Thierry. *Conversion of the Defense Industry in the Former Soviet Union.* New York, Prague: Institute for East-West Security Studies. Occasional Paper Series No. 23, distributed by Westview Press, 1992.

Mau, V. A. *Ekonomika i Vlast': Politicheskaia istoriia ekonomicheskoi reformy v Rossii, 1985–1994.* Moscow: Delo DLT, 1995.

McFaul, Michael, and Tova Perlmutter, eds. *Privatization, Conversion, and Enterprise Reform in Russia.* Boulder, CO: Westview Press, 1995.

Nolting, Louvan E. *The Structure and Functions of the USSR State Committee for Science and Technology.* Washington, D.C.: U.S. Department of Commerce Foreign Economic Report No. 16, 1979.

OECD. *Science, Technology and Innovation Policies: Federation of Russia.* Paris: OECD. Volume One: Evaluation Report, 1993.

O korennoi perestroike upravleniia ekonomiki. Sbornik dokumentov. Moscow: Izdatel'stvo politicheskoi literatury, 1987.

Ozhegov, Aleksandr. *Innovatsionnyi potentsial konversii.* Moscow: Analytical Center RAN, April 1993.

Perekhod k rynku: Kontseptsiia i programma. Moscow: Arkhangelskoe, 1990.

Prokop, Jane. "Industrial Conglomerates, Risk Spreading and the Transition in Russia." *Communist Economies & Economic Transformation* 7, no. 1 (March 1995): 35–50.

Rakitov, Anatoly. "Vyshe planku!" *Poisk*, no. 27–28 (321–322) (1–14 July 1995): 3.

Rowen, Henry S., Charles Wolf, Jr., and Jeanne Zlotnick, eds. *Defense Conversion, Economic Reform, and the Outlook for the Russian and Ukrainian Economies.* A RAND Study. New York: St. Martin's Press, 1994.

Sapir, Jacques. "Conversion of Russian Defense Industries: A Macroeconomic and Regional Perspective." In Michael McFaul and Tova Perlmutter, 137–165.

Schweitzer, Glenn E. "Can Research and Development Recover in Russia?" *Technology in Society* 17, no. 2 (1995): 121–142.

Starodubrovskaya, Irina. "Financial-Industrial Groups: Illusions and Reality." *Communist Economies & Economic Transformation* 7, no. 1 (March 1995): 5–20.

Stites, Richard. *Revolutionary Dreams: Utopian Vision and the Experimental Life in the Russian Revolution.* New York: Oxford University Press, 1989.

World Bank. *The East Asian Miracle: Economic Growth and Public Policy.* A World Bank Policy Research Report. New York: Oxford University Press, 1995 (1993).

Yaremenko, Yuri V., and Victor N. Rassadin. "The Experience and Prospects of Remodeling the Defense Industries in Russia." In Rowen et al., 99–113.

Yasin, Yevgeny G. "Market Reform and the Future of the Russian Economy." In Rowen et al., 115–133.

SECTION TWO

The Problems and Prospects of Technology Transfer

5

Inter-Organizational Technology Transfer: From Standard Technology Packages to Spin-Offs

David V. Gibson

The challenges of technology transfer now confront R&D activities at federal laboratories, research universities, and institutes worldwide as these institutions become more concerned with commercial technology applications of their research activities. Research laboratories and institutes, universities, and R&D consortia in the United States and worldwide are increasingly being judged in terms of technological applications profitably used in the global marketplace. Yet the demand for technology application is a relatively new and much debated mandate for federal defense R&D sites, in both the East and West. The experience with technology transfer in the U.S. private sector may provide a baseline model for government-funded institutions.

Technology Transfer Defined

The concepts of technology and of transfer are defined by both theoreticians and practitioners in many different ways. There is usually agreement, however, that (1) technology is not just a "thing," and (2) that transfer is a profoundly human endeavor. Essentially, technology is information that is put to use in order to accomplish some task, or the knowledge of how to do something.

Transfer is the movement of technology via some channel from one individual or organization to another. So technology transfer involves the application of knowledge, putting a tool to use. The transfer of technology is a particularly difficult type of communication in that it often requires collaborative activity between two or more individuals or functional units who are separated by structural, cultural, and organizational boundaries. Appreciation for the human component in technology transfer directs us away from thinking of simply moving technology from point "A" to point "B." Instead, successful technology transfer should be thought of as an interactive process with a great deal of back-and-forth exchange among individuals over an extended period of time.

Three major types of inter-organizational technology transfer—from research to application—directly influence industrial competitiveness (Figure 5.1): (1) spinning out technologies into start-up companies (the dashed line), (2) transferring technologies from research organizations to established firms (the solid line), and (3) spinning advanced technologies into the research activities of institutional laboratories, universities, and consortia (the dotted line). Types (1) and (2) are most concerned with shorter term capital gains and increased market share. Type (3) is most concerned with adequately leveraging R&D resources, and has important long-term implications for industrial competitiveness as well as national security.

Figure 5.1: Two Basic Forms of Technology Transfer to Commercial Applications

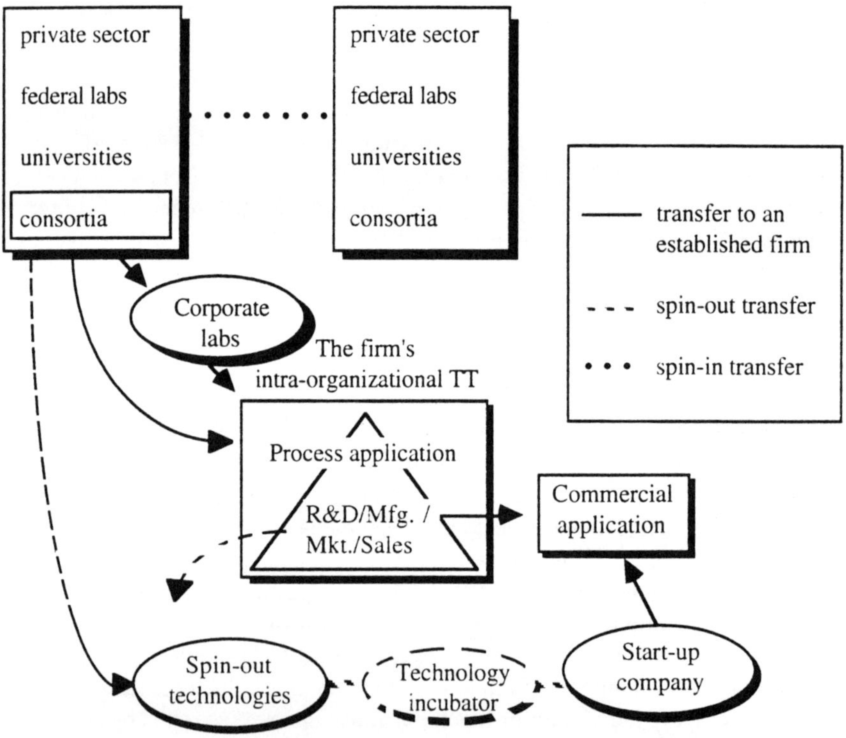

The United States is a successful role model for much of the industrialized world in spin-out technology transfer. Spin-out companies have been a vitally important if not dominant factor in the emergence and growth of the United States' two premier technopoleis or technology centers: Silicon Valley, California and Route 128, Massachusetts. Technology transfer across organizational boundaries of established firms is a seemingly more difficult challenge.

Nonetheless, there are many motivations for forming R&D consortia, including (1) to allow member firms to leverage their R&D investments, (2) to reduce the amount of duplicate research, (3) to promote long-term basic research, (4) to leverage costly and scarce intellectual resources and talent, (5) to better monitor proliferation of new technologies and the research activities of competitors, (6) to reduce risk by allowing participants to diversify their portfolio of research projects, especially given increasingly short product development cycles, (7) to increase the ability of smaller companies to compete with giants like IBM, AT&T, and NEC, and (8) to enhance corporate image by emphasizing access to state-of-the-art technology (Murphy, 1987; Fusfeld and Haklisch, 1985; Gibson and Rogers, 1988; Evan and Olk, 1990; Gibson et al., 1988).

The Serendipity of Technology Transfer

A communication-based model of technology transfer centers on information exchange between a technology source and a receptor as a two-way process (Gibson and Rogers, 1994). Such information exchange is typically not orderly or unidirectional. It is often chaotic and disorderly. Participants are "transceivers," who exchange ideas simultaneously and continuously, thereby blurring the distinction between senders and receivers. The technology being transferred is often not a fully formed idea that can be neatly packaged and forwarded. It has no inherent meaning or value: Meaning is in the minds of the participants. Accordingly, transmitters and users are likely to have different perceptions of the same technology. Feedback helps transceivers reach convergence about important dimensions of the technology. Successful technology transfer to product commercialization involves an ongoing, multilevel exchange of information.

There is a serendipitous aspect to technology transfer in that when researchers chart a course they often make unexpected discoveries. Users of a technology discover technology applications that were unintended. Researchers and users may combine in a synergistic way, one that could not be predicted nor managed, to produce unexpected results. Such technology transfer is a particular case of the "garbage can model" of decision-making (March and Olsen, 1976), which bridges the gap between the classic, rational model of organizational behavior and the frequent lack of linear rationality, efficiency, and predictability in real organizational life. A transferred technology results from an unplanned mixture of participants, solutions, choice opportunities, and problems. Both problems-looking-for-solutions (technology pull) as well as technology-solutions-looking-for-problems (technology push) are encountered.

This analysis is based on a ten-year study of the Microelectronics and Computer Technology Corporation (MCC), an important experiment in the management of inter-organizational technology transfer. MCC was the first major U.S. effort to get private corporations to collaborate in research while they competed in the marketplace. In 1983, this R&D consortium challenged U.S. antitrust law* and a long-standing belief in the value of free enterprise based on unfettered competition. In 1995, MCC is still on trial as a symbol of an emerging paradox in marketplace economies worldwide—the belief that for business enterprises to compete effectively in the global economy, they must collaborate with competitors as well as with external R&D activities in the public and private sectors.

Levels of Technology Transfer

MCC's experiences with technology transfer suggest four levels of collaborative activity, and four correspondingly different definitions of technology transfer success (Figure 5.2). At Level I, quality R&D, researchers conduct state-of-the-art, pre-competitive research, and transfer these results by such varied means as research publications, videotapes, teleconferences, and software computer tapes. Technology transfer at this level is a largely passive process that requires little collaborative behavior among the transceivers, although the researchers may work in teams or across organizational or even national boundaries. Traditionally, in the United States, technology users have not been involved at this level of the transfer process. Level I success is measured by the quantity and quality (usually through peer review) of research reports and journal articles. Technology transfer plans and processes are not considered very important. Research strength is most important. The belief is that good ideas sell themselves; pressures of the marketplace are all that is needed to drive technology use and commercialization.

During MCC's early years of operation, the work of the consortium's scientists was measured in terms of Level I activity. Indeed, these conceptions of technology transfer reflected (1) the norms and values that MCC scientists brought with them from university, federal, and corporate laboratories, and (2) the pre-competitive level of technology transfer mandated in the pre-competitive research limitation of the 1984 National Cooperative Research Act.

Level II technology transfer, technology acceptance, calls for the beginnings of shared responsibility between technology developers and users. Success occurs when a technology is transferred across personal, functional, or organizational boundaries, and it is accepted and understood by designated users. Moving from Level I to Level II technology transfer is an extremely difficult task for research organizations like MCC. Knowing the appropriate person to contact is an immense challenge as one looks at a large corporation like Lockheed or General Electric, or even at a much smaller company like Advanced Micro Devices. A Level II perspective encourages the belief that successful technology transfer is simply a matter of getting the right information to the right people at the right time.

---

* *The National Cooperative Research Act of 1984 granted antitrust protection to organizations pooling their resources for the development of new technologies. This authorized the formation of industrial consortia, which aimed to emulate the Japanese* keiretsu *system of tight industry linkages, in order to strengthen the development of strategic technologies.*

Figure 5.2: Technology Transfer at Four Levels of Involvement
Source: Gibson and Rogers, 1994

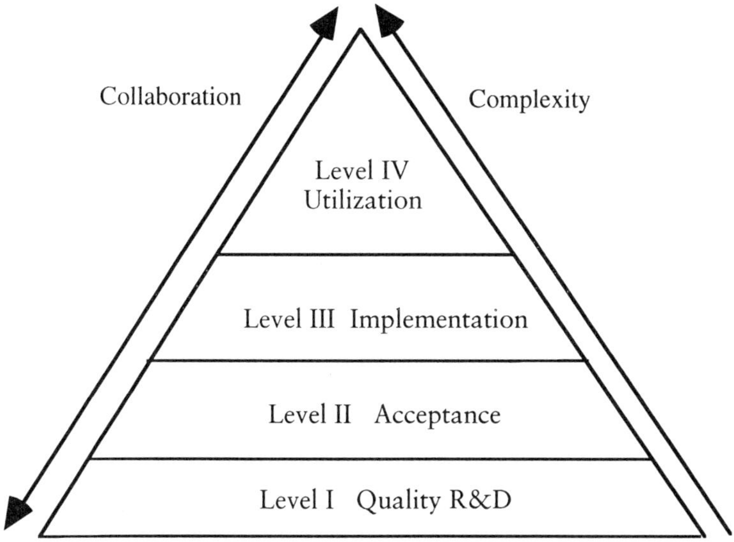

In Level III transfer, success is marked by the timely and efficient implementation of the technology. For Level III success to occur, technology users must have the knowledge and the resources needed to implement, or beta-test, the technology. Technology implementation can occur within the user organization in terms of manufacturing or other processes, or it can occur in terms of product development such as building a prototype for commercial application. Industrial-strength R&D is required.

Level IV transfer, technology utilization, centers on product commercialization. Level IV builds cumulatively on the successes achieved in attaining the objectives of the three previous stages, but market strength is required. Feedback from technology users drives the transfer process. Success is measured in terms of return on investment or market share. Here we take a longer-term view. It is for failure to achieve Level IV technology transfer that MCC has been most criticized.

Overall technology transfer success, Levels I to IV, is difficult to measure in terms of traditional cost/benefit analysis since (1) it is often difficult to quantify financial or other impacts of a technology over time, and (2) different persons involved in the process are likely to evaluate costs and benefits differently depending on their perspectives. In the case of MCC, the different member companies commonly hold different expectations as to which level of technology transfer they expect. Some member companies are happy with research reports, while others want market-strength products. Even within a member company, different managers (e.g., hierarchically) and functional areas (e.g., R&D, production, marketing, and sales) evaluate MCC's technology transfer activities by different criteria. For example, a CEO might be motivated by a new, bold concept, whereas a line manager in the same company wants technology that is of industrial strength and supported. For a scientist, success might be peer recognition or a journal publication, rather than the commercial application of his/her idea (Table 5:1).

Table 5.1: Collaborative Activity and Complexity for the Four Technology Transfer Levels

Level	Collaborative activity	Interorganizational complexity
I Quality of R&D	Technology users not involved	Trickle out over time
II Acceptance	Shared responsibility between technology developers and users	Right information goes to the right people at the right time
III Implementation	Technology users with the necessary knowledge and resources to implement or test the technology	Industrial strength and value-added ability
IV Application	Product commercialization	Market strength required

According to MCC's charter, the consortium's objective is to create and transfer generic technology that can be utilized by its shareholder companies to create new products. In keeping with this orientation, during the consortium's formative years MCC scientists generally held to a Level I perspective of technology transfer. Success was defined as (1) conducting high-quality, long-range research, and (2) making the results available in a timely fashion to the shareholder companies. The shareholder companies were to use pre-competitive MCC-developed technologies to create products of their own design and to compete in markets of their own choice. Market forces and competitive pressures were expected to drive the process.

Over time it has become apparent to MCC's managers and scientists that Level I measures of success, however impressive, would not sustain the consortium. Divisional and group managers increasingly demanded measurable technology benefits (Levels III and IV technology transfer). MCC faced a dilemma as technology transfer activities moved from a Level I to a Level IV perspective: The consortium exercised less control over events leading to successful technology commercialization as more collaboration was required across functional and organizational boundaries from a range of participants. MCC could develop a "silver bullet" technology, but if the member companies chose not to take it to industrial and market strength, MCC would be judged a failure.

There are also less tangible and less accepted criteria for judging MCC's success at technology transfer: learning from R&D failures while having the costs dispersed over a number of firms; small incremental wins; and support services provided.

Learning from failures is sometimes offered as a benefit of R&D consortia. If ten companies invest in a risky technology and it fails, then each company learns what not to do at one-tenth of the cost. Unfortunately, managers soon realized that "learning from failures" was not the way to promote the benefits of belonging to MCC; i.e., member companies were not keen on funding research, on any scale, that produced "failures" as a measure of success. MCC has also engaged in a great many collaborative research projects that resulted in seemingly imperceptible transfers of technologies to its member companies in the form of incremental improvements in member company technologies and internal processes. At times MCC researchers did not even realize they were transferring important technologies/ideas to member company engineers, researchers, and managers. Small and continuous transfers were often embedded in larger product or process technologies.

Finally, member companies have looked to MCC for support services, and particular product or process applications, that the member company could not, or would not, find elsewhere. MCC hosted teleconferences on cutting-edge research for its member companies, and it has organized seminars on important emerging technology areas. MCC's International Liaison Office provides valuable information on foreign competitors to MCC's member companies modeled on the technology scanning and utilization capabilities of their Japanese, Korean, and Taiwanese counterparts. Yet it would be difficult to place a precise value on how such information has helped MCC's member firms compete in the global marketplace.

Technology Transfer at MCC

Depending on the perspective of the evaluator, an assessment of MCC's success at technology transfer could be based on a range of criteria. The R&D consortium has published more than 2,500 technical reports; produced over 400 technical videotapes; "transferred" more than 200 technologies; published numerous research articles; and been issued or assured 117 patents and 182 technology licenses. Such quantifiable measures focus on research productivity, whereas most of the criticism of MCC has focused on research application and whether MCC's technologies have enhanced competitive activities within the shareholder companies.

Over the years, MCC used a variety of strategies and mechanisms to achieve its transferred successes. While each method has afforded unique advantages and challenges, four key issues emerge as key to successful technology commercialization across organizational boundaries: (1) interactive communication; (2) physical, cultural, and strategic proximity of members; (3) characteristics of different technologies; and (4) interpersonal motivation.

Interactive Communication

Interactive communication refers to the richness of exchange between technology transceivers—i.e., developers and users. Interactivity ranges on a continuum from passive, one-way media-based linkages such as research reports, videotapes, and computer tapes, to more interactive face-to-face linkages such as on-site research demonstrations and collaborative research projects. MCC has initiated a range of passive to interactive communication linkages between technology developers and users (Table 5.2).

Initially, MCC emphasized the use of passive modes of communication with its member companies through technology reports and videotapes. Passive communication demands less time and expense than more interactive modes of communication. Researchers can stay "at home" while transferring technologies to potential users. A large number of individual and organizational receptors can be reached at little cost. With such computer-mediated linkages as electronic mail and videoconferencing, researchers can talk directly to users about their technology. Such communication linkages diminish personal and professional risk for the technology developers, as it can be a challenging experience to go to the field to "sell" one's research. On the other hand, passive communication seldom elicits useful feedback from technology users.

MCC's various boards provided formal, structured meetings to facilitate technology transfer between MCC and its member companies. At these meetings, MCC researchers gave formal presentations, tutorials, and demonstrations about specific technologies. However, the right information was often not given to the right member company recipient at the right time for successful technology transfer to occur. Some member companies did not even send a company representative to these technology briefings.

To facilitate active technology transfer and viable communication, MCC focused particularly on the link between MCC and its shareholder companies through creating shareholder representatives. An MCC "visitor program" allowed shareholder scientists to spend time at MCC working on a collaborative research project. The shareholder representative was expected to carry MCC know-how back to his/her company and was assigned for a two to three year period to a particular research program at the consortium. These representatives were expected to participate in MCC research activities and to gain an in-depth knowledge of the available technologies. They were also expected to make periodic trips back to—

Table 5.2: Passive and Interactive Technology Transfer Mechanisms at MCC

Passive technology transfer
1. Proprietary technical reports
2. Refereed journal articles
3. Newsletters
4. Videotaped overviews
5. Videotaped demonstrations

Computer-based technology transfer
1. Video conferencing
2. E-mail consulting

Face-to-face technology transfer
1. MCC/shareholder meetings
    - Software Technology Advisory Council (STAC)
    - Technical Advisory Board (TAB)
    - Program Technical Advisory Board (PTAB)
2. Shareholder committees
    - Program Advisory Committees (PAC)
    - Technical Requirements Panels
    - Manufacturability Panels
    - Quality Assurance Panels
3. Shareholder representatives
4. Shareholders assignees
5. Visitors programs
6. Shareholder site demonstrations
7. Receptor organizations within shareholders
8. Shadow research projects within shareholders
9. Shareholder/MCC collaboration

and ultimately permanently return to—their home company in order to transfer this knowledge. Despite the soundness of the idea, MCC's conception of shareholder representatives as the premier means of technology transfer to the member companies has not worked out as planned.

First, MCC did not receive the expected number of quality shareholder representatives. Being assigned to MCC for two to three years was generally not viewed as an attractive career option for a fast-track company technician or researcher. Furthermore, shareholder companies were reluctant to give up their best researchers for a tour at MCC. As a result, shareholder representatives in residence at MCC were often not the most visible and respected company researchers, nor were they the most appropriate receptors for MCC technologies. While MCC could exercise control over the researchers it hired, the consortium had little control over the quality of a company's selection of shareholder representatives. As a result, often these shareholder representatives (1) knew little about their home company's technology priorities and product development, (2) were not able to contribute to MCC's research, and (3) had little clout in their home companies. Assigning a quality shareholder representative was perceived as a threat, by some shareholders, to proprietary knowledge and competitive advantage for the parent company. For example, if a shareholder representative was able to transfer an MCC technology to his/her parent company, and that company's research experiences were communicated back to MCC staff by the representative, other companies might gain valuable information or otherwise benefit from these technological advances. As it turned out, such concerns with protecting proprietary knowledge began to lessen in 1986. In our decade-long study of MCC, we found little meaningful scanning of a competitor's technologies for competitive advantage. In fact, many MCC researchers and managers reached the following self-evident conclusion: *Shareholder companies that participated most actively in MCC's research projects benefited the most from the technological advances.*

Over the years some very capable shareholder representatives have been assigned to MCC, and they have transferred technologies back to their home companies. But even the most capable representative found it a challenge to be knowledgeable about MCC's wide range of technologies and to keep abreast of his/her own company's vast research activities that might benefit from such MCC-produced technologies.

Physical, Cultural, and Strategic Proximity

The shareholder representative approach to technology transfer emphasized the importance of the MCC technology-based contacts and de-emphasized the importance of the representative maintaining strong, ongoing personal contacts with key employees in the shareholder company. MCC has found that the longer a shareholder representative stayed in Austin, the more out of contact the representative became with the personal networks in his/her home company that are so crucial to speeding the transfer of technology, especially to Level IV product commercialization. Similarly, occupational, or cultural, proximity to MCC technicians was also a liability for shareholders.

MCC initially thought that shareholder researchers of comparable training and technical skill would be the most appropriate receptors for an MCC-developed technology. Effective communication would be facilitated by knowledgeable experts talking with one another: Scientists talking to scientists. However, such homogeneous communication was hindered as MCC and shareholder scientists often competed for research funds and professional prestige. Over the years it has become increasingly apparent to MCC that other, more heterogeneous receptors (such as a shareholder's marketing, production, and sales personnel) may at times be the most appropriate receptors for an MCC produced-technology. MCC's successful cases of technology transfer have often involved shareholder production and marketing people who pulled the technology out of MCC.

Finally, MCC also has to contend with strategic differences among its 22 shareholders, and potential government and university receptors and sources of MCC technology. MCC shareholders differed by size, the number and location of divisions and their preferred markets, internal R&D expenditures, percentage of company annual sales going to MCC, and the particular industry represented. This diversity among MCC shareholders created significant challenges to technology transfer.

Technology Characteristics

MCC's research programs cover a spectrum from basic research to more tangible products and tools. What are the implications of such technological diversity for technology transfer? Is it more difficult to transfer basic research? Is software technology more difficult to transfer than hardware and other physical products? MCC's first chief scientist, John Pinkston, contended that "The more a technology can be encapsulated—the more the user only has to deal with the externals (as in the case of a tool), the easier technology transfer tends to be."

MCC's High Value Electronics (HVE) Division (formerly the Packaging/ Interconnect (P/I) Program) has transferred more technologies to its shareholders than all the rest of MCC's research programs combined. Packaging/Interconnect technologies are generally observable, measurable, and relatively easy to demonstrate. In contrast, the Information System Division's technologies (formerly Software Technology, Computer-Aided Design, and Advanced Computer Architecture Programs) tend to be more idea-based, with a variety of possible applications. The superiority of a tool can be demonstrated, but the superiority of an idea is more difficult to evaluate.

Secondly, risk and cost are important barriers to technology transfer. The consortium's original mission was high-risk, long-term (a 7 to 10 year time frame), pre-competitive research. Such a mandate encouraged cutting-edge research using state-of-the-art equipment. This long-term goal drove staffing and research decisions.

Thus, MCC researchers were equipped with Sun workstations years before their counterparts in the shareholder companies. While facilitating the recruitment of quality researchers to MCC, however, these advanced computer technologies also acted as a barrier to transferring research results. The hardware platforms in most MCC shareholders were DEC, IBM, or Apple computers. Even when MCC-developed technologies fit the hardware and software platforms of a particular shareholder, a lack of agreement about existing platform technologies still existed within groups of shareholders belonging to the same MCC research program.

The timing of technology transfer is also important. Because of ever shorter product life cycles and increasing worldwide competition, transfer must occur rapidly for a technology to be of maximum use to a receptor. Central to rapid technology transfer is the product planning strategy of the shareholder companies. For MCC, it has been a significant challenge to get product planners from the member companies to be committed to MCC-produced technologies. If an MCC technology is transferred after a shareholder company has budgeted for and committed to a different technology, or while the shareholder is preoccupied with other strategic contingencies (such as a financial downturn), little serious consideration is likely to be given to the MCC-produced technology.

Figure 5.3 shows four stages of idea-to-product formulation. (The arrows represent new ideas or technologies, and the solid curved lines represent project conceptualization.) During Stage 1, the research strength of a new idea has the most likelihood of effecting the process of technology development. MCC-developed technologies would receive less resistance at the user organization at this stage. During Stage 2, ideas become more focused and walls begin to form around a preferred technology. This sorting-out process is driven by product champions and organization power and politics, as well as by technological and bottom-line considerations. Such forces inhibit technology acceptance as depicted in Stage 2. During Stage 3, considerable time and effort is invested in building an industrial-strength technology and possibly a prototype, activities which correspond to the third level of technology transfer, technology implementation. In Stage 4, the technology hopefully achieves market strength (the goal of Level IV). Psychological, professional, financial, and strategic switching costs increase for the receptor site of an MCC-developed technology as one moves from Stage 1 to Stage 4.

Figure 5.3: Four Stages of Idea to Product Formulation

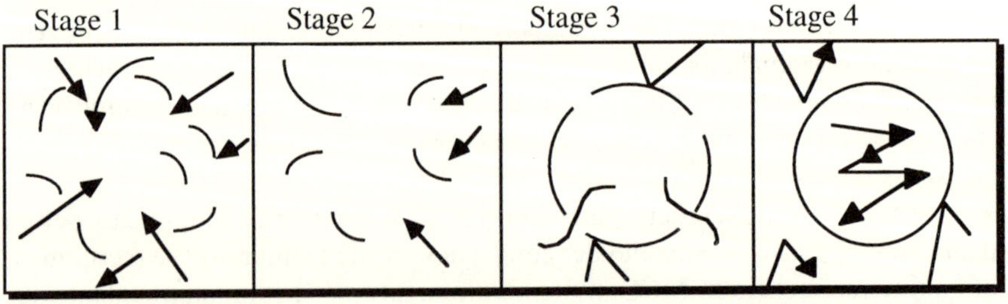

Arrows represent ideas. Lines represent project institutionalization.

Interpersonal Motivation

Researchers and managers from MCC and the shareholder companies ranked the ways to improve technology transfer. All respondents generally agreed that it was most important (1) to increase interpersonal communication while focusing on involving the shareholder researchers more with MCC; (2) to increase motivation by rewarding those involved in the technology transfer processes at both the shareholder companies and MCC; and (3) to change the culture or context of the reward structure by increasing awareness of the importance of technology transfer. Enacting such cultural and structural changes in MCC and the shareholder companies has proven to be a significant challenge.

Despite the arguments for the importance of motivating technology developers and users to collaborate in the transfer process, such motivation (from Levels I through IV) is not sufficient to achieve technology utilization. While highly motivated technology transfer participants can overcome passive or infrequent communication; physical, cultural, and strategic distance; and technology barriers, they are not a panacea for successful technology transfer. During MCC's formative years, the consortium's scientists were initially elated to locate receptive shareholder-based technology champions. However, they often spent considerable time and resources transferring technologies to these receptors, who, while being extremely motivated, lacked the political and resource clout to effectively implement the MCC technology within their shareholder company.

Successful technology commercialization between MCC and its member companies most frequently occurred in "win-win" situations. Frustrating such collaboration is the reality that the criteria for technology transfer success is often markedly different for research, product development, production, marketing, and sales personnel. One well-placed "innovation assassin" can frustrate the successful implementation of such collaborative activity.

Technology Spin-offs and the Austin Technology Incubator

Important technologies have been successfully transferred from MCC to process and product applications. However, such transfers have been relatively few in number, and they have been difficult and time-consuming. A technology transfer gap existed between MCC's research and achieving the industrial and market strength needed to successfully bring the technologies to commercial application. Partly as a result of MCC's frustration in getting its member companies to commercialize its research results, by 1990 the consortium began to foster technology use through entrepreneurship and spin-off activity.

When MCC came to Austin in 1983, many of the community's public and private leaders expected the city's economic growth to be spurred by spin-offs from the R&D consortium despite the fact that MCC officials emphasized that this was not what the consortium was about. In 1989, when the Austin economy was working its way out of a major economic recession, the idea of a regionally based technology incubator was being championed by Dr. George Kozmetsky, director of the IC² Institute at the University of Texas at Austin. To Kozmetsky, such an incubator would facilitate public/private collaboration at the regional level and would spur economic development, fill vacant office space, train entrepreneurs, and create high-value jobs. The facility, which came to be called the Austin Technology Incubator (ATI), would act as a "lightning rod," linking talent, technology, capital, and business know-how to market needs.

ATI seeks to fill the technology transfer gap between research strength and market applications. The primary drivers for the incubator are entrepreneurs and technology, which come from the private sector, universities, federal laboratories, and R&D consortia. To be accepted as a tenant company, entrepreneurs must submit a business plan that is evaluated by seasoned business professionals as well as technologists. The entrepreneurial culture of ATI fosters the linking of cutting-edge research and technology with venture financing and the realities of the marketplace. Whereas technology reports, patents, and technology licenses are often the output of R&D environments, they are considered inputs to the due diligence and business plans required at ATI. The incubator shortens the product development cycle by broadening tenant entrepreneurs' know-how in market research, finance, advertising, quality issues, management, sales, and service. ATI's culture emphasizes the importance of intangibles (e.g., business know-how and learning from each other) over tangibles (e.g., expensive office furniture) and it reinforces resource leveraging and spending hard-to-get cash on factors that added value and speed product commercialization.

The evolution of MCC came full circle—from opposing to encouraging spin-off activity—in 1990, when the consortium spun out its first company, Evolutionary Technologies, Inc., into ATI. A second company, Corporate Memory Technologies, was spun out of MCC into ATI in 1991. Three years after its founding, ATI was fulfilling Kozmetsky's vision and serving as an important bridging organization to commercialize technologies developed at MCC, thereby creating high-paying jobs, helping to decrease Austin's glut of vacant office space, and contributing to U.S. industrial competitiveness.

Since 1989 ATI has expanded its activities three times to occupy by 1993 two facilities: A suite of offices and a light manufacturing facility and wet lab, totaling 60,000 square feet. In spring 1995, ATI moved into the MCC building as the R&D consortium downsized and restructured. Twenty-seven high-tech start-ups currently reside in the incubator. The tenant companies are given three years to graduate and launch new businesses in biotechnology, telecommunications, computer, and software industries. Also resident in the incubator is the Texas Capital Network, which matches funds from private investors with entrepreneurial companies. As of 1995, ATI had graduated 18 firms and generated more than 200 technology-based jobs in the Austin area. The graduated companies occupy 51,000 square feet of office, manufacturing, or laboratory space and have attracted more than $3,000 million in capital investment.

Even with highly motivated champions, a critical gap in technology transfer from quality research to the commercial application is know-how—the ability to find and apply expertise in areas (e.g., manufacturing, marketing, finance, distributions, sales, and management) central to successful technology commercialization. The challenge is to effectively leverage academic, government, and business resources at the community level by linking talent, technology, capital, and business know-how with markets. Such value-added technology transfer can help sustain established firms, and it can foster the growth of new firms for increased economic competitiveness.

The resources of R&D organizations lie more within the technologists and researchers than with any particular set of technologies that are currently available—that is, the most valuable technology to be transferred is in the researcher's head, not sitting on the shelf. The vast array of advanced and not-so-advanced technologies that MCC developed could not simply be pushed out through superior outreach attempts and marketing. Technological capabilities had to be linked to existing and vital problems at the user level, and for new-to-the-world technologies, markets had to be developed.

The entrepreneurs that successfully spun technologies out of MCC were product champions who had a clear vision of the opportunity posed by a problem as well as an intimate knowledge of the technology they were developing. Still, they had to be adaptable and listen to business, financial, and legal advisors as they took their technologies to market strength. Transferring technology from the laboratory to successful commercial/process applications through the development of new firms requires sustained public/private collaboration at the regional level in a supportive context guided at the national level.

Conclusions

Effective technology transfer leading to successful commercial applications is an increasingly important challenge for companies, federal laboratories, and research universities in the United States and around the world. The challenges are accentuated as R&D organizations attempt to move technologies across organizational boundaries. In inter-organizational technology transfer, spatial and cultural distance impedes effective communication, which involves confidential information about complex tasks often involving high levels of competition, uncertainty, and risk. Technology transceivers who communicate across technology-producing and -receiving organizations are often not meaningfully rewarded for their boundary-spanning collaborative activities. Knowledge creators, not users, are traditionally celebrated in the technology transfer process.

From the research facilities in the nation's universities and colleges to those under federal and state jurisdiction, there is a great concern with metrics—the measuring of performance and return on investment. Such concerns increase as budgets are slashed, jobs are on the line, and established and new research missions are subject to external scrutiny. While the measures of success of R&D organizations vary considerably with the perspective (i.e., institutional affiliation and organizational position) of the evaluators, a common tendency is to emphasize only what can be counted in the short term (e.g., technical reports, journal articles, copyrights, patents, and amount of funding generated).

The Japanese tend to place a higher value on the less quantifiable and long-term benefits of R&D consortia such as information sharing and education, raising the level of competitiveness of an emerging industry, and standards setting. They work at learning from past events to improve future attempts, whether in refining a product, managing consortia, or building a science city. The Japanese credit such process technologies as being fundamentally important to their successes in competing in the global marketplace.

Washington's concern with military security—not economic security—provided the call to action in the formation of MCC. When it became clear that the U.S. microelectronics industry was threatened, the Department of Defense became alarmed, which, in turn, alarmed the White House and Congress. The 1984 National Cooperative Research Act was passed and the U.S. launched one major R&D effort after another. Federal and state research funds and tax credits have traditionally been provided for companies that invest in research and experimentation. Much less concern and funding has been directed toward understanding and facilitating technology commercialization to (1) improve the competitiveness of established firms, (2) launch and grow new firms, and (3) establish new industries. Managing technology transfer has been seen as someone else's job and relatively unimportant in the scheme of things. The traditional orientation has been that doing good science ought to be enough.

With the formation of each new major research project the federal government suggests that the way for the United States to win in global high-tech competition is big-expenditure R&D projects, plus financial incentives like tax breaks for R&D investment by private companies. We do not question the value of basic R&D, the "seed corn" of U.S. industry. Indeed, U.S. research universities and federal laboratories are the envy of the industrialized world. These knowledge resources may well constitute the best hope for increased U.S. industrial competitiveness. However, following the lead of the nation's secretary of labor and former Harvard professor Robert Reich, we label these incentives the "loud path" to international competitiveness. We recommend that more attention, research, and funding be targeted toward the "quiet path" of successful technology commercialization. The most direct route to restoring America's international competitiveness centers on improving technology transfer processes through which R&D results are rapidly transformed into high-quality, competitive products and manufacturing processes.

Businesses and governments are operating in an era where there are not national technologies, products, corporations, or industries. There are global markets for capital, labor, and technology. Regional wealth comes from world-class science and technology being quickly linked to product/process application—the timely commercialization of cutting-edge research wherever it is developed. The United States and its technologists and managers need now to become better students of global R&D capabilities so that advanced technologies flow into as well as out of the United States.

The United States' failure to compete effectively with Japan does not involve technological invention but technology application. The United States has led the world in the quantity and quality of its research and development, but this lead has not produced sufficient commercially competitive products and processes to maintain U.S. industrial strength. The importance of effectively utilizing the nation's existing research base—cutting-edge technologies and research personnel at R&D consortia, federal laboratories, and research universities—will only increase. National laboratories provided the technologies for the U.S. military to win the Cold War and for NASA to send astronauts to the moon and bring them safely back to earth. U.S. research universities are among the best in the world. Students from Japan, Taiwan, South Korea, China, Singapore, and Europe overcome strong competition and cultural barriers for the opportunity to study at these centers of educational and research excellence. The challenge is to see that America's research base is more effectively and efficiently used for shared prosperity at home and abroad.

References

Evan, William M., and Paul Olk. "R&D Consortia: A New U.S. Organizational Form." *Sloan Management Review* 31, no. 3 (Spring 1990): 37–46.

Fusfeld, Herbert I., and Carmela S. Haklisch. "Cooperative R&D for Competitors." *Harvard Business Review* 63, no. 6 (November/December 1985): 60–76.

Gibson, David V., George Kozmetsky, and Raymond W. Smilor. *Creating the Technopolis: Linking Technology, Commercialization, and Economic Development.* Cambridge, MA: Ballinger Publishing Company, 1988.

Gibson, David V., and Everett M. Rogers. "The MCC Comes to Texas." In *Measuring the Information Society*, edited by Frederick Williams, 91–115. Newbury Park, CA: Sage Publications, 1988.

Gibson, David V., and Everett M. Rogers. *R & D Collaboration on Trial: The Microelectronics and Computer Technology Corporation.* Boston: Harvard Business School Press, 1994.

March, J. G., and J. P. Olsen. *Ambiguity and Choice in Organizations.* Bergen, Norway: Universitetsforlaget, 1976.

Murphy, William Joseph III. "Cooperative Action to Achieve Competitive Strategic Objectives: A Study of the Microelectronics and Computer Technology Corporation." Ph.D. diss., Harvard University Graduate School of Business Administration, 1987.

6

A Model for Entrepreneurship Infrastructure Development in the Creation of Technopoleis

Raymond Radosevich and Gary S. Smith

Introduction

While there are many different models for the creation of technopoleis (Allen and Victor, 1986; Tatsuno, 1986; Smilor, Kozmetsky, and Gibson, 1988; Gibson and Kozmetsky, 1993), technological entrepreneurship is common to all of them. In some technopoleis, such as Boston's Route 128 area, technology commercialization originates primarily in the private sector, although local universities and government laboratories served as early sources (Roberts, 1991). In other locations, the development of an early-stage technopolis is still largely dependent on the commercialization of technology from public-sector sources.

Commercializing technology from public-sector institutions, though complicated by many strategic issues, is generally recognized as important to the health of the nation, as well as the institution and the local economy (Radosevich and Kassicieh, 1993). The two primary modes of commercialization are transferring the technology to an existing firm or working with a new entrepreneurial effort. In states like New Mexico and countries such as Russia that do not have a large private-sector industrial base to which technology can be transferred, the entrepreneurial mode may be, by default, the more important method. While embryonic firms have some limited capabilities with respect to commercial activities in general, they are frequently acknowledged to be more innovative than larger firms (Rothwell, 1989). Table 6.1 lists some of the advantages of small firms in technology commercialization.

The development of this model was partially supported by the U.S. Department of Energy under contract number DE-AP04-94AL97506. The authors acknowledge their appreciation for this support. Any representation in this model may or may not reflect the policies or views of the DOE; any errors or shortcomings of the representation are solely the responsibility of the authors.

Table 6.1: Advantages of Small Firms in Technology Commercialization

Table 6.1: Advantages of Small Firms in Technology Commercialization
Strong commitment to the technology (little of the "not-invented-here" syndrome), especially if the inventor becomes involved in the enterprise
Ability to move rapidly, both in technology development and in commercialization activities
Lower costs of development and operations
Less bureaucratic, more innovative
More efficient job and wealth creators
Entrepreneurial management for early-stage growth

Whitcomb and McEnrue (1995) argue that technology-based entrepreneurship is especially important in the conversion of centrally planned economies to free enterprise systems. In order to increase the incidence and success of technology commercialization in early-stage technopoleis, it is essential to understand the process and to dedicate resources to its improvement. If entrepreneurship is required to create the new ventures that will commercialize technology within developing areas, how is this entrepreneurial process to be further stimulated? Entirely new incentives and support infrastructure are needed to increase the number of public-sector scientists, engineers, and technicians willing to leave the security of their employment. Herlau and Tetzschner (1994) provide evidence that the highly educated unemployed and underemployed are often induced into entrepreneurship given the proper support infrastructure. Fulop's study (1994) showed that increasing unemployment in Hungary was a causal factor in the growth of entrepreneurship there.

The methodology of technology commercialization may vary among institutions and localities, but the overall process is often similar. Domicone et al. (1993) tested the validity of a U.S. derived model of entrepreneurship in Mexico and found similarities in the development stages but fewer similarities at later stages. This would suggest that infrastructure to support start-up activities may be common across international borders, but later stages of business development may be best supported by "customized" infrastructure based on local needs. Hisrich and Grachev (1993) found great similarities between present-day entrepreneurs in Russia and those in Western cultures. Similar processes of entrepreneurship certainly support the thesis that common forms of infrastructure may be appropriate even in substantially different socioeconomic contexts.

The model in Figure 6.1 presents the decisions and steps in a general entrepreneurial process for the commercialization of technology so that we can understand the resources and infrastructure necessary to support new ventures. This model should provide decision-makers at the involved institutions, and those interested in economic development within the region, with useful insights for improving the process. Gaps or weak areas identified in the analysis, such as the lack

Figure 6.1: A Generalized Entrepreneurship Model for the Commercialization of Public Sector Technology

A. Technology Source
1. Develop and mature technologies jointly with partner(s)
2. Identify and assess commercial applications
3. Protect intellectual property
4. Define commercialization strategies
5. Determine role of inventor employee(s)
6. Determine if local commercialization viable
   Select appropriate entrepreneurship mode
   Use surrogate entrepreneur alternative

B. External Commercialization Interface
Market to and interact with external organizations

Commercialization Process

C. Venture Packaging
1. Become aware of opportunity
2. Complete detailed assessment
3. Prepare business plan
4. Define venture structure and build venture team
   Expand existing firm
   Start new firm
   Inventor or surrogate entrepreneur
   Relationships with support intermediaries
5. Define relationship with technology source
   Secure intellectual property
   Joint technology maturation process
6. Secure early-stage resources

D. Venture Launch
1. Develop technology to functional prototype
2. Expand capital structure
3. Develop facilities
4. Expand venture team
5. Arrange strategic alliances
6. Progress through alpha and beta stages

E. Initial Commercialization
1. Begin manufacturing and marketing
2. Work out the bugs
3. Develop ramp-up capabilities
4. Ramp up operations

F. Steady-State Operations
1. Update business strategies
2. Begin second-generation product development
3. Work on exit issues for investors, update capital structure
4. Support local tech-based economic development

G. Recycle the Entrepreneur
1. Remove the stigma of failure
2. Support the next venture

of seed capital, experienced entrepreneurs, policies at the institution to encourage entrepreneurial spin-offs, and facilities, should be the focus of improvement.

Each of the elements of the model presented in the figure above will be described, including the resources necessary to support each. In most areas seeking to create a technopolis, there are likely to be multiple public-sector, not-for-profit, and private organizations committing resources to economic development. The resource needs identified by the model can provide a framework to survey these organizations in order to assess which, if any, can provide a given resource. This process is discussed after examination of the model.

Policymakers and providers of the resources required to build the comprehensive infrastructure described below may well question the efficacy of such an investment. In one of the few comprehensive studies of new venture creation that compared "assisted" areas (where publicly supported infrastructure is available) with "unassisted" areas, Birley and Westhead (1992) found that new firms in the "assisted" areas of Great Britain were more successful in dimensions such as employment generation than were the firms in the "unassisted" areas. Barkham (1992) also found that variations in support mechanisms provided partial explanations for varying degrees of success in entrepreneurship when comparing regions of the United Kingdom. In a study evaluating one component of infrastructure, small business counseling, Chrisman and Katrishen (1994) found that this government-sponsored activity contributed to substantial wealth and job creation and largely paid for itself through incremental tax revenues from the increased sales of client companies.

A Generalized Entrepreneurship Model for the Commercialization of Public-Sector Technology

The model presented in Figure 6.1 is described below in terms of a sequence of activities performed within the technopolis to commercialize technology from a public source through the mechanism of entrepreneurship. After each set of activities is described, the resource requirements that should be in place to support the new venture are detailed.

The model is derived from the authors' experiences of providing direct support to new technology-based firms for more than 15 years, as well as creating other support infrastructure such as an incubator and an entrepreneurs' association. This was done in the state of New Mexico; the context for this experience is presented in Table 6.2. Our experience in New Mexico may have considerable relevance for many metropolitan areas with low commercial activity in countries such as Russia.

Table 6.2:
The Technology Commercialization Environment of an Early-Stage Technopolis

A majority of technology development is publicly sponsored and early stage in the commercialization cycle
Limited headquarters of large firms
Isolated from mass markets and suppliers
A dependency culture (especially dependent upon federal government spending)
Limited institutional risk capital
A shortage of technical entrepreneurs
Fragmented but competitive technology commercialization support infrastructure (public and private)
Increasingly supportive federal legislation for technology transfer and commercialization

Technology Development and Transfer

A. Technology Source

A.1 Develop and mature technologies jointly with partners

Most technology development at public-sector research institutions is typically early stage and not directly tied to commercial products or processes. Up to now, much of it was defense related and not driven by industrial or consumer mass production requirements and cost considerations. Circumstances today are encouraging closer and more frequent interactions with commercial customers at earlier stages in the development process.

Earlier in the evolution of technology transfer, it was commonly believed that public-sector research institutions were analogous to candy stores full of opportunities to be selected off the shelf. Experience now tells us that while there is likely to be significant opportunity, realizing this potential takes much more work than walking through the aisles and filling up a cart. Also, products are typically not in a "ready-to-go" state. They are usually advanced only to the proof of concept or, at best, functional prototypes.

Continued interaction between the original developer and the commercialization entity is usually key to successful transfer. The resources at the technology source can contribute significantly to the maturation of the technology up to a beta-site or preproduction model. User facilities, technical assistance programs, and private consulting by technical staff members can be used by the technology source to contribute to the maturation.

In addition to the patent position, the intellectual property may best be protected by the use of trade secrets that evolve from the know-how generated through joint technology development between the source and recipient. The process for entering into joint development agreements must be user-friendly and timely in order to facilitate use of this important intellectual property tool.

Resource needs. The development of commercial technology requires a creative environment with sufficient resources to create a substantial inventory of technological advances. The environment within the technology source must encourage interaction with industry, and incremental funds must be available for technology maturation. The technology source should sponsor programs that encourage and facilitate interactions with small business, such as user facility and technical assistance programs of relevance to small firms, and establish incentives for technical staff to work with firms interested in establishing applications-oriented intellectual property.

A.2 Identify and assess commercial applications

In order to determine possible commercial applications of the invention, target industries must be recognized and potential industrial partners identified. Nonlocal firms identified during this process could become strategic partners for a local venture should that prove feasible.

The next step is to do a preliminary assessment of the commercial viability of the technology. This will assist in the patenting decision, in choosing a commercialization strategy, in licensing, and in helping the inventor employees determine their level of participation in the commercialization process. The assessment should be guided by the subjective judgements of experienced businesspersons, especially in determining commercialization strategies in specific industries.

Resource needs. Gatekeepers, boundary spanners, or product managers at the technology source who understand and have networks in the industries associated with the technological advances are best qualified to identify and assess commercial applications of new technology. Databases such as Corptech or Technology Targeting can be utilized to define technology areas of interest to industry, along with expert systems that match laboratory technology opportunity inventories to databases of industry needs. These commercial databases can be supplemented by the development of databases of the technology needs in local firms. And, persons who are experienced in the new venture mode of technology commercialization can assist in preparing technology commercialization opportunity assessments. Such persons may be proven entrepreneurs, venture capitalists, or those experienced in professional business support services for start-up companies.

A.3 Protect intellectual property

Protecting intellectual property is key for creating value for commercial applications (Van de Ven, 1993). Since significant additional investment is required to bring the technological advancement to market, industry requires incentives through intellectual property protection to justify and protect their investment.

Recognition of technology advancements, their patentability, and their relative value are key aspects of the intellectual property decision. Timing, effectiveness, and allocation of limited resources impact the process of establishing and protecting intellectual property. The institutional culture must also support the process. The culture should include at least the following: communication of how important commercialization is to management, a timely and user-friendly system for disclosing inventions, incentives for inventors, and recognition of the impact of intellectual property protection on publishing.

While intellectual property is widely recognized as a valuable business asset in most private enterprise economies, legal systems that establish and protect these properties need further modification in socioeconomic systems that were formerly centrally planned (Fulop, 1994).

Resource needs. Safeguarding of intellectual property requires a technical staff trained in intellectual property issues, applications identification, etc. Policies and incentives that encourage staff to create technology-based commercial opportunities and to work with industry should be in place. The technology source must employ sufficient patenting resources—attorneys, agents, filing and maintenance fees, etc.—to create a valuable intellectual property inventory. And, research of geographic markets and strategic alliance trends ensures that investments in expensive foreign patents are informed decisions.

A.4 Define commercialization strategies

The general policies of the institution and the specific decision made for each opportunity will determine how much commercialization occurs locally. If licensing to large companies takes precedence over encouraging inventor employees to become entrepreneurs and working with small or start-up firms, then areas like New Mexico will suffer because they have a small industrial base. Although fair access and maximization of rewards back to the technology source must be addressed, an active partnership with local economic development networks may enhance both local economic development and success for the technology source.

Resource needs. The technology source needs sufficient resources to determine commercialization strategies that maximize the benefits to the technology source while effecting local technology commercialization.

A.5 Determine the role of inventor employees

Successful integration of inventor and key associate employees into the commercialization process is critical to the success of commercialization. It also has an important impact on the amount of local development. In areas that have experienced significant economic development based upon technology commercialization (for example, Route 128 and Silicon Valley in the United States), spin-off entrepreneurial ventures from universities and laboratories were an important factor.

Institutions that encourage and reward inventor participation most likely increase the incidence of opportunity recognition and marketing success. Although it is possible for an inventor to be integrated into the venture effort with a large firm, most entrepreneurial cases involve the formation of a start-up firm for the specific purpose of commercializing that inventor's technology.

It is usually best if the inventor devotes full time to the new venture; however, if the inventor prefers to stay with the technology source, alternative arrangements can be made to accommodate this circumstance. A new venture can still be formed with the assistance of a surrogate entrepreneur and the inventor can participate part-time assuming the technology source has the mechanisms to allow this. This method is discussed in more detail in section C.4 below.

Resource needs. The odds that commercialization will succeed are increased if the inventors have sufficient business acumen and experience to serve as the venture entrepreneur or, alternatively, sufficient business support services are available to guide less-experienced inventors through the maze. The technology source should institute policies that permit inventors to select from a variety of participation mechanisms (consulting for equity or cash, leaves of absence, part-time employment, etc.) that which best suits their personal situation. Mechanisms should be in place that allow inventors and other new venture partners the ability to secure intellectual property rights while avoiding conflicts of interest and allow the technology source to share in the benefits of commercialization (hopefully including the holding of equity in the new venture).

A.6 Determine if local commercialization is viable

This evaluation is based on the commercialization strategies noted earlier, the technology source's awareness of the local potential, the relationship between the technology source and local economic development networks, and the viability of supporting a decision to promote the venture locally. This decision must balance the issues of fair access, the rights of inventors, value-added to the technology source and sponsors, community obligations, and the likelihood of success. To make local commercialization viable, the technology source must have the mechanisms and policies in place to support the process, and the local economic development networks (which include the private sector) must have sufficient infrastructure in place to support the venture.

Resource needs. Effective interaction by the technology source with local industry and the support infrastructure is necessary to determine local interest in commercialization. This interaction can be facilitated by formal networks and databases. The technology source must be able to assess the local potential to commercialize the technology, especially in the highly subjective aspect of management capacity. This requires extensive experience in launching new ventures in the local context.

B. External Commercialization Interface

The ability of technology producers to reach local potential users and the extent to which the interface is user-friendly is important to local commercialization. Technology producers tend to seek relationships with larger firms because it is perceived that larger firms can provide more resources for the interaction and provide a greater likelihood of success. The local small business community is sometimes frustrated with an inability to get the attention of the technology producer and the lack of resources to facilitate the interaction. Policies and programs at the technology source can mitigate this problem, and a strong local economic development network can help to provide the screening and support necessary to make this interaction more productive.

The interface with the technology source usually has two paths, the official interface through the administrative structure and the interface with the technologists and their line management. The administrative staff typically can not understand all the specific technology implications, is burdened by having to orchestrate many interactions, and does not have the same incentive as the technologists to make a particular deal work. The technologists often become personally involved in the process, understand the specific technology implications and needs, and are interested in making the deal work because the rewards are more direct and immediate. Knowing how to work with the administrative structure and developing a strong relationship with the technologists is important to potential local partners.

Resource needs. The technology source needs an adequate outreach system to reach potential customers. The system must be user-friendly in order to encourage direct interaction between the technical staff and local entities. Local infrastructure and industry should support programs that organize and facilitate this interaction, including educating local entrepreneurs and businesspersons in the methods of collaborating with technology sources.

Commercialization Process

C. Venture Packaging

C.1 Become aware of commercialization opportunity

Under ideal conditions, a substantial number of inventors at the technology sources would recognize the commercialization opportunities and form new ventures to capitalize on them. Traditionally, however, this happens with very few technical staff members. In the long term, new policies, programs, and cultural changes at the technology sources, together with more success stories to encourage and support the process, will result in more inventor-based spin-offs. In the interim, the local economic development network and private sector must interact with the technology sources to become aware of various opportunities. They must then encourage more inventors to participate in the commercialization process and provide surrogate entrepreneurial capabilities to supplement the traditional spin-off model.

Resource needs. Entrepreneurship requires a process through which the local commercialization infrastructure interacts with technology sources on a meaningful and systematic basis to become aware of opportunities.

C.2 Complete detailed assessment

Not all opportunities identified will result in a viable business. Usually, as part of the business planning process noted below, a more detailed assessment will be completed. Since resources may be very limited at this stage, the support infrastructure will play a key role in this process. A very cost-effective method which has performed well in the past is the use of graduate business classes that work with potential entrepreneurs to help them complete an assessment and write a preliminary business plan as a class project. A number of U.S. federal laboratories have developed formal relationships with local universities to provide assessments. As the support infrastructure grows and matures, pools of more professional resources should be available to guide and complement the student resource.

Resource needs. More private-sector involvement in laboratory-university programs for commercial assessments would facilitate the identification of viable commercial opportunities.

C.3 Prepare a business plan

A professionally written business plan is a necessary tool for interacting with the majority of future resource providers to the venture (this includes the technology source, investors, key management yet to be brought in, and strategic alliances). It is the primary source for understanding and evaluating the merits of the deal. The plan should articulate the commercial opportunity, the basic business strategies, the competitive advantages afforded by the proprietary technology, the resources available (including management), the additional resource requirements, and the returns realistically expected from the investment.

Resource needs. The planning process demands an experienced entrepreneurial team whose members understand and are able to address the issues that the resource providers will want answered by the plan. Entrepreneurs should utilize experienced support services for data gathering, analysis, and business plan writing.

C.4 Define venture structure and build initial venture team

Traditionally, an entrepreneur leaves the technology provider and forms a new company to commercialize the technology. Alternatively, either: (1) an existing firm with complementary resources could provide a new home for the inventor, (2) an existing firm could act as a surrogate entrepreneur, or (3) a surrogate entrepreneur can start a new firm in the event that the inventor prefers to remain with the technology source and provide only part-time support to the new venture. Under any of the above alternatives, a strong private/public local support infrastructure is important for building and nurturing start-up ventures.

The proper legal form and jurisdiction under which the firm will operate should be considered. Many states are not progressive with respect to the legal liabilities of participants in for-profit firms. The roles defined by the directors and officers of the entity should match actual capabilities and the willingness to accept risk exposure. All participants should be well informed of these risks. Since both the jurisdiction and form of entity impact the manner in which resources can be sought and accepted by the entity, professional advice with respect to capital acquisition should be sought before the firm is formed.

The new venture will likely require the services of additional entrepreneurs or early-stage managers to complement the abilities of the inventor(s). Such managers should have experience with new ventures, experience in the field of application, a willingness to take risks, experience in several dimensions including functional areas, the ability to work closely with all members of the team, and the ability to overcome difficult challenges.

Resource needs. This pool of experienced local entrepreneurs should be available to lead or support the new venture through the early stages of development, often four or five years. The technology source should develop a system to recruit potential entrepreneurs into the local situation.

Successful initial ventures start with a supportive business environment (including the perception that seed capital is available for meritorious deals) that is sufficient to induce entrepreneurs to attempt a local start-up. This environment may either exist in a dynamic private sector or be supplied through sophisticated public and not-for-profit support services. It depends on the availability of local businesspersons who are able to complement the entrepreneur's business talents and who are experienced in technology-based start-up activities (for example, to serve as members of boards of directors). Formalization of such a pool to provide a networking service to entrepreneurs could be especially valuable to surrogate entrepreneurs brought in from outside the state.

In the longer term, legislation should be passed that will foster a more positive business legal environment for participating in these kinds of entities. In the short term, a coordinated "pro-business" lobbying effort is needed to identify and promote legislation that supports new ventures.

C.5 Define relationship with technology source

The two most important aspects of the relationship between the technology source and the venture are the license for intellectual property rights and the technology interaction, which includes transferring the technology and future co-development.

To establish a viable market position and attract sophisticated investors, the new venture will need a strong proprietary position in the intellectual property associated with the invention and any improvements. A license or other assignment of the rights to the new venture should include terms that provide for access to improvements, exclusivity for initial fields of use, at least non-exclusive rights to additional fields of use should the venture be successful in commercializing the first, and most-favored status in non-exclusive fields. As suggested in the arguments in section A.1 regarding joint technology maturation, the technology source should continue to provide assistance to advance the technology toward new generations and applications, and should agree in advance to provide access by the licensee to the new technology based upon satisfactory performance to date.

Resource needs. A timely licensing process should take into account the unique needs of the nascent venture with respect to the acquisition of intellectual property (e.g., no advance payments, assistance in policing infringements, assistance in researching infringement possibilities on others' patents, verification of the relative advantage over competing technologies). The technology source should be flexible in configuring the optimal methods of recouping value in the technology, including accepting equity in the new venture instead of licensing fees.

Experienced professional legal services in intellectual property should be employed. This experience should include substantial practice in litigating intellectual property issues, securing and issuing rights (licenses, etc.) to others, and contributing to the firm's strategies relative to the use and protection of intellectual property. Given the cash flow constraints of most early-stage companies, it would be ideal if professional services suppliers would take equity in exchange for services or provide substantial discounts on normal billing rates during the firm's formative stage.

C.6 Secure early-stage resources

There are several stages of evolution in forming and growing a new venture to steady-state operations. As the venture matures, it becomes easier to demonstrate the opportunity. More and cheaper sources of capital typically become available as the venture progresses. Consequently, it is hardest to find sources of capital during the conceptual stage. The resources necessary to launch a venture in less commercially developed areas are usually limited to government sources that promote economic development (usually state and local programs, but they may also include the federal Small Business Innovation Research [SBIR] and Small Business Technology Transfer [STTR] programs), family and friends, and previously successful businesspersons. In some instances, it may not be premature, even at the formative stage, to begin seeking strategic alliances which have an interest in the technology or market applications.

Resource needs. An angels' network that is specifically oriented toward technology-based ventures is another early-stage resource. Frequently the wealth that exists in less developed areas has been earned in other forms of commercial activity such as land development or extractive industries. These wealthy persons usually are reluctant to invest in technology ventures since they have little experience in this area.

Other resource requirements are a coordinated pool of local economic development resources, including firms willing to incubate the venture, incubator facilities, loaned personnel and facilities, and access to capital; and the use of more private-sector investment criteria by public-sector and not-for-profit sources of services and funds (that is, better screening of a client's business potential before providing funds or subsidized services).

D. Venture Launch

D.1 Develop technology to functional prototype

In most cases, the technology will require significant maturation, especially from the perspective of manufacturing engineering. Manufacturability and cost considerations will likely impact product materials, fabrication techniques, tolerances, finishes, etc. If the product is consumer oriented, it is especially important to introduce product design skills that augment commercial appeal. As the functional prototype nears completion, planning and implementation of test and evaluation programs is required. These programs should also guide any required product regulatory approval processes.

Resource needs. This development depends on the availability of user facilities at universities and laboratories, especially those associated with expensive capital expenditures and short-term usage needs such as testing and certification of product designs. Commercial design skills in professional services to assure product acceptance by target customers are also necessary; these skills appear to be in short supply in areas with low commercial activity. In the final stages, manufacturing system design assistance and manufacturability studies of functional product designs are performed.

D.2 Expand capital structure

Securing capital for a venture is typically a multi-staged process. Traditional sources of pre-seed and seed capital usually have limited resources and are the most expensive in terms of equity share per dollar. Founders like to limit how much of the company they must share with investors, so these sources aren't intended to supply all of the capital needs of the venture. As the venture progresses and the burn rate accelerates, the CEO will typically spend at least half of his or her time seeking the next round of financing.

By this stage, the venture should demonstrate sufficient promise to attract alliances which may also be the main source of equity monies. In many instances, strategic alliances will be the preferred mode of capitalizing the venture. The capital pricing is often favorable, the terms are less onerous, and greater ancillary benefits are provided such as access to markets or manufacturing know-how and capacity.

Resource needs. Most desirable are local sources of early-stage risk capital that are oriented toward technology-based deals and industries in which local deal flow is likely to occur. Local sources that serve as lead investors are important to form syndicates of capital and to offset the tendency for out-of-state capital sources to encourage the venture to relocate close to them. Experienced and reputable local agents and investment bankers to assist in the search for development and expansion capital are essential. New ventures should locate a network of companies and their executives known to actively seek acquisitions and partnerships with early-stage companies. Such an inventory should include due diligence on the effectiveness of past partnerships with these companies. Finally, expanding the capital structure requires commercial bankers, commercial finance companies, leasing companies, etc., who understand the needs and peculiarities of technology-based, early-stage companies and can provide financial advice and contacts as well as limited asset-based financing.

D.3 Develop facilities

Due to capital constraints, most start-up ventures begin operations either in an incubator or in modest temporary facilities. As the firm matures and expands, it must move to a larger space to accommodate growth and begin acquiring the equipment necessary for manufacturing. This step can be traumatic: it often consumes large amounts of precious capital and time. The decision to buy or lease is important. The lack of reasonably priced space for specific applications, e.g. cleanrooms, could prove prohibitive.

Resource needs. Expansion requires the availability of a pool of office and light industrial space, including landlords willing to provide some leasehold improvements to customize the space; and lenders and leasing companies that understand leasing needs and processes for technology-based, early-stage ventures.

D.4 Expand venture team

The start-up firm often struggles through the early stages with key management doing double duty. As the venture matures, the timing of the acquisition of additional managers is critical. Often sophisticated investors will force the entrepreneurs to hire a CEO with industry experience or at least bring in experienced outsiders who were not part of the original start-up team. Salary expenses expand greatly and the transition is sometimes troubling. In areas with a limited industrial base, few executives are available locally for hiring and many must be recruited from outside of the region.

Resource needs. Successful new ventures require recruiting capability that is knowledgeable in securing management for technology-based venture firms, and a network to facilitate the search and due diligence process.

D.5 Arrange strategic alliances

Assuming the venture has significant market opportunity, it is very difficult and often not cost effective to build all manufacturing and marketing capability from scratch. Strategic alliances can provide much-needed expansion capital, access to related technologies, a second source of supply for wary customers, market recognition, and distribution and marketing resources that are especially important in foreign markets.

Resource needs. Networks, agents, and other intermediaries can help identify and arrange partnerships.

D.6 Progress through alpha and beta stages

The important aspect of this stage is finding suitable sites. Often new ventures are required to support beta systems at distant locations that present many logistical and cost drawbacks.

Resource needs. Networks can be utilized to identify and arrange beta sites. Preferably, more sites could be found in and around the local area. Often potential local sites aren't as cooperative with small firms as they might be with more influential enterprises.

E. Initial Commercialization

E.1 Begin manufacturing and marketing

Hopefully, this signifies the start of revenue generation by the new venture. Preliminary marketing and beta-site information will guide the transition into full-scale marketing. Strategies devised earlier regarding promotion, trade show attendance, brochures, distribution, geographic roll-out, servicing, etc. are all put into practice. Resources are devoted to the formalization of sales and distribution relationships, the institutionalization of market research, and the organization of service functions. Customer support is established by captive and third-party servicing functions, user support materials and advisory functions, and a highly trained, missionary sales organization. Future versions of product modifications will be defined during this process to facilitate customer use for unanticipated applications.

Resource needs. Manufacturing and marketing requires identification of local and regional early-adopter customers, the availability of marketing and advertising support services to provide promotional media, advertising copy, trade show displays, etc., and a network to third-party providers of functions such as manufacturers' representatives, product installation and servicing, and customer financing.

E.2 Work out the bugs

Resolution of unanticipated problems (and opportunities) is especially critical during the early stages of commercialization when resources are still very limited. Experienced help from strategic alliance partners is most welcome at this time; otherwise, consultants may be brought in to help resolve the problems.

Resource needs. Start-up manufacturing and marketing expertise from private consultants and public or not-for-profit support services can assist in the troubleshooting process.

E.3 Develop ramp-up capabilities

Most new ventures can't afford the optimal first facility that affords sufficient room for explosive growth. As the new firm enters the market successfully, it must begin planning for expansion and the formalization of more and more systems and functions. Real growth and its necessary requirements now put the young firm into another round of capital raising. Although it will still be difficult, success is likely to open up additional avenues for pursuing capital.

Resource needs. At this stage new ventures are aided by progressive local banking practices for technology-based ventures, local intermediaries and agents who can provide access to a broader base of financing sources, and local support services for expansion capital planning.

E.4 Ramp up operations

The dramatic, high-growth expansion of operations will cause the venture to add personnel, formalize policy and procedures, complete automation of management systems, secure banking relationships for working capital, expand distribution systems, evaluate and add suppliers, move some professional service functions in-house, and, in general, create additional capabilities which will allow rapid movement along the experience curve.

Resource needs. This requires access to resources (consultants, new employees, recruiting firms, etc.) that understand the difference between the needs of growing early-stage firms and those of large corporations.

Wealth and Job Generation within the Technopolis

F. Venture Steady-State Operations

F.1 Update business strategies

Although the initial business plan may have provided early guidance for the venture, enough changes and surprises will have taken place to require updating and revision of business strategies. Such revisions will be based upon reviews of the firm's technology position, the pursuit of internal development and technology acquisition programs, the results of marketing experience, the recognition and acknowledgement of competitor responses and customer needs, and the serendipitous opportunities for new alliances.

Resource needs. This updating will require access to a network of sophisticated private-sector industrial managers who understand business strategy and its implementation. These managers serve as directors and consultants, and occasionally replace the entrepreneurs as the firm matures to the later stewardship phase.

F.2 Begin second-generation product development

The maturing venture should be achieving efficiency of operations, cost reduction, and additional market penetration at this stage. Competitive advantage is achieved by expanding product lines, seeking new applications of the technology, developing the next generation of technology, and improving the effectiveness of the management team.

Resource needs. Competitive advantage depends on local availability of expansion resources—management, capital, facilities, and personnel. A competitive business environment does not provide disincentives to remain in the evolving technopolis. Any such disincentives should be identified and removed through the actions of "pro-business" groups.

F.3 Work on exit issues for investors, update capital structure

Investors invest for the primary purpose of making money. By now, they will want a mechanism to take their gains. This is most often accomplished by going public or selling the firm to another company, perhaps one of the strategic partners.

Resource needs. Investment bankers and other intermediaries who provide investor exit mechanisms such as IPOs and acquisition services should be located.

F.4 Support local technology-based economic development

As the replication of venture formation is the goal of technology-based, homegrown economic development, spin-offs from this venture should provide the next generation of entrepreneurs. The firm could also serve as a role model and provide resources to promote and assist in the packaging and incubation of new ventures.

An ongoing objective evaluation of the support infrastructure and services provided is needed to identify gaps, improve the services, and improve the incidence of successful ventures.

Resource needs. Network and support services that identify opportunities for local suppliers aid growing firms, as does an objective, third-party evaluation of the support infrastructure and services with feedback to facilitate learning and improvement.

G. Recycle the Entrepreneur

G.1 Remove the stigma of failure

A significant portion of all technology-based new ventures will ultimately fail because of the many risks and complexities of embryonic markets. Although the venture itself may fail, a very positive residual may survive in the form of a more experienced entrepreneur. Studies by Lamont (1972) and by Reuber and Fischer (1994) verify that greater entrepreneurial experience results in higher performance in technology-based firms. Additionally, a study by Hart (1989) suggests that a higher level of new firm formation in the Republic of Ireland (as compared with Northern Ireland) may result from greater entrepreneurial experience in the former. Therefore the technopolis that wishes to retain the enhanced entrepreneurial resource must strive to separate the failure of the business venture from the person of the entrepreneur.

Resource needs. An objective analysis should be conducted to ascertain the causes of venture failure and the lessons learned. This analysis can be of great benefit to the entrepreneur as well as to the local support services and the economic development organizations. If the analysis demonstrates a "smart" and ethical failure, economic and psychic support should be organized to rescue the entrepreneurial resource.

Policies (national as well as specific support organizations) which currently require the entrepreneur to spend all of his/her personal resources as a demonstration of commitment to the venture should be changed. These policies often result in such economic and personal devastation when a venture fails as to preclude additional attempts at entrepreneurship, thereby eliminating an essential ingredient in technopolis development. Changing institutional policies that constrain the survival of entrepreneurs will be very difficult. To do so, Van de Ven (1993) suggests that public and not-for-profit organizations that provide supportive infrastructure for entrepreneurship must themselves possess strong entrepreneurial talent to lead the change processes.

Entrepreneurial leave from technology-source jobs should be promoted, such as that granted by Sandia National Laboratories of the U.S. Department of Energy to its inventors who wish to leave the laboratory to commercialize their inventions. Leave is provided for up to two years, thus providing some security in the event of venture failure.

G.2 Support the next venture

To retain the resource embodied in the more seasoned entrepreneur, the successful technopolis will strive to provide an environment that facilitates the launching of the next venture by the entrepreneur. Assistance in locating additional technology-based venture opportunities may result in the proven entrepreneur's serving as a surrogate for the inventor who does not want to leave a public technology source (Radosevich, 1995).

Resource needs. Priority in grant programs and other forms of equity commitments in recognition of the reduced risk due to a proven entrepreneur would encourage next ventures. An integrated commitment from support organizations to provide services would induce entrepreneurs to remain within the technopolis for the launch of the next venture.

Creating a Technopolis Infrastructure Inventory

The above section details the resource requirements for a technopolis that is highly dependent upon technology-based entrepreneurship, especially if the primary source of technology is public-sector organizations such as universities and government laboratories. While other authors have implied certain resource requirements for successful technology commercialization, few comprehensive statements of requisite capabilities exist for given technopolis development strategies. (See, for example, the paper by Tatsuno [1986] which describes physical infrastructure requirements, or the paper by Smilor and Gibson [1992] which describes the use of consortia as an organizational form promoting technopolis development.)

Since the development of a technopolis requires many forms of support organizations, the resource requirements are often met by a variety of providers. This paper suggests that the development of a resource inventory can be a very useful tool for strategic planners. To develop such an inventory for an entrepreneurship-based technopolis, a format can be used that lists the resource requirements articulated above as one dimension and the support providers as the other.

The compiling of data for the inventory matrix can be done in several ways, and from several sources. A ranking scheme can be used to fill in each cell with respect to the quality or comprehensiveness of a resource represented by a specific organization. Sources of data should include the entrepreneurs who have previously been served, since support organizations frequently advertise more comprehensive services than can actually be provided. Often a technopolis will have an entrepreneurs' association or formal network which may take on the responsibility for gathering data by surveying members.

Creation of a requirements inventory matrix can provide valuable insight for technopolis developers with respect to the redundancies and gaps in services provided or the necessity for organizational change if entrepreneurs are unimpressed with the services offered by some support organizations. The inventory can also be a useful service of the entrepreneurs' association by providing information to new entrepreneurs relative to the actual quality of service provided by different sources for a needed resource.

Development of a "specialized" requirements inventory matrix may facilitate the evolution of a mini-technopolis based on the cluster theory. If unique technological resources exist in a given location—biotechnology in Boston, for example, or cryogenics in Oxfordshire, England—then special forms of support may be necessary to create the appropriate operating environment. Smith (1991) argues that the specialization required extends well beyond the technology source.

Conclusions

Many models of technopolis development suggest that technology-based entrepreneurship is a critical ingredient. In the early stages of development, dependence upon public sources of technology is a frequent necessity until a critical mass of private sector firms evolves. By examining a generalized process of technology-based entrepreneurship and then postulating the resource requirements necessary to support this activity, this paper has attempted to facilitate the development of an inventory of requisite capabilities. This inventory can be used normatively in the technopolis planning process or it can be used descriptively as a survey instrument to catalogue available resources. Both uses can facilitate the successful evolution of technology-based entrepreneurship during the development of technopoleis.

References

Allen, D. N., and L. Victor. *Nurturing Advanced Technology Enterprises: Emerging Issues in State and Local Economic Development Policy.* New York: Praeger Publishers, 1986.

Barkham, R. "Regional variations in entrepreneurship: some evidence from the United Kingdom." *Entrepreneurship and Regional Development* 4 (1992): 225–244.

Birley, S., and P. Westhead. "A comparison of new firms in 'assisted' and 'non-assisted' areas in Great Britain." *Entrepreneurship and Regional Development* 4 (1992): 299–338.

Chrisman, J. J., and F. Katrishen. "The economic impact of Small Business Development Center counseling activities in the United States: 1990–1991." *Journal of Business Venturing* 9, no.4 (July 1994): 271–280.

Cooper, A. C. "The role of incubator organizations in the founding of growth-oriented firms." *Journal of Business Venturing* 1, no. 1 (Winter 1985): 75–86.

——— "Entrepreneurship and high technology." In *The Art and Science of Entrepreneurship*, edited by D. L. Sexton and R. W. Smilor. Cambridge, MA: Ballinger Publishing, 1986.

Domicone, A. D., J. R. Montanari, A. M. Headrick, and R. L. Oldenkamp. "International dimensions of management: the trans-national efficacy of a venture development model." *Entrepreneurship and Regional Development* 5 (1993): 45–60.

Fulop, G. "Entrepreneurship, small- and medium-sized enterprises in Northern Hungary." *Entrepreneurship and Regional Development* 6 (1994): 15–28.

Gibson, D. V., and G. Kozmetsky. "Networking the technopolis; cross-institutional alliances to facilitate regionally-based economic development." *The Journal of Urban Technology* (Spring 1993): 21–39.

Hart, M. "Entrepreneurship in Ireland: a comparative study of Northern Ireland and the Republic of Ireland." *Entrepreneurship and Regional Development* 1 (1989): 129–140.

Herlau, H., and H. Tetzschner. "Regional development: who are the problem owners?" *Entrepreneurship and Regional Development* 6 (1994): 161–175.

Hisrich, R. D., and M. V. Grachev. "The Russian entrepreneur." *Journal of Business Venturing* 8, no. 6 (November 1993): 487–498.

Lamont, L. M. "What entrepreneurs learn from experience." *Journal of Small Business Management* 10 (1972): 36–41.

Radosevich, R., and S. Kassicieh. "Strategic challenges and proposed responses to competitiveness through public-sector technology." *California Management Review* 35, no. 4 (Summer 1993): 33–50.

Radosevich, R. "A model for entrepreneurial spin-offs from public technology sources." *International Journal of Technology Management* 10 (1995): 879–893.

Reuber, A. R., and E. M. Fischer. "Entrepreneurs' experience, expertise, and the performance of technology-based firms." *IEEE Transactions on Engineering Management* 41, no. 4 (November 1994): 365–374.

Roberts, E. B. *Entrepreneurs in High Technology: Lessons from MIT and Beyond.* New York: Oxford University Press, 1991.

Rothwell, R. "SMFs, inter-firm relationships and technological change." *Entrepreneurship and Regional Development* 1 (1989): 275–297.

Smilor, R. W., Kozmetsky, G., and D. V. Gibson. *Creating the Technopolis: Linking Technology Commercialization and Economic Development.* Boston: Ballinger, 1988.

Smilor, R. W., and D. V. Gibson. "Building a technology transfer infrastructure." In *Technology Transfer in Consortia and Strategic Alliances*, edited by D. V. Gibson and R. W. Smilor, 129–149. Lanham, Maryland: Rowman & Littlefield Publishers, 1992.

Smith, H. L. "The role of incubators in local industrial development: the cryogenics industry in Oxfordshire." *Entrepreneurship and Regional Development* 3 (1991): 175–194.

Tatsuno, S. *The Technopolis Strategy.* Reading, Massachusetts: Addison-Wesley Publishing, 1986.

Van de Ven, A. H. "The development of an infrastructure for entrepreneurship." *Journal of Business Venturing* 8, no. 3 (May 1993): 211–230.

Whitcomb, L. L., and M. P. McEnrue. "High tech entrepreneurship in China: developing new organizational structures and incentives for innovation." *Proceedings: Entrepreneurship and Small Business in a Changing Competitive Environment*, 21–30. University of Colorado at Boulder, 1995.

7

Defense Conversion and Restructuring in the Russian High-Technology Sector: Is There an Alternative to Uncontrolled Exports?

Jacques Sapir

The conversion of the defense industries of former Soviet-type economies is a painful and protracted process. This process is mandatory, however, for the successful adaptation to a market system in a number of countries, particularly those—like many successor states of the former Soviet Union—in which this sector was overdeveloped. Conversion is above all a giant exercise in industrial restructuring, and has both a macroeconomic and a microeconomic dimension. Traditionally, defense enterprises have been seen as a kind of blue ribbon sector.[1] However, the specific advantages these enterprises might have possessed in the older economic system do not necessarily translate into a competitive edge under new conditions. Facing a very uncertain future, some of these enterprises want to remain in their former specialized technological niche and attempt survival through a massive export push. Such a strategy at the enterprise level is influenced to a considerable extent by the macroeconomic context and by economic policies.[2] If mismanaged, it could have highly deleterious political implications, both domestically and abroad.

The problem of conversion in the former Soviet Union is compounded by uncertainties about the meaning of conversion and by the regional distribution of the military-industrial sector. The regional concentration of defense-related activities implied a potential lobbying from both enterprise managers and local authorities. It also created fear of deep regional depressions able to destabilize political authorities. The considerable level of intellectual confusion about the meaning of conversion has worsened the problem. This confusion has persisted far beyond the initial phase of the process. Signals sent to enterprises and workers alike have been mixed at best, giving birth to false hopes and creating some unrealistic expectations. This situation resulted from macroeconomic policies implemented since 1992, policies which were advised and supported by international organizations and major Western powers which bear now no uncertain responsibility. The combination of a highly destructive macroeconomic and institutional environment with a highly complex and confused microeconomic situation is producing a crisis of an as-yet unknown magnitude. The social and political consequences of this crisis are probably beyond the management abilities of the Russian government. There is left then no other option than a major export drive, both for enterprise managers, and frequently new owners, and the government. This trend is putting relations between Russia and the West at risk and has the potential to fuel antagonistic perceptions on both sides.

Regional Concentration and Diversity of the Former Soviet Defense Industrial Sector

The Soviet defense-industrial sector has been predominantly Russian. Russia accounted for no less than 73 percent of its global production. Ukraine was second, with around 15 percent, followed by Belarus (5 percent) and Kazakstan (3 percent).[3] Even in the former Soviet Union, regional concentration of military-linked activities was quite high. As a whole, employment controlled by the VPK accounted for some 25 percent of total industrial employment in the RSFSR, 18.6 percent in Ukraine, and 17.4 percent in Belarus. These very high figures do not mean that one-fourth of the Russian labor force in 1989 was engaged in arms production. The military-industrial sector was a massive producer of civilian goods and could be thought of more as a high priority sector than a purely military one. In Russia, three regions exceeded the national average with a very high percentage of VPK-controlled employment: the Northwest (the St. Petersburg industrial basin as well as military production in Pskov and Narva), the Urals, and Volga-Vyatka. However, at the subregional level, we find some local concentration of military-industrial facilities even in regions usually thought to be less directly affected. A case in point is the concentration of defense-related factories near the city of Tambov in a region generally assessed as agricultural. In Ukraine, too, some regions had a concentration of VPK-controlled employment far superior to the republic average. VPK-controlled employment neared 72 percent in Nikolaevsk, and was significantly above 30 percent in regions such as Ivano-Frankovsk, Kiev, Crimea, Lvov, Kherson, Khmelnitski, and Cherkasy.[4]

This regional repartition led to two distinct problems. First, pure conversion is prone to generate massive regional unemployment pockets, with extremely severe social consequences, as is happening now in the Volga-Vyatka, Ural, and Western Siberia regions of Russia. Exacerbating this problem is the fact that part of the social system was managed by enterprises and not through a federal or local public sector. Second, the military-industrial sector was a Soviet creation; inter-republic trade was essential to its operation. The collapse of the Soviet Union, and much more so the collapse of inter-CIS trade, created its own set of economic and social difficulties. It makes sense to think about conversion at the level of the CIS and not as a Russian or Ukrainian policy. However, such a strategy is extremely remote if not only because Russian liberals savaged inter-CIS economic relations for the sake of the struggle against inflation. Conversion must be viewed as simultaneously a microeconomic and a macroeconomic issue, a national and a regional undertaking.[5]

The Soviet military-industrial sector should not be regarded as a homogeneous entity. It was not producing weapons alone, but many raw or semi-processed materials as well, and large quantities of civilian products.[6] One of every four workers in this sector was directly engaged in arms production. This heterogeneity also owed itself to the military-industrial sector's organization in concentric circles. The first circle was made up of enterprises belonging to ministries of defense-industrial branches, or MOOP. After the 1988 administrative reform this nucleus housed eight production ministries and the State Committee for Technical and Scientific Information.[7] The second circle was comprised of civilian ministries that oversaw enterprises manufacturing products with important military implications, such as machine tools, nonferrous metals, and atomic energy. A third circle was made up of various enterprises that belonged to other ministries and acted as subcontractors or suppliers for the two first circles.[8] These circles were under the control of the Military-Industrial Commission (VPK), the defense industry section of the Gosplan, and of ad hoc committees in the Central Committee of the CPSU. Furthermore, this sector lacked homogeneity as far as priority levels were concerned. More than a true military sector or military-industrial complex in the Western sense, it could have been described as a high priority sector.[9] However, its very size precluded the application of highest priority levels to the whole sector. Even if the most modern equipment was to be found in VPK-controlled enterprises, all enterprises belonging to this sector were not equally well equipped. Because it was much more profitable in terms of political leverage and status to create new plants rather than to modernize older ones, a structure developed in which very new factories existed alongside much older ones. Looking at the defense sector capital structure, we would find that different historical capital layers—early thirties, war years, late forties/early fifties, and so on—were still active but in separate plants. As a matter of fact, the global efficiency of this sector was not far better than that of the rest of the economy; it was no island of efficiency in a sea of waste.[10] At the same time, however, it was a highly capable sector, and certainly one with very strong administrative and political leverage, enabling it to frequently gain the upper hand against the user—that is, the Ministry of Defense.[11]

Is There a Specific Advantage for Defense-Related Enterprises?

One could argue that defense-related enterprises in former Soviet-type economies, because they enjoyed a very high position on priority lists, are in better shape than "pure" civilian enterprises and can thus adapt more easily to the new conditions. It is beyond doubt that industrial equipment is usually better and more modern in defense-related enterprises, and that this is the case even if the defense sector in the former Soviet Union was never a system completely insulated from deficiencies in the civilian sector.[12] However, this does not translate automatically into a greater flexibility. If priority rankings are assumed to have played under the Soviet system the same role as relative prices in a theoretical market economy, that is a good indicator of how difficult, and ultimately costly, it was to obtain resources.[13] To be higher on such a list would have been the systemic equivalent of paying a lower price for a given resource in a market economy. And a lower price translates more easily into waste than into productive efficiency, as users can be supposed to be more prone to misuse what they got for free or for a reduced amount of effort.[14] Hence we can assume that defense-related enterprises were certainly more capable than civilian ones, but not necessarily more efficient.[15]

A second drawback is the degree of technological specialization of industrial equipment. Overly specialized equipment is a liability when conversion toward completely different products is required. This point underscores the need for a microeconomic approach to the conversion process. At the microeconomic level, conversion potential could be better assessed by analyzing what assets are deployed by enterprises, and how are they deployed, than by focusing on what enterprises actually produce. As explained by Hilton,[16] any industrial enterprise can be broken down into four groups of assets. Competitive assets provide the enterprise with a given advantage on the demand side; efficiency assets lend advantage on the supply side by enabling enterprises to reduce costs. Complementary assets are those which enable any given enterprise to deploy in the most effective form other assets—that is, organizational skills—but also an ability to monitor and efficiently forecast what development could happen either in the supply or in the demand side. Knowledge of specific markets could then be critical. Lastly, specialist assets are those used by the enterprise to exploit the division of labor in the most efficient way. To some extent, the policy debate has focused so far on competitive assets; that is, what the would-be converted enterprise could produce for the civilian market. However, the complementary nature of all asset groups, and the very fact that they have to be used in a coordinated way, is a critical point in understanding the conversion potential of any given enterprise. What is more, an enterprise's competitive assets are usually linked to its technological base, normally assumed to be the strong point of military-industrial enterprises. But the technological base matters only when civilian products are already technologically close to military ones, or when the enterprise was producing dual-use components. And, military-related enterprises have been notoriously lacking in marketing skills and the ability to provide customers with specific services (maintenance, training, etc.).

A third important problem is the very nature of the technological culture that had developed in specific enterprises.[17] This is a particularly significant point when one keeps in mind the collective nature of the knowledge acquisition process, and the fact that a large part of economic and technological knowledge is both tacit and procedural—that is, not only the knowledge of something but the knowledge of procedures needed to improve the already existing stock of knowledge should unforeseen events arise.[18] Technological culture here means the core of tacit and collective knowledge accumulated over the life of the enterprise;[19] it amounts mainly to an unwritten internal history of an enterprise's record of success and failure.[20] The Soviet technological culture in defense-related enterprises appears to have been geared toward capabilities at any cost, or toward achievements in production not necessarily linked to achievements in efficiency.[21] More often than not the foreign observer touring defense enterprises today will hear about "what we can do"; very infrequently does one hear about "what we can sell." This engineer-driven culture has certainly been the key to some Soviet success, but it must recede in order for true entrepreneurial behavior to develop. Enterprise culture also has an organizational dimension; each enterprise can be looked upon as an organization resulting from diverse trade-offs between transaction and organization costs.[22] It is obvious that the very nature of the informational environment has played a major role in the development of defense-related enterprises in former Soviet-type economies. These enterprises had to achieve capability levels similar to those of Western firms in a poor, shortage-prone environment. Such a situation led to the development of specific organizational skills that are probably redundant in a transition economy where constraints and uncertainties belong to another logic.[23] What is more, the culture of secrecy which was so important in the defense sector can put an enterprise at a disadvantage when openness and ability to develop relations with new partners is needed.

A fourth important point is linked to formal and informal networking, as it was practiced under the Soviet-type economy.[24] It is important here to make a clear distinction between formal and informal networking. Formal networking is usually associated with subcontracting. The history of industrial development in the twentieth century shows a shift from the vertically integrated firm model (so obvious in the actual structure of Krupp or Ford[25]) that was dominant until the big depression to a new model in which one can identify a "leader" surrounded by suppliers. This shift was probably linked to a relative reduction in transaction costs and the simultaneous rise of competition, which necessitated cutting as many organizational costs as possible.[26] In the traditional Soviet economy, supply disruptions and shortages led to a very high level of transaction costs, and competition was restricted if only because the general excess of demand on supply would have led, as in any war economy, to a seller's market even without State regulations and planning. State property has at the same time considerably restrained the process of new firm creation even if some defense-related Soviet enterprises helped engineers to set up nearly autonomous firms well before the beginning of perestroika.[27] It was then logical for Soviet industry to follow the pattern of early-century Western industry. To some extent, the situation of the military-industrial complex was quite different from that of the rest of the economy. Because of its high priority, disruptions and shortages were considerably less acute. Competition, even if not at the same level as in the civilian sector of market economies, was nonetheless more present than elsewhere, either because of direct competition between two potential suppliers (like MiG and Sukhoi for fighter aircraft, or in tactical missile design),[28] or because of extremely demanding requirements and specifications (Taktiko-Tekhnicheskoye Zadaniye or TTZ, and Tekhnicheskiye Usloviya or TU) aimed at pushing enterprises to the very edge of their technological capabilities. It is no wonder that Soviet defense-related enterprises have been less affected by transaction costs and more sensitive to organizational rather than civilian ones. For the very same reason as in market economies they responded to such a situation by developing subcontracting when civilian enterprises were still strictly committed to the vertical integration model[29] and were clearly conservative in design.[30] Interfirm cooperation has thus been much more developed in the military-industrial sector, a fact helped by the greater political status achieved by some military-industrial designers. However, formal networking is not the last word as far as networking is concerned. Allocation system rigidities and shortages have induced plant managers to develop an informal, and illegal, barter system stressing complementarities either in production or in abilities to get specific inputs. The level of this trade in the final twenty years of the Soviet system seems to have been quite considerable, reaching perhaps as much as 50 percent of the official trade.[31] Because of its informal and illegal nature this trade system relied upon the development of a high degree of confidence between partners. This probably explains why it was mainly a local phenomenon, leading to barter networks at the city or the *krai* level. Because of their relative advantage in acquiring some inputs, defense-related factories have generally achieved a dominant position in these networks, using them not primarily to overcome actual shortages (with the exception of consumption goods needed by the workforce) but to gain more flexibility in responding to new or shifting requirements. One can then assume that networking, formal and informal, developing along both vertical (i.e., technological complementarity) and horizontal (i.e., local) lines, had been much more prominent for defense-related enterprises than for civilian ones.

During transition, where the usual information-gathering channels are blurred or inoperative, and under a high degree of institutional uncertainty, bilateral relations are of the utmost relevance. The ability to maintain formal and informal networks could be a decisive advantage for defense-related enterprises. On the other hand the nature of transitional change is probably lessening the value of some of the accumulated knowledge of defense-related enterprises, particularly that of administrative bargaining and priority-hunting skills. The balance between advantage and disadvantage is then far from obvious.

To sum up, defense industries in former command economies have frequently been highly capable. However, they have also been—and remain today and probably will remain for some time to come—highly inefficient. They are unable to immediately compete effectively with companies from the West or from the newly industrialized countries without some major restructuring and/or a massive undervaluation of the national currency. It has to be understood that restructuring as touched on here does not end with reducing the number of employees. Reducing the adjustment problem to overstaffing alone would be a tragic mistake, even if overstaffing is certainly present.

Blurred Conversion Perceptions and Policies before and after 1992

The conversion policy was officially instigated by Gorbachev in 1988. However, one can find some earlier examples of conversion attempts in the sixties.[32] What was different then was the scope and the context. But if it was clear even then that conversion would become a major aim of the new policy in the future, it was far from clear what conversion meant.

Uncertainties surrounding conversion from 1988 up to now were linked to many factors. First, and most obvious, the Soviet leadership was divided about the role of the conversion process. Soviet economists have argued that the VPK sector could be used as a reserve of advanced technology to achieve modernization of Soviet industry, enabling it to attain the level of competitiveness necessary in the course of marketization.[33] Conversion was first understood simply as a means to channel more resources toward civilian production. Then it was argued that the military-industrial sector could be an example for the civilian industry. Some of its organizational patterns were applied in the "civilian" sector. Second, there was an obvious gap between a consensus understanding of conversion, the subsequent ability to formulate policies based on this, and the fact that military spending was rapidly declining. Whatever the exact role of conversion in the initial blueprints, as military contracts were vanishing it became suddenly urgent to think about a global policy.[34] Part of the former reasoning was then devalued by the very speed of the contraction of the military budget. A third factor here was the doctrinal debate about structural policy. For Russian liberals, at least in 1991 and 1992, the very idea of any industrial policy was a symptom of conservatism, a relic from the bygone central planning period.[35] During the first half of 1992, no conversion policy at all was discussed in the Russian government. Only the increasing potential for economic and social problems ensuing from the possible collapse of a major part of the industrial infrastructure roused some of the young liberals. But once the idea of a structural policy was revived, the difficult issue arose of determining who would develop and implement it. Debate then became less doctrinal and much more political.[36] Various administrations clashed on competence issues in the second half of 1992 and 1993. In 1994, it was still far from clear if conversion would be run by the Ministry of the Economy, the Ministry of Industry, the Ministry of Defense, or some kind of inter-ministerial committee. One has here to consider the perverse effect of the political and administrative instability that developed after mid-1993. Even if the presidency formally has the upper hand, confusion over political and administrative agendas has resulted in constant infighting between the major ministries and the administration. Except for the president's declarations, which to date have been far from coherent, there are no rules to manage these conflicts and bureaucratic competition. The new Russian system is repeating an old Tsarist tradition which has historically not been very helpful in establishing economic stability.

The hesitations and fluctuations that characterized the conversion policy nonetheless led to some learning. By 1993, more realistic assessments of what could realistically be achieved were evident. Conversion prospects would have certainly been better but for the industrial crisis of the first half of 1994, which has been worrisome. Official declarations have stressed three main points of what is yet to be called an industrial policy program. First, the old Soviet practice of maintaining extensive mobilization capacities will be abandoned, enabling the Russian government to focus its efforts on a reduced number of purely military enterprises. A nucleus of some two hundred to four hundred firms is to be preserved at any cost, as they are seen as the core of the industrial-technological potential of the country. These enterprises would concentrate on military contracts but simultaneously maintain civilian production capacities. This is a kind of diversification, with priority nonetheless assigned to the military production. Such a policy certainly entails the concentration of production in a few plants and the closure of redundant facilities.[37] A process of dismantlement is seen as inevitable, and the government will try to concentrate orders on the most efficient producers. Organizational changes are the second feature of this program. Many Russian decision-makers have urged the creation of big industrial-financial conglomerates on the South Korean *chaebols* model.[38] Partial privatization would then be possible, even if the government kept strict controls on these groups. Specific financial infrastructures are to be developed to help investment and ease transactions, but such a program would require a general stabilization of the banking and financial sphere, something which does not seem to be at hand in 1995. Cooperation with Western companies is seen here as extremely important to enable Russian companies to both improve their technological abilities and to gain a foothold in Western markets.[39] Some results have already been obtained, including the development of a new MiG trainer aircraft in cooperation with the French, and extensive agreements between the Klimov jet engine company and U.S. and Canadian companies.[40] The third emphasis is on identifying priorities. Military contracts will be concentrated on research and development, with some small-scale production for testing and keeping production facilities operational. But the technological potential of military-industrial companies is to be used also in rebuilding some civilian infrastructure in areas including transportation, oil and gas production, telecommunications, and health. Such a program is certainly more realistic than what has been previously put forth. It does not try to convert the whole sector, seeking instead a balance between conversion and diversification, and between microeconomic and macroeconomic conversion, through closing redundant facilities.

Many problems are still evident, however. First, Russian decision makers have tacitly assumed a demand level, both for military hardware and civilian products, that was not compatible with the budget policy of early 1994, as the budget debate of April and May 1994 amply demonstrated, and was not consistent with the pledge of a reduced deficit made to gain IMF standby loans in 1995. Diversification and conversion also need an important investment effort which clearly is lacking. Whatever industrial and restructuring policies emanate from the government are closely dependent on the macroeconomic context. Here, conversion interacts with macroeconomic stabilization.

Macroeconomic Constraints on Conversion: The Problem of Disappearing Investments

Whatever the kind of conversion envisioned, be it through restructuring and adaptation of existing firms, or through a creative destruction process wherein new enterprises rise from the ashes of older ones, investment is the key. But investment is not only a microeconomic decision; it is clearly linked to the macroeconomic context. Obviously, no one will invest if no returns can be seen in the short to medium term. There is a clear link between demand and investment levels. Further, demand fluctuations could play an extremely destabilizing role whatever the demand level, because of specific investment characteristics. This is why it is important to assess the effects of stabilization policies implemented from 1992 to the present on the investment process.

Investment decisions are usually linked to four factors. Profitability[41] can be seen as the main incentive to invest, even if computation is quite difficult.[42] We must nonetheless take into account the fact that stability of profit could be as important as the actual profit level. If the recoupment time of an investment is any but the shortest possible, and in industry it is frequently quite long, particularly when investments are focusing not on expanding the production basis but on changing the technology, relative prices and production are decisive factors. When profit parameters are unstable, rational economic agents will reduce their risk by concentrating on extremely short-term activities, with a low capital intensity.[43] But profit parameters could also be linked to demand in a traditional Keynesian view. This explanation emphasizes the fact that investment is a two-step decision.[44] First, economic agents have to make up their minds about what level of additional production capacity they want. Second, they must then decide the capital intensity and the technology to be used. The Keynesian approach highlights an extremely important fact: investment decisions are irreversible.[45] Any mistake would entail a significant loss of welfare. Uncertainties about the future state (quantitative and qualitative) of the aggregate demand are thus an extremely important factor restraining investment.[46] Here the nature of the information and the degree of uncertainty are the two main factors to be taken into account. Even when profit expectations are good in the short run, uncertainty about the medium-term aggregate demand could, because of the irreversibility factor, lead enterprises to invest less than otherwise expected. Then, if we admit that at least some economic agents are able to implement strategic actions—that is, individual strategies able to significantly affect the situation of other agents—coordination could be a critical factor as well. This is particularly important if increasing returns (as in emerging industries) are to be expected, or if external finance (either bank credit or stock markets) is to play a significant role in investment. External finance is frequently linked to too low a profit ratio. But even enterprises with a very high profit ratio could need external finance, as capital assets are not infinitely divisible. Investment could have to take the form of a step-by-step process, where enterprises are spending discrete amounts of money. Internal and external finance are then not completely substitutable.[47] Coordination failures, mainly in markets where increasing returns can be expected, could lead to an insufficient investment trap. If all agents are too cautious (or risk-averse, in economic jargon), they will opt for insufficient investment levels, which will reduce their profitability and, in the second phase, justify caution, and more restraint, and cause another deterioration of profitability.[48] If enterprises have to rely upon external finance, and if the financial and banking system is disorganized and itself a cause of uncertainties, which is exactly the situation in Russia, the cost of external finance will be too high. Enterprises would then have to use internal finance and, as it is not perfectly substitutable for external finance, both the aggregate level and the composition of investment could be inefficient. A fourth factor influencing private investment decision is the nature and level of public investments. Private and public investment are complementary in many cases, and public investment is instrumental in creating conditions that favor and encourage private investment.[49]

Usually, stabilization programs focus on restricting public expenditures, tightening monetary policies, and accessing the international economy through convertibility. These programs have been found to deepen the economic depression and considerably slow the adjustment process.[50] If this happens, transition from stabilization to sustainable growth could become a goal out of reach. Even if stabilization can be achieved, belated restructuring would induce a resurgence of inflationary pressure, increased depression, and growing political discontent.[51] Among the worst problems linked to stabilization programs is the tight budgetary policy. It would first of all translate into a sharp reduction of public investments. But in former Soviet-type economies, including Russia, public and private investments are complementary, especially because much of the infrastructure necessary for a market economy is lacking.[52] Second, the tight monetary policy, if implemented through interest rates as has been done since mid-1993, could induce a kind of perverse crowding out, where privately owned enterprises would be ousted from the credit market[53] and, at the same time, tight money would induce firms to rely upon internal finance, a situation leading to both insufficient quantity and composition of investment.[54] Last but not least, tight monetary policy is here generating a much greater depressive effect than on a market economy, as, because of the lack or underdevelopment of financial markets, credit plays a much more important role for enterprises. The reduction in aggregate demand, induced by tight budgetary and monetary policies, combined with a brutal increase in competition through openness, could, in an oligopolistic environment, generate the kind of coordination failure already described. Enterprises would overreact by reducing both investment and production levels, which would induce a greater-than-anticipated depression which would then validate and consolidate this overreaction for the next period.

Too sharp an initial devaluation, usually in stabilization programs using a nominal anchor (that is, a fixed exchange rate), could make imported capital goods too expensive for private investors. The question then is to what degree economic restructuring is dependent upon imported capital goods, a question obviously related to the former industrial development history of the country. The export boost induced by such a devaluation would also be limited to raw materials or semi-processed products, as enterprises are lacking both knowledge of and means of sale to foreign markets to make decisive progress in exporting industrial, high value-added goods. The devaluation could then restrict, both through limitation on imports and distortion of exports, a move toward high value-added production, leading to the kind of perverse restructuring we experienced in Russia. In the second phase, if inflation has not been kept under control, as in Russia, any attempt to use the exchange rate as an anti-inflation device would lead to a creeping reevaluation of the national currency, as inflation would outpace devaluation. Then enterprises starved of investment in the first phase would be unable to compete with their foreign competitors because they would lack any protection while undergoing restructuring.

The indirect negative consequences of stabilization programs are no less important. One can first point to the very combination of a quick move to convertibility with widespread price liberalization when neither financial markets nor trade system infrastructures exist. Such a move could become a kind of massive "noise" creation, distorting all explicit signals upon which the stabilization program relies to achieve a more efficient resource allocation. Relative prices and other profit factors would then become highly unstable, with wide variations across the national territory.[55] Trading activities, too limited in scope and carried out under massive uncertainties, would then increase the instability of relative prices.[56]

Even if internal relative prices are evolving predictably, a combination of a nominal anchor with a non-negligible residual inflation would lead to actual exchange rates becoming unstable.[57] Such instability is highly detrimental to any mid- to long-term decisions on resource reallocation. When this happens, results of the stabilization programs become uncertain, government credibility is eroded, and a new uncertainty is created.[58] Economic agents wonder if the government will loosen the belt a little, with a resulting increase of demand, or if it will try to tighten it. Medium-term evolutions of interest rates (or credit allocations), demand, and prices also become uncertain. By itself, the erosion of credibility induced by aiming at a too-sharp and too-quick stabilization, a common mistake in stabilization programs, is extremely self-destructive for any economic policy.

Currently, the investment problem translates into internal demand levels, as Russian exports are still more or less unwelcome in Western markets, and in particular a high level of government-induced demand. Health, transportation, and telecommunication infrastructures are to be financed through government spending, but even when funds are allocated for these programs, they are more frequently than not "sequestrated" by the Ministry of Finance to keep the deficit down. Another side of the internal demand level is trade protection. As the purchasing power of the ruble has been fast increasing in 1993 and early 1994 (the exchange rate drift is slower than inflation), internal production has been facing tougher competition from imported goods. This is particularly detrimental in the delicate restructuring phase, when Russian companies are far from being as efficient as Western firms. Taxes on imports or manipulation of exchange rates are obvious necessities, not only on products that the converted or diversified industry is supposed to make but also on goods produced by users of converted industry production. But by mid-1994, for doctrinal and political reasons, the Russian government remains extremely cautious on this issue. As a matter of fact Moscow and St. Petersburg authorities have threatened not to pay federal taxes if import taxes are raised. The current behavior of the government casts doubt on its ability to make choices and to implement a true industrial policy.

Tentative Enterprise-Level Adjustment to the So-Called "Shock Therapy"

The macroeconomic policy implemented first in the beginning of 1992 and again after June 1993 obviously marked a new stage in the conversion process and has significant implications at the enterprise level. The military-industrial sector had to face a three-pronged shock.

First was a demand shock, linked to the accelerated contraction of defense spending and restructuring with a much greater than proportional contraction of procurement. Simultaneously there was a supply shock, induced both by the price liberalization and the collapse of inter-republic trade. The critical situation of interregional trade in Russia exacerbated this shock because the military-industrial sector, as the beneficiary of a greater priority status than other sectors, had developed internalization at the network level more than at the local or plant levels. Third, this sector had to face a major organizational or administrative shock with the formal dissolution of the VPK, of ad hoc CPSU committees, and of various command and information links instrumental in getting relevant information.[59] This was particularly important as this sector was probably the most dependent on information not carried through the media of prices. Change in administrative structures was another side of this organizational shock. In the former Soviet Union networks of large companies with multiple plants did not exist. When we heard about MiG, Sukhoi, or Tupolev, we heard about two different things, design bureaus with limited production facilities, and networks of main production plants and subcontractors actually producing the hardware. Links between design bureaus and networks were real but loose, at least initially. A main production plant could switch from a MiG type to a Tupolev type according to VPK plans. With possible privatization, and mandatory "commercialization" (that is, transformation to joint-stock status), military-industrial enterprises had to strengthen their links to achieve better cooperation and reduce overhead and transaction costs. This process led to the creation of associations, which can be seen as harbingers of large industrial groups. The whole process actually was a blessing for the Russian industry in the long term. However, during 1991 and 1992 it induced painful negotiations and bargaining, which consumed part of the time that management could have used to devise internal restructuring programs.

In spite of a critical situation, economic discipline was quite good in this sector in 1992 and the first half of 1993. The big arrears crisis of 1992 was not linked to military enterprises.[60] Enterprise managers genuinely tried to develop a kind of entrepreneurial conversion, feverishly seeking Western partners and new organizational patterns. It was only when they perceived that they were not being rewarded for their efforts that discipline began to weaken, and exports and an increase in the Russian military budget began to be seen as alternatives to conversion. As a matter of fact, the former VPK sector was initially less affected by the economic depression in 1992 than pure civilian industries.[61] This was partly because of the better equipment, greater material reserves, and good connections between managers of these enterprises, which enabled them to ride out part of the supply shock affecting Russia's economy. The situation began to worsen in 1993 with the new round of shock therapy that began in June. Problems arose not only from the nature of the economic policy but from how it has been implemented—that is, through budget sequestrations and a brutal rise in interest rates. It was mainly with the beginning of State non-payments on a large scale, after September 1993, that a true crisis began to develop in this sector. Part of the rapid decline in industrial production at the end of 1993 and first half of 1994 could be linked to this crisis. Significantly, regions where the military-industrial sector is dominant that supported Yeltsin and reforms in the April 1993 referendum voted against Gaidar and reforms in the general elections of December 1993.[62]

Changes both in the macroeconomic and the institutional context have been considerable and far-reaching. They had a considerable impact on all Russian enterprises, but particularly on defense-related enterprises. Four strategies employed by Russian defense-related enterprises have been identified: real adjustment, fragile real adjustment, sophisticated rent seeking, and crude rent seeking.[63] Real adjustment is obviously the most promising and implies an ability to find new products, new ways of producing, and new markets. However, the combination of demand contraction and instability has considerably restricted opportunities to develop such a strategy. Whatever the inclination of the local manager, the macroeconomic context and the institutional and legal environment (or the very absence of the latter) are doing a lot to push enterprises away from realistic real adjustment strategies. Either enterprises have found it impossible to secure the investment needed to modernize already existing civilian production capacities, or new products were unable to find a market even when they were clearly needed because of a lack of money. Financial structures were frequently too unstable to support enterprise restructuring and at the same time too closed to new entrants to promote enterprise creation. Relative price variations induced a large profit uncertainty in manufactured goods production, leading what has been left as far as investment was concerned to be used in trade and speculation. Such a situation was not only materially detrimental to any industrial restructuring but also psychologically discouraged long-term investment by making nonproductive activities much more rewarding than productive ones. One has also to add that the crumbling of the legal basis, and the loss of State legitimacy resulting from some aspects of the official liberal policy, led to a considerable rise in transaction costs. Only bilateral links can be used to enforce contracting. Such a situation not only limits opportunities for existing enterprises to find new partners but also dramatically restricts the possibility of new entries. Destruction is then moving quickly, but without any creation. In some cases rent seeking, particularly if connected to exports, can be seen as a more realistic option at the company level than any other adjustment strategy.

The Roots of the Export Drive

The export drive can be seen as the first consequence of the lack of a clear and sensible conversion policy and the preeminence of a macroeconomic context adverse to any attempts at large-scale industrial restructuring.

Exports were first seen as a way to finance conversion, not as an alternative to it. In a sense one can view the massive Russian exports of raw and semi-processed materials in 1992 (which nearly destroyed the world aluminum market) as a response to a vanishing internal demand.[64] The military-industrial sector was a massive producer of these commodities. Exports here were not only a way to survive but were also seen as a way to raise funds for retooling factories and adapting them to a market environment. However, such an export drive quickly ran into opposition from the West. The aluminum market is a typical example. The reactions of Western governments, especially those already faced with internal depression, are understandable. Nonetheless, restricting Russian exports of raw materials can only exacerbate the problem of arms exports.

The legal framework and control system in this field was established quickly. From February to May 1992, four decrees were aimed at the creation of an organizational system clearly inspired by what is operating in some Western countries.[65] Two special bodies were created, the Commission for Military and Technical Cooperation (KVTS) and the Commission on Export Controls (CEC). They were intended to give the State a means to distinguish itself from industrial interests and to organize cooperation between the various bodies, from ministries to companies, which were to be involved in arms or high-technology items. These decrees were accompanied by a published list of sensitive materials subject to specific restrictions.[66] However, until November 1993, at least three different organizations were empowered to promote arms exports: Oboronexport, Spetsveshtekhnika, and GUSK. The fact that they all emanated from the Ministry of Foreign Trade did not help to reduce the competition between these organizations, competition that submitted the system of controls to manifold pressures. Compounded by the fact that results have not been successful, this led to a process of partial deregulation of arms exports, with licenses given to a large number of enterprises and even regional authorities. By November 1993, these three organizations were merged as *Rosvooruzhenie*, a monopolist company under strict presidential control. It appears, however, that *Rosvooruzhenie* is more likely to be active in finalizing contracts. A large part of the marketing business still seems to be handled either by individual companies, or by special emissaries of the General Staff. As with the rest of the Russian administration, arms export controls suffered from the same process of blurred command and hierarchical lines, with those able to claim favor with the president or his closest advisors often bypassing the system. Bureaucratic competition and the multiplication of vested interests entail a drastic reduction in oversight and actual control abilities. Exports are then seen as the only means to assure the survival of the high-technology sector. One can assume that many attempts to sell "hot" high-technology items, even when officially labeled as irresponsible local actions, have met with the tacit approval of some policymakers. Certainly such efforts had some support from the General Staff if one can extrapolate from a paper on conversion that was written by an instructor at the General Staff Academy. In this document, conversion is viewed as a long and arduous, but ultimately needed, process. Conversion is also linked to economic national security, which is understood as the ability of Russian industry to provide needed equipment, and the creation of an internationally competitive industry as well.[67] In such a context every means must be employed to prevent the technological decline of key enterprises; in the absence of a buoyant civilian market for high-technology products, exports are and will be for a time the only solution.

Arms exports were also sought, but at first not very aggressively. This progressively changed when it was realized that no government subsidies were forthcoming and that internal demand was spiraling downward. But then Russian arms producers had to face a world market in which demand had contracted and competition from the West was fierce. The exhibition of advanced Russian military hardware at international fairs (like in Abu Dhabi) helped a little, but results were far from satisfactory: arms sales in 1992 generated between 2.7 and 4 billion U.S. dollars, versus 11 billion in 1989.[68] But even if the global amount has been considerably restricted, it is clear that exports have helped some Russian weapons manufacturers to survive. It would be a mistake to infer from a reduced amount of money a reduced relevance of these exports. On the contrary, with current internal procurements just a drop of what they were only five years before, these contracts are giving manufacturers vital breathing room. Because of this it is easy to foresee not only the continuation but quite likely an increase in Russia's willingness to export at any cost.

This raises some political problems, however, due in no small part to the identity of the customers. Western governments, and particularly the U.S. government, have voiced strong concerns about Russia's export policy. Here, however, there is an obvious contradiction in Western behavior toward Russia. On one hand, G7 countries are troubled by Russian exports, but on the other hand they support the very economic policies advised by international organizations they dominate which lead to a massive contraction of internal demand, making conversion nearly impossible. This at the very least ambiguous attitude has generated strong anti-Western feelings in Russia. Some Russian leaders view the Western policy as an attempt to destroy Russia's industry, and in particular high-technology fields linked to the military-industrial sector.[69]

Is There Really an Alternative to Export?

There is as yet no indication of a significant change either in economic or institutional policy in Russia. If some limited successes in conversion can be noted, usually near Moscow, still it can be reasonably forecasted that the present industrial crisis will not recede in the next two to three years. The Russian government, and some of its Western advisers, have idealized marketization and have been unable to address correctly the industrial restructuring problem, either at the macro or the micro level.[70] Unless a drastic change occurs, closing plants and implementing bankruptcy law will become the only possibility, but will give rise to severe regional and social problems. The most obvious would be a quick rise in unemployment. If by October 1994 official statistics were showing only 1.69 million unemployed people,[71] a national survey in December 1993 showed that the actual number was probably four million.[72] Researchers from the International Labor Organization (ILO) have convincingly argued that by the end of 1994 actual figures on unemployment were five times higher than official ones.[73] But the unemployment problem does not stop there. A large number of workers are left without work and without pay even if they are not officially dismissed. As much as 35 percent of the national labor force may be in such a situation, according to ILO surveys.[74] Such a situation results directly from the way social benefits, and the social infrastructure more globally, is managed in Russia. A large portion of social benefits are managed through enterprises; the closure of any large enterprise threatens to destroy the local social infrastructure. Making things worse is the fact that military-industrial enterprises are highly regionally concentrated. This leads obviously to the problem of regional lobbies. Economic choices translate in Russia into regional choices.[75] Such choices are extremely difficult to manage due to the near impossibility of having large-scale labor mobility because of the lack of housing, and the quasi-destruction of the federal tax system resulting from Boris Federov's policy in 1993.

Any industrial policy, and we assume that conversion is basically a case in industrial policy, must be matched in Russia by a regional development parallel. Today, if some decision-makers have clearly understood this link and its related problems, there is still no administrative structure and no financial or economic means to implement such a regional policy. Here economic and political problems merge; the lack of a true federal structure, reliance on presidential power, and neglect of the political consequences of public budget contractions are making implementation of a sensible conversion policy, or for that matter of any industrial policy, an extremely difficult undertaking.

A second problem is the large income differentiation process taking place in Russia. This problem has to be seen from two different perspectives. From an economic point of view, too high a concentration of income in too small a segment of the national population is usually linked to speculative or rent-seeking activities. Historical studies make clear that people who are benefiting from such a process are usually aware of its instability. They are much more prone to spend their money in conspicuous consumption or to engage in capital flight than to invest. The actual level of invested savings is then lower than what it could have been with a different income distribution. The second perspective is clearly political; to what extent is income differentiation politically acceptable when not linked to economic fairness? The very fact that the new income differentiation combines special "links" (or *svyazy*), which translates into the notorious corruption and insider problems faced by many economic activities in Russia, and a premium on activities which are not perceived as productive (banking, hot money speculation), is giving rise to considerable discontent and even anger. This reaction is nowhere more palatable than among engineers or highly skilled technicians previously working in the defense-related sector.

Gini's curve for Russia at comparable times in 1993 and 1994

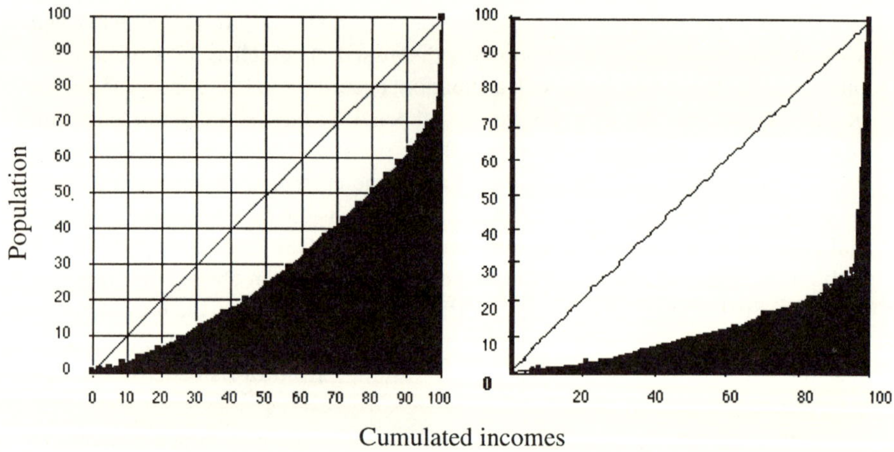

Source: GOSKOMSTAT data.

The combination of these two problems is making it highly improbable that the Russian government will show any willingness to reduce the current export drive, assuming there is still a way to implement an efficient control, which is increasingly doubtful. Furthermore, it offers little hope for any agreement, be it implemented or not, on either customers or on products exported to Western countries.

Conclusion

As discussed previously, Russia is currently committed to an economic policy that places considerable strain on the internal market, because of a deflationary budget and monetary policies and an overly rapid opening to world markets, without any insurance that the world market can compensate for the constrained internal demand. To make things worse, most of the administrative tools which could have been used together with a market approach to help the restructuring process are already tainted or made inefficient because of political manipulation for the sake of a limited number of operators usually closely connected with the presidential structure. Institutional chaos is abetting mistaken macroeconomics. The very legitimacy of the state has been weakened considerably, and relations between the center and the regions have been strained. Such a context is making conversion, by whatever means, an extremely improbable achievement. The export drive, and particularly arms export, is the only way for Russian decision-makers to avoid a complete collapse of the military-industrial sector with all its attendant social, political, and strategic consequences. Such a situation gives rise to two primary concerns.

Nuclear proliferation is the obvious first. However, the focus on nuclear proliferation could very well be a case of the trees masking the forest. Up to now, nuclear proliferation brought about by the disintegration of the Soviet Union has been negligible. This fact has been officially acknowledged by the U.S. intelligence community.[76] The making of a viable nuclear arsenal is a long, difficult, and costly undertaking, and not something which can be dramatically sped up through the smuggling of materials. This issue would certainly be better handled by addressing the national security problems of the countries involved and offering them assurances of regional strategic and economic stability rather than through negative sanctions.[77] But behind nuclear proliferation is a real problem, the diffusion and proliferation of conventional weapons. New and advanced military or dual-use technologies could be used to develop advanced weapon systems or to upgrade existing ones to a level that would create regional strategic instability. Not only the combination of highly potent but simultaneously highly vulnerable weapon systems could make military preemption a viable strategy in a number of cases, but the very power of some conventional weapon systems, mainly ballistic missiles, could cause the regional expansion of what was initially a localized conflict. Escalation, both vertical and horizontal, would become relevant not only for the superpowers, but also for small or regional powers. The second problem is directly linked to what we have seen in previous sections of this paper. It is clear that Russia's future course concerning exports is of serious concern. To argue that Russia will become what could be called a rogue exporter is not without meaning, but it is important to define more accurately what such a notion means. A rogue exporter can be seen as any country that refuses to comply with international agreements on international trade. However, such a definition would be broad enough to include even the United States, as the U.S. agreed to be part of the new World Trade Organization under the condition that the organization not adopt policies contradictory to U.S. interests. A second possible definition could be any country breaking UN-endorsed embargoes. Here again, this definition, if narrower than the first, is not without problems. If UN resolutions need a positive majority, any member of the Security Council with veto power can block a majority attempt to overturn a previous decision. Then an embargo can begin as a majority decision and survive as a minority one. Breaking the embargo would then have quite different meanings depending on the situation. And, even if an agreed-upon definition of a rogue exporter can be achieved, the reasons behind rogue behaviors can be far from clear.

Government-level decisions to supply some countries with arms or critical technologies are behind some "rogue" behavior. But there is also a second situation where a government is unable to control the export process because it lacks either the tools or the legitimacy to enforce control. And then there are countries where the governmental administration is divided and part of it is turning a blind eye to export practices either for political reasons or because of corruption. In the first situation the country is a rogue by intent, in the second a rogue by default, and in the third a rogue by consent.

Today Russia is running the risk of becoming a rogue by default. The weakness of the State and the near disintegration of the administration have reduced the degree of control over some critical flows. However, this situation could devolve into that of a rogue by consent, and then a rogue by intent. Such a bleak evolution would probably be interpreted by some Western analysts as the result of conservative Russian imperialist or neo-communist thinking. This would be a tragic mistake. If Russia becomes a rogue exporter it will be the result of pro-Western, liberal-minded policies. No one could deny a country the right to have a share in high technologies. But international organizations, like IMF or the World Bank, have backed Russia into an economic and social corner where exports, at any political price, are appearing to be the only solution for the survival of its high technology base. If Western countries look upon Russia as a rogue exporter, a large part of the Russian elite is already looking upon Western countries as double-dealers. The combination of these two perceptions does not bode well for the future of international relations.

Notes

1. J. Sapir, *The Soviet Military System* (Oxford: Polity Press, 1991).

2. J. Sapir, "Conversion of Russian Defense Industries: A Macroeconomic and Regional Perspective," in *Privatization, Conversion, and Enterprise Reform in Russia*, edited by M. McFaul and T. Perlmutter (Boulder, CO: Westview Press, 1995); idem, "La conversion des industries de défense de l'ex-URSS: dimensions, convulsions, perspectives," in *Mondes en Développement*, Tome 21, no. 83/1993, pp. 99–114.

3. J. Sapir, "L'avenir des forces armées en Russie (I)", in *L'Armement*, no. 38 (July 1993), pp. 8–17.

4. J. Sapir, *Les bases futures de la puissance militaire russe*, GSD-CIRPES, Cahiers d'Etudes Stratégiques no. 16, Paris, January 1993

5. J. Sapir, "Conversion of Russian Defense Industries: a Macroeconomic and Regional Perspective."

6. Including 100 percent of Soviet production of radio sets, 94 percent of television sets, 100 percent of sewing machines, 77 percent of vacuum cleaners, and so on; see *Vestnik Statistiki*, no. 5/1989, May.

7. "Pravitel'stvo strani: ego funktsii, polnostiya, otvetsvennost'," in *Pravitel'stvenniy Vestnik*, no.18, 1989, p. 23.

8. GERSS, "Reconversion Industrielle?" in J. Sapir, Gen. R. Ernould and D. Pineye, *La Décomposition de l'armée soviétique*, Presses de la Fondation pour les Études de Défense Nationale, Cahiers no. 46, 1992.

9. C. Davis, *Interdependance of the Defense and Civilian Sectors in the Contemporary Soviet Economy: Concept, Problems and Reforms*, paper presented to the AAASS convention panel on Major Defense Economic Trends under Gorbachev, Chicago, November 1989.

10. L. Badgett, *Defeated by a Maze: The Soviet Economy and Its Defense Industrial Sector* (Santa Monica, CA: RAND, 1988).

11. J. Sapir, *The Soviet Military System*. P. Almquist, "Soviet Military Acquisition: From a Seller's Market to a Buyer's?" *In The Soviet Military Factor in a Changing World*, edited by S. Clarck (Boulder, CO: Westview Press, 1991).

12. C. Davis, *Interdependance of the Defense and Civilian Sectors in the Contemporary Soviet Economy*, paper presented to the AAASS 1989 conference, Chicago.

13. See J. Sapir, *L'Économie Mobilisée*, Paris, la Découverte, 1990.

14. J. Sapir, *Les bases futures de la puissance miltaire russe*, Cahiers d'Études Stratégiques no.16, Paris, GSD—École des Hautes Études en Sciences Sociales, 1993.

15. See L. Badgett, *Defeated by a Maze: the Soviet Economy and Its Defense Sector*. T. Malleret, *Les transferts de technologie non négociés de l'ouest à destination de l'URSS et leur contribution au potentiel militaire soviétique*, Ph.D., École des Hautes Études en Sciences Sociales, Paris, 1992.

16. B. Hilton, *Defence Conversion or Diversification, East and West: An Overview of the Literature and the Arguments* (Oxford: Templeton College, Management Research Papers, 1992).

17. J. Sapir, *La culture technologique: le cas des matériels militaires soviétiques*, Compiègne, Université de Technologie de Compiègne, Cahiers de Recherches no. 3, 1990. idem, "Culture économique, culture technologique, culture organisationnelle," in *Cahiers du Monde Russe*, no. 1–2/1995.

18. F. A. von Hayek, *The Sensory Order* (London: Routledge & Paul, 1952).

19. On the tacit dimension of the technological knowledge, see D. C. Mowery, "Economic Theory and Government Technology Policy," *Policy Sciences*, no. 1 (1983).

20. See S. Hooker, *Not Much of an Engineer* (Warrendale, PA: Society of Automotive Engineers, 1991).

21. J. Sapir, *The Soviet Military System*.

22. E. Penrose, *The Growth of the Firm* (Oxford: Basil Blackwell, 1968). O. J. L.Ravix, "L'émergence de la firme et des coopérations inter-firmes dans la théorie de l'organisation industrielle," in *Revue d'Économie Industrielle* 51, no.1, 1990.

23. B. Hilton, *Defence Conversion or Diversification.*

24. L. Freikman, "Shaping a market environment and analysis of the Enterprise Interaction Mechanism," in *Studies on Soviet Economic Development* 3, no. 2 (1992).

25. Ford's organization was closely copied in the early thirties by Soviet engineers building the Gorki car plant. See J. Sapir, *Organisation du travail, classe ouvrière et rapports sociaux en URSS de 1924 à 1941*, Thèse de 3e cycle à l'EHESS, 20 May 1980, vol. 2, p. 570.

26. M. Aoki, "Horizontal vs. vertical information structure of the firm," *American Economic Review* 76, no. 5 (December 1986). B. Guilhon, "Technologie, organisation et performance—le cas de la firme-réseau," in *Revue d'économie politique* 102, no. 4 (July–August 1992). W. Powell, "Neither market nor hierarchy: networks form of organization," *Research in Organizational Behavior* 12, no. 1 (1990).

27. L. Chaiko, *Helicopter Construction in the USSR* (Falls Church, VA: Delphic Associates, 1986), Ch. 5.

28. See A. Steinhaus, *The Beginnings of Soviet Military Electronics—1948–1961* (Falls Church, VA: Delphic Associates, 1986).

29. This is not to say that there has been no tendency toward vertical integration in defense-related design bureaus and enterprises. But the very complexity of tasks demanded by TTZ and TU and the relatively more competitive environment than in civilian production frequently allowed for no other option than subcontracting. A good example is given in A. Steinhaus, The *Beginnings of Soviet Military Electronics—1948–1961*, when comparing the K-5 missile program to the K-13.

30. N. Bogorad, *The Soviet Machine Tool Industry* (Falls Church, VA: Delphic Associates, 1989). See also V. Peysakhovich, *Economics of Automation in the Soviet Machine-Building Industry* (Falls Church, VA: Delphic Associates, 1987).

31. L. Freikman, "Shaping a market environment and analysis of the Enterprise Interaction Mechanism."

32. GERSS, "Reconversion Industrielle?" in Sapir, Ernould, and Pineye, *La Décomposition de l'armée soviétique.*

33. The debate can be read in V. K. Fal'tsman, "Konveriya i ekonomicheskaya reforma," in *MEiMO*, no. 10/1990, October. E. Rogovskiy, "Ekonomicheskaya orientatsiya konversii," in *Ekonomicheskie Nauki*, no. 8/1990, August. M. Spekler, A. Ozhegov, and V. Malygin, "Konversiya oboronnykh predpriyatiy: vybor strategii," in *Voprosy Ekonomiki*, no. 2/1991, February. A. Ozhegov, Ye. Rogovsky, and Yu. Yaremenko, "Konversiya oboronnoy promishlennosti i preobrazovanie ekonomiki SSSR," in Kommunist, no. 1/1991. The most integrated understanding of how conversion could be understood in the 1989–1990 context is to be found in the economic program published in *Problemy Prognozirovaniya*, no. 2/1991.

34. J. Cooper, *The Soviet Defence Industry, Conversion and Reform* (London: RIIA and Pinter Publishers, 1991). T. Malleret, *Conversion of the Defense Industry in the Former Soviet Union* (New York, NY: Institute for East West Security Studies, Occasional Paper series, no. 23, 1992). J. Sapir, *Les bases futures de la puissance militaire russe.*

35. J. Sapir, *Feu le système soviétique?*, Paris, La Découverte, 1992.

36. J. Cooper, *The Conversion of the Former Soviet Defense Industry* (London: RIIA, Post-Soviet Business Forum Publications, 1993).

37. Interview of A. Kokoshin by P. Felgengauer, in *Nezavissimaja Gazeta*, 18/02/1994, p. 9. Interview of A. Kokoshin in *Morskoy Sbornik*, no. 11/1993, November, pp. 3–7. Interview of A. Kokoshin in *Rosiiskie Vesti*, 23/07/1993, p. 7.

38. Interview of Vice Premier O. Soskovets in *Kommersant'*, 31 July 1993, p. 5.

39. A. Martini, "La coopération franco-russe: l'hélicoptère Mi.38," in *L'Armement—revue de la Délégation Générale à l'Armement*, no. 38, July–August 1993.

40. E. Kogan, "Is there a future for Russia's defence industry? Conversion in the aircraft industry," in *FOA Lectures and Contributions* , no. 6, 25 July 1994, Stockholm.

41. What is called marginal Q is the increase in the value of the firm linked to installation of an additional unit of capital less its replacement cost, at market value. James Tobin, "A General Equilibrium Approach to Monetary Theory," in *Journal of Money, Credit and Banking* 1, no. 1 (1969). Hiro Yoshikawa, "On the Q Theory of Investment," in *American Economic Review* (September 1980).

42. J. Chan Lee and Raymond Torres, "Q de Tobin et taux d'accumulation en France," in *Annales d'Économie et de Statistiques*, no. 5, 1987.

43. Edmond Malinvaud, "Profitability and Investment Facing Uncertain Demand," INSEE, Document de Travail no. 8303, Paris, 1983.

44. Edmond Malinvaud, "Wages and Unemployment," *Economic Journal* 92, no. 1 (1982).

45. Kenneth Arrow, "Optimal Capital Policy with Irreversible Investment," in *Value, Capital and Growth: Essays in Honor of John Hicks*, edited by John Wolfe (Edinburgh: Edinburgh University Press, 1968). Ben Bernanke, "Irreversibility, Uncertainty and Cyclical Investment," *Quarterly Journal of Economics*, no. 98 (1983).

46. Edmond Malinvaud, "Capital Productif, Incertitudes et Profitabilité," in *Annales d'Économie et de Statistiques*, no. 5, 1987.

47. Glenn Hubbard, *Asymetric Information, Corporate Finance and Investment* (Chicago: University of Chicago Press, 1990).

48. Nobuhiro Kiyotaki, "Multiple Expectational Equilibria under Monopolistic Competition," *Quarterly Journal of Economics* 103 (1988).

49. Joshua Green and Delano Villanueva, "Private investment in developing countries: an empirical analysis," in *IMF Staff Papers* 38, no. 1 (1991).

50. Luis Serven and Andrés Solimano, "Private Investment and Macroeconomic Adjustment: a Survey," *The World Bank Research Observer* 7, no.1 (1992).

51. This is more or less the current Polish situation, where investment is lacking and the reformist government was ousted in the 1993 general elections, even if a kind of limited static stabilization has been achieved. See Jérôme Sgard, "Pologne, la transition achevée?" in *La Lettre du CEPII*, no. 122, March 1994.

52. Michael Bruno, "Stabilization and the macroeconomics of transition," *The Economics of Transition* 1, no. 1 (1993): 13.

53. Joseph E. Stiglitz and Andrew Weiss, "Credit Rationing in Markets with Imperfect Information," *American Economic Review* 71, no. 3 (1981).

54. Which is exactly the Polish situation, as forecasted in Yves Zlotowski, "Politique monétaire et thérapie de choc en Pologne: les limites de l'expérience 1990–1991," *Economie Prospective Internationale*, no. 53, 1993, 1st quarter.

55. As experienced in Russia. Jacques Sapir, "Formes et nature de l'inflation," *Économie Internationale*, no. 54, 1993, 2nd quarter.

56. S. Salop and Joseph Stiglitz, "Bargains and Ripoffs: A Model of Monopolistically Competitive Price Dispersion," *Review of Economic Studies* 44, no. 138 (1977).

57. Eliana Cardoso, "Macroeconomic Environment and Capital Formation in Latin America" (paper presented at the World Bank conference on Private Investment and Macroeconomic Adjustment, Washington, D.C., March 21–22, 1991). Felipe Larrain and Rodrigo Vergara, "Investment and Macroeconomic Adjustment: the Case of East Asia," ibid.

58. This perverse circle is described in Shafiqul Islam, "Conclusion: Problems of Planning a Market Economy," in *Making Markets*, edited by M. Mandelbaum and S. Islam (New York: Council on Foreign Relations, 1993).

59. J. Sapir, "Conversion of Russian Defense Industries: a Macroeconomic and Regional Perspective," in M. McFaul and T. Perlmutter, *Privatization, Conversion and Enterprise Reform in Russia.*

60. The author was able to consult the *CARTOTEK 2*, where documentation relative to non-payments was then stored.

61. J. Sapir, "Conversion of Russian Defense Industries: A Macroeconomic and Regional Perspective," in M. McFaul and T. Perlmutter, *Privatization, Conversion and Enterprise Reform in Russia.*

62. C. Gaddy, "Economic Performance and Policies in the Defense Industrial Regions of Russia," in M. McFaul and T. Perlmutter, *Privatization, Conversion and Enterprise Reform in Russia.*

63. K. Gonchar, Ye. Kuznetsov, A. Ozhegov, *Conversion of the Post-Soviet Defense Industry: Implications for Russian Economic Development*, Brief 1 (Bonn: Bonn International Center for Conversion, February 1995).

64. N. Arutyunov, "Tyazhelo s samym legkim zhelezom," in *Kommersant'*, no. 49, 6–12 December 1993, p. 11.

65. No. 179 (2/22/1992), No. 312 (3/27/1992), No. 388 (4/11/1992), and No. 507 (5/12/1992).

66. *Comparative Strategy* 13 (1994): 231.

67. Col. A. Piskunov, in *Voyenno-Ekonomicheskiy Zhurnal*, no. 9/1993, September, pp.16–18.

68. The first estimate was given by A. Kokoshin, the Russian defense vice-minister, in *Rossiiskie Vesti*, 23 July 1993, p. 2; the second by V. Lukhin, in *Delovye Lyudi*, no. 32, April 1993, p. 14.

69. Personal discussions in 1993 and 1994 with high-ranking officials.

70. J. Sapir, "Conversion of Russian Defense Industries: A Macroeconomic and Regional Perspective," in McFaul and Perlmutter, *Privatization, Conversion and Enterprise Reform in Russia*, 137–168.

71. See G. Standing, *Developing a Labour Market Information System for the Russian Federal Employment Service*, Budapest, ILO Central and Eastern European Team, Policy Report, August 1993.

72. See G. Standing, "Why is measured unemployment in Russia so low? The net with many holes," *The Journal of European Social Policy* 4, no. 1 (February 1994).

73. G. Standing, *Enterprise Restructuring in Russian Industry and Mass Unemployment*, Labour Market Papers no. 1, Geneva, ILO/BIT, 1995, p. 7.

74. Ibid., p. 31. Such a proportion is compatible with results of monograph studies made on cities like Perm, Tambov, and Tver.

75. See C. Gaddy, "Economic Performance and Policies in the Defense Industrial region of Russia," in McFaul and Perlmutter, *Privatization, Conversion and Enterprise Reform in Russia*, 103–136.

76. CIA, *Rough Road to Markets in Russia and Eurasia.*

77. J. Sapir, *Feu le système soviétique*.

8

Commercializing State-Owned R&D: A Russia–United States Comparison

Judith B. Sedaitis

The end of the Cold War left the defense industries in both the United States and the Russian Federation scrambling for new funding and a redefinition of their institutional role. The growing trade deficit and alarming decline in the U.S. share of global markets have increased pressure to find ways to utilize defense R&D investments to increase the competitiveness of U.S. industry (McTague, 1988). Financial pressures in the Russian Federation are much more severe than those faced by U.S. state researchers, as the dramatic loss of state subsidies threatens the very existence of most Russian research institutes. Their disintegration would not only mean the loss of badly needed growth potential, but it could also destabilize control over the massive Russian military-industrial complex with potentially serious consequences for democracy in Russia as well as for Western and international security. The great infusion of funds by the two powers into federal research over the last forty years fuels the perception of the states' R&D sectors as "sleeping giants" with inherently great innovative potential. This is especially true for Russia, where defense production received the lion's share of resources and skilled personnel and now accounts for most of those goods with the greatest competitive potential (Shlykov, 1995). As global conflict shifts from military to economic competitiveness, however, it remains no small task to commercialize the defense-based research establishment in either country.

State ownership, regulation, and monopsony created a non-market U.S. federal research system that has not been able to interface easily with the private sector. In the Russian Federation, private industry is only beginning to develop. In both countries, therefore, federal research establishments have great innovative potential, but face a "critical gap" on several levels in communicating with the private sector (Geisler & Rubenstein, 1994). As a result, they can miss the commercial potential of their new innovations or are unable to find interested producers or investors. At the same time, the relative seclusion of the U.S. national R&D system and its staggering bureaucracy make the cost of bridging the gap by the private sector too high. Thus the current attempts by U.S. national labs to commercialize can provide important lessons to Russia in its nascent process. Drawing on studies of technology transfer and commercialization in U.S. federal labs, this paper analyzes survey data from 100 Russian R&D enterprises to compare the commercializing process of both national systems.

A Basis for Comparison: Some Shared Fallout of Structural Isolation

The United States and the Russian Federation continue to support the two most militarized state research systems in the Western world. Currently, both the United States and Russia allocate about 4 percent of their gross national product to research expenditures (OECD, 1994, p. 65) of which roughly 60 percent in the United States and 45 percent in Russia is spent on defense-oriented R&D. Furthermore, in both countries, the state-supported research structures are embedded in an organizationally secluded military procurement system, unlike state R&D in other countries. In France, for example, another big military R&D spender, approximately 37 percent of state research funds are spent on defense research agencies, but state industrial policy integrates the defense research programs with the aims of the private sector. In contrast, both the U.S. and Soviet systems were honed in the era of superpower rivalry during which policymakers purposefully sought to seclude and isolate the defense sector from the economy at large.

In the former USSR, virtually all scientific research was funded and monitored by the state. The military-industrial complex was historically administered by nine vertically integrated ministries, including the Ministry of Heavy Machine Building, Ministry of Electronics, and Ministry of Aviation, wherein each was responsible for the production of different aspects of military output and a wide array of civilian production as well. The Russian Federation inherited the lion's share of the Soviet system, which in 1990 included roughly 3,000 individual production firms and roughly 920 military R&D organizations (see Table 8.1). Shrouded in secrecy and isolated even from one another, contractual relationships were dictated from above. Outside of military competition with the West, Soviet research took place largely in the absence of market forces and consumer demand. Ironically, the U.S. military research establishment came to resemble its Soviet rivals in this seclusion from consumer demand.

While only about 150 labs are of national scope, there are in total more than 700 federal labs in the U.S. system, which belong to one of seven different federal departments or agencies including the Department of Defense (DoD), the Department of Energy (DOE), the National Institutes of Health (NIH), and the National Aeronautics and Space Administration (NASA). The labs receive about 35 million dollars annually, the largest part of which goes to the Department of Defense, which gets about 50 percent, followed by the Department of Health and Human Services and NASA. Central agencies determine the "mission" or substan-

Table 8.1: Key Characteristics of U.S. and Russian Federation State R&D

	Allocation of Total Federal R&D Funds, 1991		Performance of R&D, 1991[a]		Type and Number of Organizations Involved	Scientists Employed	Type of Research	
United States	Private sector	41.6%	Private sector	70.2%	729 federal research labs.	1,090,922	1993	
	Intramural[b]	25.8%	Intramural[b]	10.5%	In 1992, 100 big private firms	(in 1990)		
	Academic R&D	16.8%	Academic R&D	12.2%	received 89% of R&D		Development	60.0%
	U&C FFRDCs[c]	8.6%	U&C FFRDCs[c]	3.6%	contracts, and 31,300 private		Applied	19.6%
	Industry FFRDCs	4.6%	Non-profits	3.5%	firms received federal funds.		Basic	20.4%
	Non-profits	1.9%						
Russia	Industrial ministries	69%	Industrial ministries	78%	900 military R&D sites;	949,300	1991	
	Academic R&D	21%	Academic R&D	12%	3,128 industrial R&D	(in 1989)		
	Higher education	7%	Higher education	5.5%	institutions including 400		Development	57.3%
	Enterprises	3%	Enterprises	4%	scientific sub-units of		Applied	33.4%
					industrial enterprises.		Basic	9.3%

Sources: OECD, *Science, Technology, and Innovation Policies* 2, 1994; *Science & Engineering Indicators*, 1994.

a In terms of total R&D funds for the United States. The Russian funds are federal, but represent ~98 percent of total R&D spending.
b In-house R&D by research labs and centers that are solely government funded.
c Federally owned research centers (FFRDCs) administered by universities or colleges.

tive research focus of the different agencies and regulate their overall production specifications and funding processes.

Structurally, both federal research systems are vertically integrated and dominated by large firms and central state agencies that play a key role in determining their funding, mission, and internal procedures. While the U.S. private sector is the largest recipient of all federal research funds, it also came to be dominated by large and monopolistic firms that in some cases command an entire industry. The output, for instance, of the entire Soviet aviation ministry in 1989 was estimated to be about $23–28 billion, roughly equal to the combined sales of McDonnell Douglas and Boeing alone (Kuznetsov, 1994). Funding of defense research was particularly concentrated. In 1988, for instance, the top 10 research grant recipients received nearly half of the moneys allotted for large defense research contracts and only 100 research sites accounted for 86 percent of the total budget (Branscomb 1993, 167). Responding to the demands of a secretive, largely non-market environment, the research establishments in both the United States and Russia have come to share some common cultural and technical obstacles to technology transfer.

In the context of having to respond to only one customer, the U.S. state-supported research system developed an organizational culture and set of practices distinct from the market culture around it (Adelman & Augustine, 1992; Dorf & Worthington, 1990). In lieu of a consumer-driven culture of close price monitoring, cost accounting, and active searching for new markets, U.S. defense R&D in both federal and even private firms developed a bureaucratic culture instead. Internal procedures had to meet the staggering demands for detailed reporting and specifications made by various and sundry state agencies. In particular, their monopolistic structure served to mitigate the important role that competition plays in stimulating innovation. The large and experienced U.S. defense researchers often took advantage of their relative concentration to undermine the formal competitive bidding procedure and collude on prices and proposals.

Finally, the research focus in both countries on producing weapons systems leaves the state R&D structures with technical disadvantages for commercialization. Federal R&D in both countries was dominated by long-term, highly formal, development-oriented research. As such, state R&D was less involved with process technologies and general innovation than with clear and specific products and tasks, making subsequent discoveries more difficult to adapt to other applications. Whereas Russia shares both the cultural and technical basic obstacles to defense conversion, its difficulty is compounded by the sheer scale of the problem and the more severe consequences of failure.

Soviet R&D permeated the whole economy, from academia and research and design institutes to research divisions in industrial associations and enterprises, so that even in 1990, when the flight from the research sector had already begun, Russia employed approximately 126 scientists per 10,000 labor force in comparison with 76 scientists in the United States (OECD, 1994). In many regions, defense work accounted for more than 50 percent of total employment. In addition, more than 50 closed science cities existed across the USSR that were essentially military R&D company towns. Thus, the legacy of gross dependence on the Russian military-industrial complex could lead to the economic devastation of whole regions in the event of their inability to actively transfer technologies to viable commercial production. At the same time, they are crippled both by the lack healthy private sector R&D and the reluctance of some military officials to create one from the ashes of their defense establishments.

For state funded research to find fertile soil, equally sophisticated R&D needs to exist in the private sector (Mowery, 1994). Yet in-house R&D facility in Russian civilian production has never been as strong as defense research, and current studies show that it is declining even further (*Kommersant-Daily*, March 28, 1995, p. 3). In addition, most defense officials would prefer that R&D commercialization be limited to dual-use applications which meet both military and commercial demands (Erokhim, 1995; *Delovie Lyudi*, April 1993, pp. 22–23). While they want to see Russia regain a large share of the global arms trade, they are nonetheless unable to finance the necessary demand themselves and will keep only a fraction of the former defense establishment under state control. In lieu of viable domestic users, therefore, both privatized R&D firms and those enterprises that remain part of the state system must find demand for their technology in the global market. Yet Russian R&D managers have little past experience in meeting consumer demand or Western technical and quality standards. As a result of their isolation from market forces and consumer demand, they face a critical gap between their past practices and the current demands of commercializing their technical capabilities similar to that already confronted by their U.S. counterparts.

Technology Transfer Models and the Critical Gap

Perhaps the single most important lesson from the U.S. technology transfer experience is that market forces alone are insufficient for bridging the institutional isolation of military R&D. The first wave of technology transfer legislation focused on strengthening incentives for individual technologists to license or patent their new discoveries, assuming that the market itself would pull out the innovations it needed (Branscomb, 1993; Bozeman & Coker, 1992). The obstacles outlined above, however, proved too costly for the private sector to bear alone, and prompted a second wave of legislation that encouraged federal R&D organizations to provide more active, institutional support for technology transfer. This included relaxation of antitrust regulation that would allow labs to partner with firms in the private sector and follow the successful Japanese model of innovation based on strong, inter-organizational linkages. Three levels of analyses, therefore, define the U.S. experience of technology transfer and frame our study of the Russian case: individual entrepreneurship, organizational strategies, and inter-organizational networks. Aspects of these three levels that have been flagged as key by the literature on technology transfer in the U.S. lab system will be defended below and linked to the measure of technology transfer success in Russia.

The Technologist Entrepreneurs

In Russia, the high barriers to new market entrants have made the spin-off mechanism the most common form of new firm development, with former scientists and academics, especially in the natural sciences, taking a leading role. The best and the brightest researchers were among the first to leave their impoverished state institutes to start new businesses in all spheres, including manufacturing (Webster and Charap, 1993) as well as in the new fields of commodity trading and commercial banking (Sedaitis, 1995). The R&D defense industry has also supported internal venturing, but not equally among all former defense R&D firms. In our sample of 100 firms, 69 percent created no new spin-offs or subsidiaries. The 31 firms which did spin off new businesses, however, were much more active than any U.S. federal labs have been and created more than 200 daughter firms. On the one hand, this exceptionally high degree of fragmentation reflects the loss of funding and general chaos in the state sector that prompted scientists to seek new or additional employment. However, our data suggest that these spin-offs were not less likely to engage in scientific work any more than their parent organization (see Table 8.2). Very often parent organizations would give the spin-off access to the core technologies in order to perform the same service as the parent would, but at a lower rate, free of the overhead costs of the parent. State R&D managers generally construe these arrangements as temporary ones that allow their scientific personnel to supplant their low incomes while the parent firm readjusts to new conditions. At the same time, other, generally more commercial spin-offs were clearly intent on maintaining their independence, such as the shoe factory located on the premise of TsAGI, an aerodynamics testing facility similar to a U.S. NASA site.

Table 8.2: Production Activity by Organizational Form

Production Activity	Spin-off (n=43)	Privatized (n=37)	State-owned(n=20)
Scientific-technical	49.8	34.8	61.2
Consumer	21.8	18.9	3.8
Light industry	0.35	1.5	0.0
Equipment	17.3	22.9	16.2
Defense	4.4	15.1	14.0

A quite extensive literature debates the relative merits and drawbacks of the technical entrepreneur. For many, the inventor's knowledge of the technology and, in particular, his or her commitment to the innovation and its successful application are key to the commercialization process (Radosevich, 1995). A study by Battelle Laboratories (1973) found that in nine of the ten major innovations of the twentieth century, the tenacity and dedication of technical entrepreneurs played a crucial role in the realization of new products. The development of photocopying technology provides a good example. The faith in its potential helped managers of Haloid, a small manufacturer, maintain persistence in light of initially negative market analyses to uncover an enormous new market and become the giant Xerox corporation (Zanan & Kim, 1988).

While some point to the value of the technologist's intimate technical knowledge in forecasting new applications, however, others charge that technologists are generally too wrapped up in the technical details of their work to see beyond it. Studies on the personal attributes of entrepreneurs find the skilled technologist less market-oriented and more risk-averse than entrepreneurs in other industrial sectors (Roberts, 1991). Those technologists employed by the state in particular are found to be even older, better educated, and with longer tenure at one employer and less industrial experience than their private sector counterparts (Kassicieh et al., 1995). The implication is that the technologist personality lacks the energetic, risk-taking traits necessary for successful entrepreneurship, even when he uncovers commercial applications for his innovation. The lack of aggressive, entrepreneurial qualities might explain the relatively poor entrepreneurial showing among technologists at U.S. labs. Although technology transfer was singled out for funding and legislative support, only about 50 firms have spun off from national labs over the last decade (Carr, 1994).

An alternative explanation for low levels of entrepreneurship among lab technologists has been to focus on structural rather than personality variables such as the state technologist's lack of access to venture capitalists and other elements of the entrepreneurial infrastructure. Kassicieh et al. (1995) found that the number of successful new spin-offs was much higher at those DOE labs which provided access to local networks of marketing specialists, financiers, and others. Similarly, other studies suggest that one of the most important ways organizations can promote technology entrepreneurship is simply to allow employees to retain the security of their job while they use its infrastructure and social network to develop their new spin-off firm (Seawright, 1988; Autio, 1995). As a result, some of the key institutional concerns in the United States have been relaxed, such as prohibitions against letting public goods result in personal wealth. Federal labs, especially in the DOE, are now offering their employees more active on-site support for those technologists interested in creating spin-off firms and becoming entrepreneurs. Drawing from Western literature, we pose the following hypotheses to study the alternative explanation for Russian R&D entrepreneurship by contrasting the strength of personal characteristics to environmental supports:

H1a: *The younger the director of a firm, the more likely the firm will actively engage in technology transfer activity.*

H1b: *The more actively management supports entrepreneurship, the more likely the firm will actively engage in technology transfer methods.*

Organizational Flexibility and Cross-Pollenization

Organizational coordination was the only strong suit of the Soviet command economy. Its U-form structure was coupled with a long cultural tradition of one-man rule (*edinochalie*) to promote historically strong, vertical lines of authority at the enterprise level (see Granick, 1954; Berliner, 1983). Managers have retained the potential to exercise control even in privatized firms by acquiring large holdings of their firm's shares. However, case study evidence suggests that successful commercialization attempts are linked with those firms which have decentralized decision-making and information flow, often by giving separate divisions control over their finances and thus the ability to monitor their individual profit or loss (Bernstein, 1994). An active, if small group of approximately 10 to 12 firms allowed internal subdivisions such control by making them independent profit-loss centers. Our sample also finds that R&D firms that are moving away from traditional, centralized authority also tend to be those firms that have chosen to privatize (see Table 8.3). These firms also tend to decentralize their approach to marketing in particular, by engaging in this new activity energetically but without much focus or development strategy.

Table 8.3: Organizational Strategy

Very strongly or strongly agree

Organizational Strategy	Spin-offs	Privatized	State-owned
We place a strong emphasis on decentralized rather than centralized search for new clients and markets.	76.2%	71.4	65.0
We restrict managerial access to important information on financial and strategic issues as commercial secrets.	74.4%	58.3	70.0
We operate with heavy dependence on expertise and market forces instead of formal authority.	83.3%	75.0	38.9

Many of the new and privatizing enterprises, therefore, are also strongly diversifying their production lines. VNIEFF, for example, the state physics institute in the closed city of Arzamas-16 responsible for nuclear warhead design, has moved from its constricted and militaristic core activities by branching out into conversion projects related to energy production, environmental protection, and textile manufacturing (Zisk, 1995). Impuls, a scientific production organization in Moscow, has moved from the research and design of optical guidance systems of bombs to the adaptation of its technology for innovative measuring tools, such as the first non-invasive pipeline measurement technology (Bernstein, 1994). In these cases, firms diversified from the strengths of their core competency; however, other studies indicate that wide and undirected diversification is more the norm among formerly state-owned defense firms. In lieu of marketing research, Randall (1995, this volume) finds that R&D managers follow a hit-or-miss approach based on pursuing several unrelated product lines simultaneously. Our sample supports these findings. Managers at privatized parent firms were the most likely to favor a strategy of streamlining their production, but it was not a popular perspective by any means (see Table 8.4). Instead, Russian R&D managers place high expectations on the infusion of funds from outside investors, much the same as they had relied on guaranteed funding from the state in the Soviet era.

Table 8.4: Product Strategy

Product Strategy	Spin-offs	Privatized	State-owned
We plan to decrease the assortment of products produced.	0.0%	17.9	0.0
We plan to shrink one or more production divisions.	8.7%	15.4	0.0
We plan to find outside investors.	57.5%	63.9	66.7

The debate over which type of organizational form best facilitates innovation has a long and illustrious history, beginning with arguments by Schumpeter in the late 1950s for the greater capacity of large, monopolistic organizations to absorb the high risk of innovating. More recent studies, however, document the sharp rise of small organizations as innovation generators and show how their flexible decentralized structures allow for rapid change and inter-organizational networking (Mowery, 1994). A flexible organizational structure is also more compatible with the emerging dynamic model of technology transfer as cyclical and iterative, in which feedback from potential users refines the direction of the research and its practical applications (Alic et al., 1992).

A key example of organizational flexibility is the important role of spin-offs in creating new products from biogenetic research, much of which sprang initially from laboratories of the National Institutes of Health (Powell et al., 1994). By 1990, the NIH had signed 114 cooperative agreements, while the DOE weapons laboratories had established only a handful. While both sponsored broad research agendas, the NIH operated a diverse, decentralized research program with relatively short time horizons. In contrast, the DOE lab projects tended to be more integrated, lengthy, and managed by a large team of researchers. Its formal, large-scale and centralized organizational management provided fewer opportunities for private sector collaboration and hindered the diffusion of its innovations into commercial markets (Branscomb, 1993). Unfortunately, legislation designed to fuel technology transfer at government labs extended this organizational tradition by mandating the formation of a separate internal Office of Research and Technology Application (ORTA) at each lab. This neatly segregated the responsibility for technology transfer into its own niche and perpetuated organizational divisions instead of decreasing or creatively rearranging them. Lines of authority stayed clear as a result and ORTAs acted as important clearinghouses of information and encouragement for invention disclosures (Wessel, 1993), but their effectiveness was largely clerical and limited to facilitating licensing agreements and other legal transfers. The lack of flexible organizational boundaries accounts in large part for the relatively dismal failure of past conversion attempts even by private sector defense contractors. As in the former USSR, civilian and defense operations in U.S. firms were generally kept segregated from one another. By separating these two spheres, managers effectively limited the free and serendipitous flow of information between experts in civilian and military operations that helps conversion into new product markets stay focused on core technical competence (Gansler, 1995; Alic et al., 1992). Instead, firms or divisions that concentrated exclusively on military production would often diversify into new products unrelated to their core abilities only to find themselves quickly bankrupt. There is a price to pay for the benefits of decentralization, however.

While flexible structures increase communication and "cross-pollinate" specializations, these benefits come at a cost; the most noted are the weakened mechanisms of coordination and control that result. For this reason, the first for-profit R&D consortium (MCC) was structured along formal lines of centralized authority. MCC brought together a group of diverse and often competing firms to share the high cost of innovation. Hierarchical organization was viewed as the most effective form for coordinating the complex process of meeting the needs of a diverse membership especially when sensitive and proprietary information was often involved (Dove, 1992). Thus,

H2: *The more centralized enterprises are less likely to be successful at engaging in technology transfer.*

H3: *The less diversified the product strategy at an enterprise, the more likely it is to be associated with active technology transfer attempts.*

Inter-Organizational Cooperative Networks

The cornerstone of the second wave of legislation in the United States was the National Cooperative Research Act of 1984, which allowed organizations to pool their resources for the development of new technologies by granting them antitrust protection. So began a wave of industrial consortia aimed at emulating the Japanese *keiretsu* system of tight industry linkages in order to share the cost of developing strategic technologies (Lieberman, 1992). National labs now strongly promulgate cooperative R&D acts (CRADAs) with industry and their numbers have risen dramatically over the last few years, from a total of about 60 in 1990 to more than 900 in 1991 (*Science and Engineering Indicators*, 1994). Despite the dramatic rise of partnering by U.S. labs, however, Japanese government labs remain the leaders with more than twice as many cooperative agreements with private firms (Bozeman & Pandey, 1994).

In lieu of restrictive state regulation, the Russian R&D sector may actually be in a better position to emulate the Japanese model than U.S. labs are. As Russian firms move beyond privatization, they have sought close links to banks by forming financial industry groups (FIGs) which, like the Japanese *keiretsu*, are based on mutual shareholding. For the moment, it appears these arrangements are less geared to providing long-term financing required for technology commercialization than to quick profit-seeking (Gorbatova, 1995). Most of them involve trade in highly marketable commodities and natural resources, especially metals. Despite formal pronouncements by the Ministry of Defense in favor of their creation, only three or four defense R&D enterprises, such as the Sukhoi aircraft design bureau, have captured any substantial interest among banks. Secondly, shareholder relations are limited in importance largely to privatized parent firms, since spin-off firms have few assets to speak of (see Table 8.5). In general, ties to their customers, both state ministries and industrial producers, were the primary interorganizational links of any great importance to our sample of R&D firms. Certainly the tie to private sector users of their technology has had real benefits for the U.S. national labs in both tangible ways, such as development of new products (Bozeman, 1995), as well as in the intangible acculturation to private sector pace and concerns (Gomes, 1995, this volume). Thus,

H4: *The greater the interorganizational integration, the more successful the attempts at technology transfer.*

Table 8.5: Inter-Organizational Ties

Organizational Type	Spin-off (n=43)	Privatized (n=37)	State-owned (n=20)
State Administration	32.6%	40.5	75.0
Local Administration	11.6%	18.9	30.0
Banks	16.3%	43.2	20.0
Industrial Producers	62.8%	40.5	45.0
Foreign Investors	11.6%	10.8	25.0
Political Organizations	4.7%	2.7	15.0

The Survey

After completing a review of technology transfer literature with a special focus on the U.S. national lab system, a questionnaire was devised and administered in conjunction with a Russian research team working in the Moscow region only. The sample consisted of 100 former defense enterprises or subunits thereof. All of the enterprises had worked largely in R&D which included design, prototype production, and/or testing activities. Although most of the enterprises had some production facilities, industrial production was not the major activity at any of the surveyed enterprises. Of the total 100 enterprises, 43 were new entrants spun off from older existent enterprises; 57 were older firms, 37 of which were already privatized.

The average age of the 100 respondents surveyed was 50.4 years old, which was substantially older than the average age of technologists (46.1) at the DOE labs studied by Kassicieh et al. (1995). At the same time, the older average age is expected considering that all the respondents were top managers. The respondents were all well educated: all had completed university training and 29 percent had graduate degrees. Their level of education was correlated to the propensity of their firms to have inter-organizational ties in general and with banks and financial institutions in particular.

The key independent variable of technology transfer success was operationalized by two variables: NSUB, or the number of spin-offs created, and TTWSCORE, a composite score based on twelve actual methods of technology transfer as complied by Gibson and Rogers (1994). They ranged from passive methods of transfer attempts such as publishing, which were scored lower than various forms of active, interpersonal methods, such as arranging special programs and cite visits for potential users and investors.

General background of the respondents was measured in terms of their age and educational background delineated by subject matter studied. Managerial support and centralization variables were measured by a five-point Lickert-type scale asking about the degree of concrete support and formal procedures present at the enterprise. The diversity of production was gauged by a creating a Herfindahl measure of concentration based on the proportion of enterprise output in terms of eight industrial categories, including heavy industry, consumer goods, and R&D-intensive civilian products. Inter-organizational integration was measured as an indicator variable distinguishing those enterprises that reported reliable ties to other producers from those that did not. Finally, the effect of the size of the enterprise was controlled for by the natural log of current staff size excluding temporary workers. The number of personnel in the sample ranged from fifty workers to about 12,400 and would present heteroscedasticity without the control.

Survey Results

Our hypotheses testing yielded a number of interesting and contradictory results that often suggested relationships in the opposite direction to the one predicted. The regression model (Table 8.6) indicates all those relationships that emerged as statistically significant in relation to the dependent variable, active technology transfer methods.

Technologist Entrepreneur

In terms of entreprenuerial behavior, personal traits proved less important than social structures. Neither the respondents' age nor education was linked to spin-off creation. However, managerial support was significantly related to the propensity to actively pursue technology transfer, suggesting that contextual elements are crucial for the success of promoting entrepreneurial activity.

Organizational and Inter-Organizational Variables

The organizational variables proved fairly strong and surprising. Only the measure of internal centralization proved no relation to active technology transfer methodologies. While centralization was expected to be negatively related, the Russian case may also involve the positive aspects of centralization as well, such as effective resource mobilization, and these diametrical effects would cancel each other out. Production diversification was noteworthy and in the direction predicted,

Table 8.6: Regression of Enterprise Variables on Technology Transfer Methods

Age	.08[†] (.07)
Managerial support	.60* (.35)
Centralization	.28 (1.1)
Product diversity	2.2[†] (2.1)
Inter-organizational ties	2.1* (1.2)
Firm size	.71** (.22)
Adjusted R squared	.12

[†]≤ .1, *≤ .05, **≤ .01

suggesting that, as in the West, less diversity is linked to greater attempts at commercializing one's technology. Finally, inter-organizational ties proved statistically significant. Enterprises with the stronger links to other producers were more likely to engage in more active technology transfer attempts.

Conclusion

As still largely state-owned, defense-oriented systems, the substantial state R&D establishments in both the United States and Russia share some key features in their attempts to interface with commercial markets, such as their noncommercial culture and the particular benefits and drawbacks of the "technology-entrepreneur." On the basis of the preliminary findings here, some important initial similarities in nascent Russian R&D commercialization to the more developed U.S. process can be summed up as follows:

(1) The focus on scientific-technical work. More than half the sample was still engaged in scientific work despite the dramatic loss of funding experienced by the R&D sector. The majority also considered their scientific skills the greatest potential on which to base future growth. However, few could afford to continuing working in R&D alone but diversified into the production of machinery or consumer

goods. Scientific work was the sole focus of activity for only about 15 percent of the sample which were mostly still state owned.

(2) The level at which technology transfer obstacles are perceived. Russian firm managers, like many of their colleagues (Gibson, 1994), find it most difficult to apply their innovation to the creation of a desirable new product. Secondly, they share a lack of experience in marketing any new product that they may develop. This in part accounts for the strong tendency among Russian managers to not create a separate marketing department and centralize the activity. Marketing has not yet developed the level of professionalism found in the Western private sector, nor do Russians have the necessary tools, such as centralized data on product consumption or the distribution of domestic market shares. Instead, one common method for deciding new product lines was simply to send out scouts to surrounding department stores to find out what products customers had been requesting.

(3) The importance of horizontal ties to potential users. Russian firm managers view as most important those interorganizational ties to potential users, or buyers of their products. In light of current policy debates to link firms more closely with banks in order to create financial-industrial groups, the relative lack of connection to financial institutions among the Russian managers is particularly notable. Instead, the ability to discuss common vital business issues with one another was one of the most strongly cited technology transfer methods. At the same time, the importance of both state organs and political groups suggests that expectations remain strong for the Russian state to play an important role in the maintenance and further development of its high-technology base.

Despite commonalities in the attempts between Russian and U.S. R&D managers to commercialize, however, the breakdown of state controls has clearly given Russian R&D enterprises greater need to find alternative sources of funding as well as the greater freedom with which to do so. Unfortunately, while they have more opportunity to pursue less fettered private use of public goods, they lack the resources of a developed market system necessary to make effective use of this opportunity. Their undeveloped system of marketing practices and information in part explains why broad product diversification is a working strategy at the moment. To develop a long-term commercialization strategy that will bridge the critical gap they face in accessing marketing knowledge, Russian R&D managers will eventually need to follow the example of several U.S. national labs to act as high-tech incubators. The DOE INEL and Sandia laboratories have drawn on the direct assistance of private sector supports.

In lieu of domestic partners, however, this role falls to the international community, whose assistance and influence is critical to many of the particular needs of high-tech incubation. In particular, partnering with the West could support the acculturation of Russian R&D to Western standards and quality. At ground zero, the success of commercializing technologies begins with the nature of the research itself, both basic and applied, and the extent to which worldwide standards, language, and procedures are understood and applied. Secondly, U.S. funds have been used to help Russian R&D managers find and contract with technology receptors. There are a few U.S. consulting organizations that connect commercializing Russian firms with the specific industry niche that might have a use for their technology. In this way, Russian firms can move to profitable civilian production, as U.S. companies access low cost technologies and new market entry. Given the paucity of market knowledge in Russia, funding for such third-party assistance work should continue. Finally, adopting the U.S. technology incubator model would require work on securing international financial and technical assistance. A number of U.S. and international funds exist which are earmarked for conversion efforts in Russian defense and high technologies. Our data indicate that Russian R&D managers that seek foreign investment are also among the most actively engaging in technology transfer. Failure to help them access financial and marketing support would signal the inability of transferring lessons from our own conversion experience, with consequences that could return to haunt us.

References

Adelman, Kenneth L., and Norman R. Augustine. "Defense Conversion: Bulldozing the Management." *Foreign Affairs* (Spring 1992): 27–47.

Alic, John, Lewis Branscomb, Harvey Brooks, Ashton Carter, and Gerald Epstein. *Beyond Spinoff: Military and Commercial Technologies in a Changing World.* Boston: Harvard Business School Press, 1992.

Autio, Erkko, and Annareetta Lumme. "Does the Innovator Role Affect the Perceived Potential for Growth? Analysis of Four Types of New, Technology-Based Firms." Paper presented at the Babson Conference on Entrepreneurship, London, April 1995.

Berger, Beverly J. "Technology Transfer in a Time of Transition." In *From Lab to Market: Commercialization of Public Sector Technology*, edited by S. Kassicieh and R. Radosevich. New York: Plenum Press, 1994.

Berliner, Joseph, A. Bergon and H. Levine, eds. *The Soviet Economy in the Year 2000*, pp. 350–90. London: Allen and Unwin, 1983.

Bernstein, David. *Defense Industry Restructuring in Russia: Case Studies and Analysis*. Stanford, CA: Stanford University Center for International Security and Arms Control, 1994.

Bozeman, B., M. Papadakis, and K. Coker. "Industry Perspectives on Commercial Interactions with Federal R&D Laboratories." Report to the National Science Foundation, Research on Science & Technology Program, 1995.

Bozeman, B., and S. Pandey. "Cooperative R&D in government laboratories: comparing the U.S. and Japan." *Technovation* 14, no. 3 (1994): 145–159.

Bozeman, B., and K. Coker. "Assessing the effectiveness of technology transfer from U.S. government R&D laboratories: the impact of market orientation." *Technovation* 12, no. 4 (1992): 239–254.

Branscomb, Lewis M. "National Laboratories: The Search for New Missions." In *Empowering Technology*, edited by Lewis M. Branscomb. Cambridge, MA: The MIT Press, 1993.

Buck, D., and K. Hartley. "The Political Economy of Defence R&D: Burden or Benefit?" In *Defence Science and Technology: Adjusting to Change*, edited by R. Coopey, M. Uttley, and G. Spinardi. Chur, Switzerland: Harwood Academic Publishers, 1993.

Carr, Robert K. "Doing Technology Transfer in Federal Laboratories." In *From Lab to Market: Commercialization of Public Sector Technology*, edited by S. Kassicieh and R. Radosevich. New York: Plenum Press, 1994.

Dorf, R., and K. Worthington. "Technology Transfer from Universities and Research Laboratories." *Technology Forecasting and Social Change* 37 (1990): 251–266.

Dove, G.A. "Encouraging Innovation and Excellence in Technology Transfer." In *Technology Transfer in Consortia and Strategic Alliances*, edited by D. Gibson and R. Smilor. Lanham, MD: Rowman & Littlefield Publishers, 1992.

Erokhim, V. M. "Voennaya Ekonomika i Tyl." *Voennaya Mycl'* 3 (May–June 1995): 61–70.

GAO. *Technology Transfer: Implementation Status of the Federal Technology Transfer Act of 1986*, May 1989.

GAO. *Technology Transfer: Improving the Use of Cooperative R&D Agreements at DOE's Contractor-Operated Laboratories*, April 1994.

Gansler, Jacques S. *Defense Conversion: Transforming the Arsenal of Democracy*. Cambridge, MA: The MIT Press, 1995.

Geisler, E., and A. Rubenstein. "The Role of the Firm's Internal Technical Entrepreneurs in Commercializing Technology from Federal Laboratories." In *From Lab to Market: Commercialization of Public Sector Technology*, edited by S. Kassicieh and R. Radosevich. New York: Plenum Press, 1994.

Gibson, David V. "Inter-Organizational Technology Transfer: From Standard Technology Packages to Spin-Offs." This volume.

Gibson, David V., and Everett M. Rogers. *R&D Collaboration on Trial.* Boston: Harvard Business School Press, 1994.

Gorbatova, Larisa. "Formation of Connections between Finance and Industry in Russia: Basic Stages and Forms." *Communist Economies & Economic Transformation* 7, no. 1 (March 1995): 21–34.

Granick, David. *Management of the Industrial Firm in the USSR*. New York: Columbia University Press, 1954.

Holloway, David. *The Soviet Union and the Arms Race*. New Haven: Yale University Press, 1984.

———"Innovation in the defence sector." In *Industrial Innovation in the Soviet*

Union, Vol. 2, edited by R. Amann and J. M. Cooper. New Haven: Yale University Press, 1982.

Kassicieh, S., R. Radosevich, and J. Umbarger. "A Comparative Study of Entrepreneurship Incidence among Inventors in National Laboratories." Unpublished report, 1995.

Kuznetsov, Yevgeny. "Adjustment of Russian Defence-Related Enterprises in 1992–1994: Macroeconomic Implications." *Communist Economies and Economic Transformation* 6, no. 4 (1994): 473–513.

Lieberman, M. "The Importance of Technology Transfer and Commercialization to U.S. Competitiveness." In *Technology Transfer in Consortia and Strategic Alliances*, edited by D. Gibson and R. Smilor. Lanham, MD: Rowman & Littlefield Publishers, 1992.

McTague, John. "Technology: Wielding a Three-Edged Sword." In *Federal Lab Technology Transfer: Issues and Policies*, edited by G. Bopp. New York: Praeger Press, 1988.

Mowery, David C. "The changing structure of U.S. industrial research: Implications for R&D organization in the Russian Federation." *International Journal of Technology Management* 9, no. 5–7 (1994): 89-102.

National Science Board. *Science and Engineering Indicators 1993.* Washington, D.C.: U.S. Government Printing Office, 1993.

OECD. *Science, Technology and Innovation Policies: Federation of Russia.* Paris: OECD, 1994.

Powell, Walter, Kenneth Koput, and Laurel Smith-Doerr. *Technological Change and the Locus of Innovation: Networks of Learning in Biotechnology*. Dallas: Academy of Management, 1994.

Radosevich, Raymond. "A model for entrepreneurial spin-offs from public technology sources." *International Journal of Technology Management* 10, no. 20 (1995).

Roberts, Edward B. *Entrepreneurs in High Technology*. New York: Oxford University Press, 1991.

Seawright, Gary L. "Entrepreneur-Driven Technology Transfer." In *Federal Lab Technology Transfer: Issues and Policies*, edited by G. Bopp. New York: Praeger Press, 1988.

Sedaitis, Judith B. "Investing in Russian High Technology . . ." Paper presented at the Babson Conference on Entrepreneurship, London, April 1995.

Shlykov, V. V. "Economic Readjustment within the Russian Defense-Industrial Complex." *Security Dialogue* 26, no. 1 (1995): 19–34.

Von Hippel, Eric A. "Users as Innovators." *Technology Review* (January 1978): 31–39.

Webster, Leila M., and Joshua Charap. *The Emergence of Private Sector Manufacturing in St. Petersburg*. World Bank Technical Paper 228. Washington, D.C.: The World Bank, 1993.

Wessel, V. William. "Technology transfer and intellectual property: an analysis of the NASA approach." *Technovation* 13, no. 3 (1993): 133–146.

Zanan, Peter, and Won Yong Kim. "Market Research for Innovative Products: The Delphi Method." In *Federal Lab Technology Transfer: Issues and Policies,* edited by G. Bopp. New York: Praeger Press, 1988.

Zisk, Kimberly Marten. "Arzamas-16: Economics and Security in a Closed Nuclear City." *Post-Soviet Affairs* 11, no. 1 (January–March 1995): 57.

Section Three

Managerial Strategies and Strategic Alliances

9

Defense Conversion in the Former Soviet Union: The Influence of Culture on the Strategic Planning Process

Lori A. Coakley and Linda M. Randall

In the former Soviet Union (FSU), the managers of many defense companies face a difficult defense conversion process (Lipsitz, 1995) complicated by systemic changes in economic and political institutions. Defense companies in the United States also faced with the need to convert are employing strategic planning to aid in the successful transition from military to commercial production. The general question in this study is whether the strategic planning process can be applied in the FSU context.

The profound influence of the Soviet command system on the values and practices of enterprise managers continues to affect their decisions today. This paper focuses on the attitudes toward defense conversion of managers from six enterprises in Russia and Belarus, and examines the factors that influence their decisions regarding the transfer of military technology to commercial production. First, a brief discussion of the strategic planning process and its role in defense conversion is offered, followed by a discussion of Russia's economic environment and the importance of defense conversion in the FSU to successful market transition. Finally, the six FSU defense enterprises are profiled. This information provides a context in which to assess the current decision-making processes of management and the effect of cultural influences in response to changes confronted by each of the six enterprises.

Strategic Planning

The strategic planning process is crucial to business performance in the commercial market environment of the United States. A sound strategic plan details what managers believe is the best use of an organization's resources and talents, and prescribes how they will deploy those to the mutual benefit of the organization and its customers (Bounds, Dobbins, and Fowler, 1995). Strategic planning requires the ability to assess the company's core capabilities within dynamic business environments. Several studies have found a positive relationship between the use of strategic plans by management and improved company performance (Barney, 1986, 1991; Hansen and Wernerfelt, 1989; Hart and Banbury, 1994; Powell, 1992).

American defense companies have been slow in recognizing the need for strategic planning compared with their private sector counterparts, however. This is largely because competition among defense companies was confined to who best understood the government's procurement process (Coakley and Segovis, 1995). Because of their lack of experience in more competitive markets, managers at U.S. defense companies have had and continue to have a difficult time learning the analytical skills needed to successfully convert their companies. Some scholars and practitioners fault in particular the incompatibility of the defense industry with the dynamic market environment. In one study of American defense conversion, those companies that have incorporated strategic planning are converting from military to commercial production more successfully than the companies that have failed to do so (Coakley and Segovis, 1995).

In the FSU, strategic planning is even more foreign to defense enterprise managers. The primary reason for this is the conflict faced by these former Soviet managers as they attempt to reconcile their entrenched managerial culture, past training, and work experience with the new demands of the market. Besides technological advancement, production and the provision of jobs for the people in the region were paramount concerns for Soviet defense companies. In a command system with non-market determined costs and prices for goods and services, promotions and bonuses for managers were determined by the attainment of production quotas. To meet the unfamiliar demands of a market economy, managers of defense enterprises in the FSU must not only adopt an entirely different set of values, skills, and goals, but they will also need to develop strategic plans to address these demands.

Strategic Planning and Defense Conversion

American defense companies facing conversion have chosen to either shrink their defense operations, maintain their existing operations, expand through acquisition of other defense companies (for example, Martin Marietta's acquisition of General Electric's defense operations and Lockheed Corporation), or pull out completely from the defense industry.

The strategic choices of defense companies in Russia and Belarus are limited in comparison with their American counterparts primarily because of insufficient capital. Managers from the Russian and Belarussian defense companies emphasized their lack of money and potential investors to fund their conversion projects, and complained of commercial bank interest rates as high as 200 percent.

However, even at these high rates, many defense companies were still seeking to borrow funds, but the commercial banks were reluctant to lend them money. Instead, banks could gain higher returns on their investment in a shorter period of time by participating in foreign exchange trading. The amount of capital needed to rebuild the infrastructure of these defense companies is tremendous; and the potential payback period for many of these defense company investments is too long for bank consideration (i.e., many are projected to have a 10 to 15 year payback period, and Russian banks typically lend funds with a six-month maturity). Furthermore, hyperinflation and the depreciating ruble in both Belarus and Russia have diminished the companies' retained earnings, resulting in insufficient reserves to pay salaries or to invest in plant and equipment to improve their competitiveness in the commercial market.

With estimates of the decline in military orders ranging from 65 percent to 90 percent since 1992 (manager interview, February 1995), successful defense conversion in the FSU is of critical importance. There is a tremendous concentration of national resources—both capital and labor—in the defense sectors. According to Lipsitz (1995), in the 1980s, 20–25 percent of the gross national product, and one-third of income generated for the country, came from the Soviet Union's military-industrial complex. Furthermore, nine to ten regions were economically dependent on the manufacturing of defense-related products. And, Lipsitz (1995) claims that approximately eight million people, plus another 1.8 million scientists and engineers, were employed in defense-related activities. In St. Petersburg, Russia's second-largest city, one in four workers still works in a defense company. Russia's dependence on its military-industrial complex, therefore, makes the need to successfully convert its defense companies more urgent for the general viability and health of its economy.

Faced with the need for greater capital and the desire to maintain employment of defense workers, managers of defense companies in the FSU have considered the following options: (1) expanding their military sales to foreign governments, (2) retrenching military production and seeking foreign assistance to develop commercial applications directly tied to their technological competencies, and (3) manufacturing products that are not necessarily limited to their existing technological expertise. The limitations imposed by a lack of resources and a strictly technological focus have negatively influenced their approach to the development of a rational strategic plan for successful defense conversion.

Strategic Planning and the Rational Model

A number of strategic planning models exist, and there is no real consensus among scholars as to which model is best. For the purposes of this study, the rational model comes closest to providing a framework for assessing defense conversion strategy in the former Soviet Union. In brief, the rational model approach requires managers to outline a comprehensive strategy for the company that (1) sets a clear direction for the company, (2) determines the company's strengths and weaknesses, (3) identifies factors in the political and social environment that require careful monitoring, (4) recognizes which competitors need critical attention, and (5) devotes its resources to projects that employ the company's core competencies (Certo and Peter, 1994).

The key to this model is the use of "SWOT" analysis, a management tool that identifies a company's strengths (e.g., core competencies), weaknesses (e.g., lack of capital), opportunities (e.g., demand for the company's product), and threats in its business environment (e.g., competitors), and uses the information in the development of a strategic plan. This model is appropriate for the assessment of defense conversion in the six companies in this study: all of them have implemented parts of the rational model (most identified their company's strengths), although the overall process remains incomplete by American standards. These companies have yet to focus on the ultimate goal of gaining market share and maximizing profits given the characteristics of the industry, the product positioning, and the available resources of the company. Instead, the primary objective of the six has been to remain technologically focused. They have failed to adopt a broader strategic approach to defense conversion that addresses not only the change in technological application, but also the need to change organizational structures and management orientation and to develop workers' and managers' skills to meet the requirements for successful operation in a market economy.

The Six FSU Defense Companies

Data for this study were collected at two defense companies in each of the following three cities: Minsk (in Belarus), and Nizhny Novgorod (NN) and Samara (in the Russian Federation).

The Nizhny Novgorod and Samara regions were once closed to foreigners because of their heavy concentration of sensitive defense technologies. The two companies studied in Nizhny Novgorod are referred to as NN A and NN B. NN A develops and manufactures products that utilize microwave technology for MiG aircraft. NN B is a radio electronics manufacturer. In Samara, Samara A manufactures products that use radio electronic technology for the defense industry, and also produces black-and-white television sets. Samara B manufactures satellites and rockets, and is the only company in the study that is 100 percent government owned.

Minsk, the capital of the Republic of Belarus, contains a large number of defense companies but was never a closed city. Belarus became an independent country in 1991 with the creation of the Commonwealth of Independent States.

Summary of Six Defense Companies in the Former Soviet Union

Company Name	**Minsk A**	**Minsk B**	**NN A**
Company Location	Belarus	Belarus	Russia
Ownership	100% private	40% Belarus govt. 30% Russian govt. 15% collective workers 15% foreigners	50% govt. 50% private
Principal Activity	Super-thin fiber for insulation in defense-related products	Wafer-grinding equipment	Microwave technology, Radio transmitters
Number of Employees	45	3,500	1,800
Conversion Plans	Insulation for commercial equipment	Commercial use for wafer-grinding equipment	Microwave ovens, transmitters for civilian radio stations
SWOT Status	Identified strengths only	Identified strengths and weaknesses	Identified strengths

Company Name	**NN B**	**Samara A**	**Samara B**
Company Location	Russia	Russia	Russia
Ownership	38% govt. 62% private	38% govt. 20% banks 42% private	100% govt.
Principal Activity	Radio electronics	Radio electronics	Manufacturing of satellites and rockets
Number of Employees	Not provided	2,600	Not provided
Conversion Plans	Audio recorders, Medical equipment, Ultrasound	Black & white televisions, Diabetes kits	Low-cost space launches, medical equipment distribution
SWOT Status	None	Identified opportunities	Identified strengths and opportunities

However, it is currently in the process of realigning itself more closely with the Russian Federation, both politically and economically. Minsk A, a manufacturer of equipment that produces super-thin fibers for insulation in defense-related products, is the only totally private company in the sample. Minsk B, jointly owned by the Belarussian and Russian governments, manufactures integrated circuits and wafer-grinding machines.

Minsk A was founded by scientists from a larger defense company that produced fibers for insulation. After their split from the larger company, these scientists developed a more effective and less expensive super-thin fiber. In an interview, the director discussed the extensive market research conducted to determine the uniqueness of their product. The scientists raised sufficient funds to develop and manufacture a prototype, and intend to patent their technology throughout the key industrial areas of the world. These managers assumed that receiving a patent for their technology would mean near-completion of their conversion. According to the management of Minsk A, their last phase in the conversion process was to search for Western partners to provide capital and, essentially, to purchase the new super-thin fibers. All resources—capital and labor—were focused on the achievement of this goal.

Minsk A management did not consider the strategic, marketing, manufacturing or workforce implications of this goal for their company. In essence, they determined through discussions and research what the product should be, without an assessment of consumer need or potential competition. They believed that by improving their technology, they would make their product sufficiently attractive to the world markets. Realizing that the Belarus market would be insufficient, managers looked to penetrate first the Russian market and then the West. When asked for more detail concerning their export strategy, they conceded that they did not know what the potential demand for their product would be in Russia or in the West. Nevertheless, they believed that their product would be half as expensive as any Western competitors' products. On the whole, Minsk A was disciplined in targeting its resources toward the one technology that was the company's core competency. However, management has remained technologically focused, and has yet to include in its strategic plan an assessment of the consumers' demand for its product or the company's potential competitors.

After sending a number of managers to training centers in Minsk to learn to write strategic plans, Minsk B, a large integrated circuits manufacturer, concentrated its efforts on developing its core technological competencies. Minsk B based its strategy on developing a technology that would surpass that available in the United States and Germany. Furthermore, as part of the company's conversion process, management stressed the importance of receiving a patent in several key countries, including Germany, Japan, and the United States. Managers realized the importance of recognizing the company's strengths and weaknesses in relation to their competitors. And, Minsk B did recognize the needs of its potential customers, although it failed to adequately address pricing and quality concerns. Minsk B also failed to consider other factors of strategic planning, such as industry and competitive analysis. Technological considerations remained the company's primary focus. Additionally, while the managers of Minsk B emphasized that they wanted a Western partner to contribute capital and equipment, they were not focused on the need to find a Western partner who was also well informed about consumers, and knew how to enter a market dominated by the Japanese and Americans.

NN A produces microwave ovens for the Russian market, as well as radio transmitters and receivers for both Russian and Ukrainian markets. It has also begun to export to the United States and to participate in Nizhny Novgorod's tax-free zone program. As part of this program, the company offers factory space and underutilized employees to entrepreneurs to start companies. These small companies receive a five-year tax deferral. The creation of these tax-free zones is one of the regional government's attempts to assist the conversion process of companies in the Nizhny Novgorod region.

NN A did not develop a strategic plan. It has lost 90 percent of its government sales, and the company claims that it is operating at 20 percent capacity. In the United States, this company would be bankrupt. But because the Russian government owns half of the company, it is not. NN A's management is basically trying to market its skilled labor force, whom they believe can produce any product, as an inexpensive alternative for foreign manufacturers. For example, when asked if management wanted to remain within the company's core technological competencies (i.e., microwave technology), they replied, "Not necessarily." As one manager asserted, "Our scientists and engineers are smart. They can produce anything; just give us the specifications, the equipment and supplies and we will produce it." NN A's technological focus overshadowed its need for a more strategic approach to competing in Russia's newly emerging market economy.

The management of NN B, a radio electronics manufacturer, hired a consultant to write a strategic plan that would help the company focus on using its core technology in the development of commercial applications. The consultant suggested the company manufacture audio recorders for the general communications industry. NN B manufactured these products at a lower price than its foreign competition. The strategic plan prepared by the company's consultant was incomplete, however; it failed to consider external threats to NN B, mainly Sony, which had also entered the Russian and Belarussian markets. Western- and Japanese-made products were of much higher quality. Therefore, if consumers could afford the foreign recorders, they would pay more for the better quality equipment. Because of this, NN B has not been successful in exporting its recorders to the West.

Because of their lack of commercial success with the consultant's recommendations, management of NN B decided to seek alternative uses of their core technology by entering another market: ultrasound and heart monitoring equipment. Their decision to enter the medical equipment industry was based on the knowledge that Western medical equipment was overly expensive, and that the regional government needed to upgrade the equipment in the hospitals. It is too early to determine the success of this new product line since management has yet to conduct a SWOT analysis, or to determine how the shift from defense to commercial production will impact their entire organization or labor force.

Samara A, which used radio-electronic technology to manufacture defense products, has experienced a 90 percent drop in orders from the military. Currently, the company's entire defense conversion effort is centered on the production of black-and-white televisions. Production has been expanded as much as possible given available equipment at the plant. Samara A's biggest constraint is supplies. Because of hyperinflation, the cost of supplies has increased more than the price of the television. For example, the prices of televisions sold in February reflected the cost of production prior to February. However, the cost of supplies in February has increased by more than the revenues received from the February sales. Therefore, the company does not earn enough revenue from sales of televisions to pay for supplies for the manufacture of the next batch of televisions.

Like NN B, Samara A wants to provide home health care products (e.g., diabetes detection kits) for the medical industry. The regional government is willing to fund the acquisition of new equipment for the regional hospitals and clinics, and so is providing a local market for these products. Based on discussions with the government, the company wants to shift its focus to the medical industry, and away from the increasingly foreign-dominated domestic consumer market. Samara A is in the process of developing several prototypes of medical equipment and is seeking a partnership with a Western company to contribute hard currency or up-to-date equipment. While developing their prototype and seeking funding, however, management is focusing their strategy away from exploiting their technological expertise, and more toward the manufacturing of any product, not exclusively televisions or medical equipment. Once again, management neglected to focus the company's conversion strategy on its core competencies.

Samara B, a 100-percent government-owned manufacturer of satellites and rockets, is not facing the crisis-like situation of many defense companies. Management admitted that the government considers their company to be crucial to the advancement of scientific knowledge in Russia, and that there is no effort, at the present time, to privatize its operations. Samara B is receiving less government funding, but it has not been cut to the dramatic levels that have prompted other Russian defense companies to seek alternative ways to remain financially viable. Nonetheless, Samara B has put some effort into its conversion process and has focused on finding direct commercial applications for its technology. For example, the company provides low-cost, commercial space launching and puts other nations' satellites into space. Management has identified NASA as the company's primary competitor. During the interview, they acknowledged that they would not be able to earn enough funds from this activity to make up for the loss in revenue from the government, so they are seeking other commercial applications, both related and unrelated to their core technological competencies.

Samara B believes that its rocketry and space technology, specifically heat-sensitive material, is its most marketable commodity. However, the company is reluctant to apply its knowledge of heat-sensitive materials to commercial applications, because it is unsure how to apply this technology without revealing proprietary information to competitors. This attitude limits Samara B's efforts in defense conversion. Although it does not have the necessary resources to go beyond its core competencies, it continues to seek products for the commercial market (e.g., distribution of medical equipment) that do not use its "proprietary" knowledge, thus creating the need for additional resources: a catch-22.

Managerial Culture: The Foundation for Managerial Action

Based on the information from these six case studies, the strategic decisions of managers in the FSU in their companies' defense conversion processes differ from what the rational strategic model would dictate. An argument can be made that these decisions are grounded in the managerial culture as evidenced in each of the six enterprises. "Culture" is the catchphrase through which behaviors, attitudes, and patterns of people's life experiences are explained (Benedict, 1934). Through extensive cross-country studies, Hofstede (1980) argued that culture is the persistent variable that explains managerial decisions. Managerial culture in the former Soviet Union, as exemplified by the decisions made by managers of defense companies there, is still greatly influenced by the values of the previous command system. The managers from the six enterprises gained most of their professional experience working under the old planned economy. Therefore, even when some employed a rudimentary SWOT analysis to assist them in their company's conversion process, their decisions reflected the ideologies grounded in a production-reward system. In particular, three distinct cultural variables were observed that create challenges for these managers as they attempt to convert their operations in the emerging market environment. These variables are attitudes toward technology, labor, and production. They are examined below.

Russia's View of Technology

During the Soviet era and before, Russia's desire was to be at least a technological equal of the West, if not to surpass it. Not only did the Soviet Union have the largest number of engineers in the world, but the way to top management positions was to be an engineer (and, of course, a member of the Communist Party). It is interesting to note that 80 percent of the Politburo members in 1980 were engineers (Balzer, 1990).

Soviet managers viewed the solutions to their problems as technical in orientation, and not organizational, strategic, or managerial (Randall, 1994). The centralized Soviet suggestion system called the rationalization proposal program is a good example. Briefly, a suggestion system is a way of continuously improving the operations of a company. In the United States and in a vast majority of companies in Japan, valid suggestions from employees generally are not technical, but usually involve organizational, manufacturing, marketing, and product development proposals. However, in the former Soviet Union, the only suggestions accepted were those technical in orientation and which required an engineer to fill out the necessary documents so that the worker could receive the reward (Randall, Robinson, and Tolstaya, 1994).

This heavy engineering influence in the management ranks, as well as the historical desire for technological preeminence over and respect from the West, has deeply affected the actions of defense company managers in the former Soviet Union. This attitude is most evident in the approach taken by the six defense companies to sacrifice sound strategic decision-making in favor of retaining their technological expertise.

Management's Perspective on Labor

The views of managers in the FSU regarding labor differ from their American counterparts because, under a market system, American managers view labor as a cost item needing to be managed in order to maximize a company's profitability. During the late 1980s and 1990s in America, the phrase "lean and mean" came to describe the strategy of many American companies seeking to maintain their competitiveness by reducing their workforce. As American defense companies are grappling with conversion and entering commercial markets, one of their primary concerns is that their cost structure (i.e., the cost of manufacturing a product) is too high to price their products competitively. Many American defense companies are laying off workers in order to get their cost structures in line with those of their commercial counterparts.

As the economy in the former Soviet Union moves slowly toward a market-like system, most of the managers' attitudes and goals are still greatly influenced by the "old" command system. The Russian managers commented that employment of the workers was a very important goal. As one Russian manager remarked, "I have heard that the goal of the American manager is to get rid of as many workers as possible. Isn't it the job of a manager to seek ways to provide employment for their workers? How can people buy all of the goods produced in America, if no one has jobs and therefore cannot afford to buy them?" (Interview, Nizhny Novgorod, February 1995).

Layoffs are occurring in the FSU, but not quite in the same manner as in the United States. For example, five of the six case study companies, even with 30 percent to 40 percent reduction of labor, are still practicing overemployment. Samara A's military production was approximately 85 percent of total capacity. The company lost 90 percent of its military orders and only made up 10 percent of the lost revenue with its television production. Yet, with a 76 percent loss of production and no prospects for future business in sight, it only provisionally "laid off" 40 percent of its workers—i.e., employees did not work, but still received benefits, and expected to be "rehired" when work became available. No additional firings occurred. When asked why, Samara A managers said that entire families were dependent on them for a salary and benefits, and the managers hoped to find projects for their workers.

A second example that illustrates the cultural differences in management's goals of employment stems from an interview with management from Minsk A. The managers of Minsk A asked several Western business people for a recommendation regarding the best strategic move for their enterprise. The Westerners recommended licensing its super-thin fiber technology to a Western company as the best way for Minsk A to enter the world markets. Minsk A managers concluded, however, that such licensing would mean they would not be able to employ as many people, and decided that they would rather have a co-production agreement with a Western company. This predicament was outlined by Nizhny Novgorod and Samara government administrators. They said that "dynasties" of families have worked at particular plants for decades, some even before the Revolution of 1917. Managers have a social obligation to these families to provide employment and services that were part of the value system of the planned economy. This economic system's purpose was to provide full employment and social services to the population. These officials commented that if these defense companies laid off workers in the numbers that occur in comparable companies in America, then the regional (oblast) governments would be placed under much more pressure. In the United States, workers and their families can move from one region to another to seek employment opportunities. In the FSU, workers do not have the same amount of mobility because of the lack of housing. Therefore, it is imperative for Russian managers to employ, despite economic hardship, as many workers as possible.

Unfortunately, the enterprises are facing tremendous cash deficits and do not pay well. And, in many instances, these companies cannot pay their workers for several months at a time. The unwritten rule is that workers remain officially employed with the large enterprise but find other jobs, usually a number of part-time jobs. It is not unusual to find a computer engineer still officially employed by a defense enterprise moonlighting as a taxi driver, for example. His missed time at the company is not punished, because everybody does it to "make ends meet."

The cultural impetus for maintaining full employment conflicts with the goals derived from the rational strategic planning process. Specifically, the need to ensure that all workers are employed impedes the ability of the six enterprises to completely exploit their strengths (i.e., gain a technological competitive edge), minimize their weaknesses (i.e., cost of production), and fully engage in a market economy.

Production versus Market Orientation

The FSU managers in this study considered the achievement of production goals paramount, often retaining excessive capital and equipment that no longer were required to manufacture the newly (converted) commercialized product. This production mentality overshadowed the demands of the emerging market economy, weakening the prospects for successful conversion. For example, the management of NN B, the radio electronics manufacturer, not only was willing to manufacture any product, but their decision to do so was based on idle production capacity and not on costs, consumer needs, or competitive advantage. Again, this behavior stems from the dictates of a command system. In the previous command system, managers of these defense enterprises were trained and compensated by achieving certain production quotas determined by the central government administrators in the ministries and the State Committee for Planning (Berliner, 1988). These production goals reflected the Soviet Union's drive for military superiority and technological preeminence during the Cold War. With the ending of the Cold War and the emergence of the market system, production goals need to be realigned with the demands of a competitive business environment.

Challenges and Conclusion

This study examined the efforts of six defense companies in three regions of the former Soviet Union to convert their operations from a defense to a commercial orientation. These case studies emphasized several critical prerequisites for the achievement of successful defense conversion in Belarus and Russia. Each of the six enterprises focused its conversion efforts on promoting technological abilities, not developing strategic plans. Therefore, they were not fully employing the rational model of strategic management: identifying their strengths, weaknesses, opportunities, and threats. This impeded their ability to fully exploit their core competencies and recognize the needs of their customers, resulting in the failure to position their enterprises in the most advantageously competitive situation.

Much of the companies' behavior may be attributed to the fact that their strategic decision-making is grounded in the old command system mentality, which affects their approach to issues of labor, production, and technology. Therefore, one of the greatest challenges to successful conversion facing these enterprises is reconciling the demands of defense conversion in a market economy with the cultural values of the old Soviet system. A change in culture can occur during times of dramatic crisis, such as that facing these Russian and Belarussian managers. But such a change requires tremendous commitment and effort from top management.

This may be problematic for the enterprises in our sample, given that many of the top managers were trained and indoctrinated in the old system, and therefore will find it difficult to adapt their socialist-based values to a market orientation.

A key step to any successful defense conversion is the training of managers and employees to think strategically. As discussed earlier, the defense conversion process should encompass more than the transference of technology from military to commercial means. The process should include a change in the way business is conducted and a change in the types of skills required of workers and managers.

The need to train managers to think strategically is becoming more evident in U.S. defense companies. The product needs of consumers are no longer given to companies in detailed specifications. Companies now must assess the market's needs, have a sales force, advertise their product lines, and develop and refine new products to match the changing needs of the consumer. This requires managers to start thinking of conversion as a strategic process involving the entire organization, and not just as a transfer of technology.

It is critical, though, that as scholars, we do not fall prey to framing a solution solely within the boundaries of a Western perspective. The SWOT analysis is based on an economic context that is wholly market driven. The SWOT framework has not been tested in any of the markets-in-transition. It is imperative that differences in the cultural and economic systems of the West, particularly the United States, and Russia and Belarus be recognized, and a greater understanding of the role culture plays in the defense conversion process be assessed. This will enable researchers to develop a framework for assisting Russian managers in their endeavor to become competitive.

References

Balzer, H. "Engineers: The Rise and Decline of the Social Myth." In *Science and Soviet Social Order*, edited by Loren Graham. Cambridge: Harvard University Press, 1990.

Barney, J. "Organizational Culture: Can it Be a Source of Sustained Competitive Advantage?" *Academy of Management Review* 11, no. 3 (1986): 656–665.

———"Firm Resources and Sustained Competitive Advantage." *Journal of Management* 17, no. 1 (1991): 99–119.

Benedict, R. *Patterns of Culture*. Boston: Houghton-Mifflin, 1934.

Berliner, Joseph. *Soviet Industry from Stalin to Gorbachev.* New York: Cornell University Press, 1988.

Bounds, G. M., G. H. Dobbins, and O. S. Fowler. *Management: A Total Quality Perspective*. Cincinnati, Ohio: South-Western College Publishing, 1995.

Certo, S., and P. Peter. *Strategic Management: Concepts and Applications*. Chicago: Irwin Press, 1994.

Coakley, L., and J. Segovis. "Successful Management of Retrenching,

Reengineering, and Downsizing: Examining Best Practices in Industries Confronting Defense Conversion," article forthcoming.

Hansen, G., and B. Wernerfelt. "Determinants of Firm Performance: the Relative Importance of Economic and Organizational Factors." *Strategic Management Journal* 10, no. 5 (1989): 339–411.

Hart, S. L., and C. Banbury. "How Strategy-Making Processes Can Make a Difference." *Strategic Management Journal* 15, no. 4 (1994): 251–269.

Hofstede, G. *Culture's Consequences*. Beverly Hills, CA: Sage Publications, 1980.

Lipsitz, I. "Problemi Rossiskoi Konvercii." *Ekonomist*, January 1995.

Powell, T. "Strategic Planning as Competitive Advantage." *Strategic Management Journal* 13, no. 7 (1992): 551–558.

Randall, L. *Determinants of Joint Venture Success in Russia*. Michigan: UMI, 1994.

Randall, L., A. Robinson, and A. Tolstaya. "What U.S. Managers Can Learn from the Former Soviet Union's National Continuous Improvement System." *Academy of Management Proceedings*, 1994.

10

Learning to Learn: Emerging Patterns of Enterprise Behavior in the Russian Defense Sector, 1992–1995

Yevgeny Kuznetsov

Introduction

Learning is a process involving repetition and experimentation which enables tasks to be performed better, and, at the enterprise level, new production opportunities to be identified. On the basis of case studies of 24 enterprises, this paper explores the Russian defense sector's process of learning to restructure. Since learning is cumulative, it can be viewed as an accumulation of competencies, or more generally an accumulation of intangible capital.

In an economy with rudimentary market institutions, enterprise learning also involves significant start-up costs to create the sort of quasi-market institutions[1] required for the adjustment. From such a perspective, the adaptation, or adjustment, of an enterprise in a transitional economy is somewhat similar to the growth of a firm in an economy with antiquated physical infrastructure or pervasive government regulation. In both cases, an entrepreneur must provide capital to establish at least part of the required infrastructure, time and financial resources to get through the tangle of regulations to set up a firm,[2] and intangible capital and financial resources to create a new institution.

The Economic Situation in the Defense Sector

The Russian defense sector is defined here as enterprises under the auspices of the State Committee for Defense Industries, which includes all military-related plants

The author is grateful to Ksenia Gonchar (Bonn International Center for Conversion), Lev Freinkman (World Bank), Victor Rassadin, Maya Shcuhglater (Institute of Economic Forecasting, Moscow), and Aleksandr Vitebski'i (Russian Committee on Defense Industries) for stimulating discussions which provoked many of the ideas of this paper. The opinions expressed in this paper are those of the author and should not be attributed in any manner to the World Bank, to its affiliated organizations, or to the members of its board of directors or the countries they represent.

with the exception of nuclear facilities belonging to the Ministry of Atomic Energy. By the end of 1994, 80 percent of the output of the nearly 2,000 enterprises of the complex was civilian output.[3] Output level in 1994 was 61.4 percent of what it was in 1993, and the level of employment was 83 percent. Compared with 1991, military output in 1994 was 19.9 percent, civilian 52.6 percent, the aggregate output was 39.2 percent, and industrial employment 68 percent. Compared with other industrial branches, military hardware ranks first in decrease in output, followed by light industry (20.5 percent of the 1991 output level) and machine building (approximately 35 percent of the 1991 level). The average wage in 1994 was 64 percent of the average wage in industry, and 87 percent of the average wage in the machine building sector. There is a substantial intrasectoral gap. The lowest average wage is in the electronics industry (44 percent of the wage in industry), while the highest is in the shipbuilding industry (89 percent of the wage in industry).

As these figures indicate, the economic situation in the defense sector parallels the situation in the machine building sector. Yet the following circumstances illustrate a difference. First, the financial situation of defense enterprises is unusually volatile due to the dependence of defense procurement on state funds. In 1994 the government procurement debt amounted to 38 to 50 percent of accounts receivable and payable of the defense sector. Second, the maintenance of the so-called mobilization capacities remains a sizable financial burden. While the government is supposed to reimburse enterprises for the full cost of the maintenance, this never happens. As a result, overhead costs remain prohibitive, reaching at times 1,000 percent of the wage bill; the typical figure is in the magnitude of 200 to 400 percent. Because of the government-induced illiquidity and high overhead cost, the scope for adjustment is smaller in the defense sector than at the majority of civilian enterprises.

The enterprises in the sample (24 companies in the radio, electronics, aviation, and shipbuilding industries: see Table 10.1) reflect this general picture, with two important caveats. First, the official employment figures appear to overestimate actual employment. Enterprises have to pay the excess wage tax if their average wage exceeds a certain threshold level. In order to avoid this, enterprises in the sample inflated employment figures by "hiring" labor "just for the record." In the sample in 1994 the average wage ranged from 25 percent (at a radio plant in Siberia) to 150 percent of the industrial wage (at an export-oriented manufacturer in Moscow). Second, the official output figures appear to underestimate the actual output level by 5–10 percent by not taking into account unrecorded economic activity.[4]

Table 10.1: Regional, sectoral, and size distribution of enterprises included in case studies

Regional distribution	
Voronezh region	3
Novosibirsk region	4
Moscow and metropolitan (Aleksandrov, Fryazino, Zelenograd) area	7
Tula	1
Eastern Siberia and Far East (Tomsk, Komsomolsk na Amure)	2
St. Petersburg and the region	4
Urals (Izhevsk, Ekaterinburg)	2
North (Severodvinsk)	1
Sectoral distribution	
Aircraft and space	4
Shipbuilding	1
High-precision mechanics	4
Radio equipment, communications electronics	12
Ammunition, armored vehicles	3
Size distribution (1993)	
Fewer than 5 thousand employees	3
5–10 thousand	14
More than 10 thousand	7

Although the sample does include enterprises from all branches of defense industry and all major regions of Russia, it is biased at least in three ways. First, half of the sample is made up of the radio, communication, and electronics branch, which is less asset-specific compared, for instance, with tank manufacturing, and thus has more favorable conditions for conversion and diversification. A useful way to think about the diversity of defense enterprises is to collapse it into a matrix (Table 10.2) which combines asset specificity and the demand orientation of enterprises (export vs. domestic demand). Second, almost half of the sample belongs to Moscow and St. Petersburg and their metropolitan areas. Third, and most importantly, the sample represents more active (but not necessarily more successful) adjustment than appears to be the case in the industry as a whole. Information important to the study on non-adjustment cases has been collected through regional administrations and less-detailed interviews which did not cover the entire range of issues included in case studies.

Table 10.2: Demand Orientation and Asset Specificity of the Sample

Export-oriented, large system manufacturer with specialized capital stock (1)	Domestic demand, large system manufacturer with specialized capital stock (7)
Export-oriented, relatively universal capital stock (3)	Domestic demand, relatively universal capital stock (13)

A Tale of Two Adjustment Paths

Puzzle 1: There are growing differences even between companies in the same industry, often located in the same town, and having similar factor endowments, that cannot be explained by differences in economic fundamentals.

Compare two medium-sized (about 5,000 employees) military enterprises which used to produce electronic control equipment. Both are located in the same town close to the Finnish border. Civilian production used to be marginal for both of them. In 1992, military procurement was decreased by a factor of 10, relegating defense-related manufacturing to a marginal role. There is an adjustment cost to transfer to civilian manufacturing; because use of expensive ferrous and precious metals is critical, a direct application of technologies to the civilian sphere is not possible.

Tangible endowments of both enterprises can be divided into illiquid (fixed capital) and relatively liquid (input inventories such as ferrous and precious metals, and funds to facilitate adjustment, provided by the federal government). One enterprise chose to adjust by engaging in transactions with liquid assets, using the proceeds to maintain high income in top management and modest wages for employees, with no effort invested in enterprise restructuring. Its adjustment strategy included five elements:

(1) Organization of a number of spin-offs owned by the top management to deal with the real estate belonging de facto to the company. Profit of these firms is the income of its owners, and is not reinvested.

(2) Manufacturing of simple consumer durables (such as unsophisticated electronic games) to justify credit from the government.

(3) Investment of the government funds, however small,[5] as well as any proceeds from the plant operation, in the financial market. Financial experts are hired so that the company focuses mainly on rents derived from financial transactions.

(4) Leasing or selling portions of the plant's real estate to foreign or wealthy domestic customers.

(5) Allocation of a part of the proceeds from trade operations and financial and real estate transactions to maintain a modest wage for company personnel, in order to prevent any unrest.

Management interests are thus outside the plant, which is left to disintegrate: fixed assets that are not maintained deteriorate and skilled labor leaves the plant because of low wages.

The manager at the second plant focused on adjustment in the real sphere: he believes that eventually the plant will export low-cost electronic microscopes for certain market niches such as educational institutions. He pursues a strategy consisting of the following steps:

(1) "Primary accumulation"
He engages in transactions similar to those pursued by the manager at the first enterprise in order to obtain "breathing time" to readjust technologies with the expectation of becoming a subcontractor to Western (Swedish and Finnish) companies.

(2) Unrelated diversification
To ameliorate the defense demand shock and to avoid consequent labor separations, the manager acquires sewing machines: female employees make intermediate parts for Finnish clothing producers. Note that it is not social responsibility which motivates the manager to avoid layoffs, but rather an awareness that unless social stability is maintained, the company's economic advance will be inhibited.[6]

(3) Learning to export on the basis of transitional comparative advantage
The plant starts to focus on relatively unsophisticated subcontracting and manufacturing of intermediate output (electrical equipment for cars) for export with the emphasis on quality of output rather than cost. By the end of 1992 the average monthly wage was $25, and even though it approached $80 in 1994, labor cost has diminished to 10–15 percent of output cost. Because energy prices in 1992 were still a fraction of world market prices, output costs were by definition much less than across the border. Output quality and organizational ability to find and maintain relationships with suitable foreign partners mattered.

(4) Investment in reputation
The manager realizes that his advantage in low-cost manufacturing is partly transitional, and will disappear with an increase in energy prices. Modest investment is made to move into more sophisticated manufacturing and most importantly to ensure timely and reliable delivery of the output for the foreign partner. Sources of investment are government credit and retained earnings from the previous stage.

(5) Increase in the foreign partner's commitment
The reliability of the Russian manufacturer, against a backdrop of institutional turmoil and costly contract enforcement, convinces the foreign partner to disre-

gard the otherwise uninviting Russian business climate and make investment commitments, starting with the long-term leasing of equipment.

(6) Measures to retain valuable employees

Because of the cost structure in which the labor share is small and the energy and materials share is substantial, the manager does not focus on the decrease of redundant employment. But in order to preserve the core group of employees he must maintain large wage differentials. This is where installation of sewing machines comes into the picture. This is perceived as a symbol of the manager's loyalty to the employees, which helps him undertake unpopular measures without inciting unrest.

(7) Export of microscopes

The plant enters into servicing and marketing agreements with a Swiss partner, which permits exporting microscopes. (In 1994 export revenues provided 60 percent of all revenues of the plant, the remainder falling on increased military procurement.)

Why did two plants with similar endowments pursue different adjustment trajectories? Why did the first manager choose private rent-seeking, focusing on financial transactions, while the other took advantage of the enterprise's proximity to the border to raise its export? Intuitively, it is clear that the longer planning horizon of the second manager accounts for the difference, but what explains the disparity in planning horizons? Before proceeding to the second puzzle let me add two important caveats which reflect observations relevant to all case studies in this paper. First, when comparing these adjustment patterns I am not implying that one manager is more entrepreneurial than the other. Both of them exploit all profit opportunities they see (the second manager, even while being a quite successful exporter of civilian output, seizes the opportunity of getting a large defense contract). It is the area of search for these opportunities that is different. Second, in the long run it is not clear which adjustment strategy is superior. The first plant continues to exist only as a social protection unit while the second exports high-tech output without investing in further R&D. Its export of microscopes proceeds entirely from a stockpile which is so large that it will maintain exports for years. Once extensive R&D comes to a halt, however, a high-tech export strategy is clearly unsustainable. What is sustainable is the lower labor costs of the maquiladora regime, which provides employment but generates little added value.

Puzzle 2: With the accumulation of adjustment experience more firms abandon efforts to adjust, while firms that continue adjustment are more likely to succeed.

On the basis of case studies one can suggest the following sequence of enterprise adjustment. The first phase is the waiting (winter and the first half of spring 1992) which immediately followed the demand shock and price liberalization. The second phase is experimentation (spring of 1992 to the end of 1993), when every enterprise (subject to the range of its competencies and entrepreneurial qualities of the management) began searching for restructuring options. By 1993 it became clear that two major factors are responsible for the differences in adjustment strategies. The first is the planning horizon of the manager, and the second is the manager's organizational focus in restructuring: whether she/he aims to revitalize part of the plant or the whole enterprise. In combining these two factors one can observe four types of adjustment strategies (Table 10.3):

- sustainable real adjustment (long-term oriented turnaround)
- fragile real adjustment (the attempt to maintain all parts of the enterprise)
- sophisticated rent-seeking (short-term extraction of rents from enterprise assets[7])
- traditional rent-seeking (reliance on the government)

Table 10.3: Managerial Incentives and Types of Adjustment

	Planning horizon of the management: long. There is a high probability that the current management will remain the effective owner of the enterprise after its privatization.	Planning horizon of the management: short ("fly-by-night" strategy).
Focus of the management's attention: certain viable segments of the enterprise. Manager is more a turn-around specialist than a conventional manager.	*Sustainable real adjustment* Management perceives that (1) certain segments of the enterprise are potentially quite competitive; (2) because of the pervasive scarcity of managerial expertise, it is bound to remain the effective owner of the enterprise even after privatization.	*Sophisticated rent-seeking* Motivation to extract high personal rents from certain lucrative segments of the enterprise and then either retire or set up a brand-new private venture not necessarily related to the production line of the enterprise.
Focus of the management's attention: the whole enterprise rather than its segments.	*Fragile real adjustment* Strategy to maintain all technological and human capabilities of the enterprise, which is not financially feasible.	*Traditional rent-seeking* "Fly-by-night" with exclusive reliance on government assistance and favors.

In 1993 about two-thirds of the enterprises in the sample pursued fragile real adjustment; i.e., they were actively searching for restructuring technologies and ways to finance them. The remaining enterprises were evenly spread between the other three adjustment types. By the middle of 1994 the situation had changed. Some enterprises (about a quarter of the sample) had become social protection units producing little and channeling all revenue of the "cash cows" within the enterprise to maintain social infrastructure and (although substantially reduced) employment. The management of those plants had practically become part of the local government.

Private rent-seeking became more flagrant because of an emerging alliance between banks interested in enterprise assets and the managers who were waiting for a takeover.[8] Since they were certain of being replaced, the managers became very keen to extract all rents from their position while they could. For instance, there was a shift in the allocation of the federal conversion credit. If in 1992 and 1993 half of the sample had channeled at least part of these funds toward investment purposes,[9] in 1994 and 1995 the approach to utilization of those credits had changed. Enterprises started to enter into agreements with banks that use those funds for high-return projects (such as trade operations) and receive in return a negotiable interest rate which is often higher than the going market rate. Cash, if needed, is obtained from commercial banks through market lending. In both cases (social protection and private rent-seeking), all searches for restructuring alternatives have been abandoned.

At the same time, sustainable adjustment has also become somewhat more widespread and more evident. Managers became more active in ordering layoffs, and began devoting a major share of their time to the search for strategic investors. Thus, the experimentation stage is also the bifurcation phase of adjustment: it is during this stage that the quite distinct adjustment trajectories pursued during the subsequent shake-up stage are revealed. The shake-up stage is characterized by the struggle for control over the privatization process and redistribution of company stock.

The puzzle is that the changes in enterprise behavior cannot be explained by a variation in economic policy and the macroeconomic environment. If anything, financial transfers to defense enterprises were steadily diminishing beginning in 1993. It appears that adjustment dynamics are based largely on the ability of managers to create new institutions and on their ability to learn how to learn.

The Start-up Cost of Institutional Creation

As the examples in the previous section illustrated, learning to adjust can be viewed as an accumulation of competencies or an accumulation of intangible capital. The latter notion includes the firm's technical and organizational know-how, its reputation as embodied in trademarks and otherwise, its organizational and customer networks, its culture (e.g., its ability to change), and its trade secrets. Accumulation of the intangible capital may be incremental or, when the new institution is being created, may involve investment which implies fixed cost in terms of financial resources and/or the firm's competencies. For instance, firms seeking to attract foreign capital invariably find it necessary to switch to Western-style accounting systems. This requires not only relevant investment in subcontracting with Western accounting firms but non-negligible learning on the part of the firm's employees to master the new system as well.

In the case of the transition to a new accounting system, the start-up cost of learning can be reduced by resorting to foreign expertise, though this option is not always available. For instance, beginning in 1993 funds for defense procurement in Russia were allocated to the final producer of weapons systems. In the past funds were routinely allocated by the military-industrial commission to every enterprise, irrespective of whether it was a subcontractor or final manufacturer. Similar to a switch to a new accounting system, organizational resources should be expended to make the procurement headed by final producers (rather than the central authority) function smoothly. Why should the final producer pay on time? What happens if it cancels the order after inputs have been already produced? Because the lack of an enforceable business code in Russia makes it impossible to settle these issues in court, an organizational network of subcontractors was created to deal with these issues.[10]

In a few cases we also observed the creation of institutions that provided information about potential suppliers and customers. A defense enterprise in St. Petersburg, for instance, contributed space for the exhibition of civilian products manufactured in the defense sector. The exhibition has a database of specific outputs which customers in the region (St. Petersburg and the area) require. The exhibition is open to everyone, but in order to join, the interested enterprise needs to include in the database the range of its output, along with relevant price and delivery data. A number of matches between customers and supplies have been accomplished via the database. What is noteworthy here is that the database and exhibition have emerged as a cooperative effort on the part of defense enterprises frustrated by the wasteful way they enter the civilian market,[11] rather than as the creation of an entrepreneur who perceived and seized an opportunity. Similarly, new institutions are invariably created when a firm starts exporting. All relevant firms in the sample had to establish agencies (with the help of Western consulting firms) for the preparation of export contracts, service and customer networks abroad.

Serious adjustment involves the creation of new institutions; clearly, however, the cost of such organizational innovation may be prohibitive. There are two reasons for this. The first is that the firm may be unable to raise the capital necessary to undertake investment. The deficiency of the revenue stream because of the lack of demand, for instance, is a common obstacle to recouping the fixed cost. As a result, adjustment normally proceeds as incremental learning rather than along the more costly path of creating new institutions.

One can think of a range of restructuring and learning options varying according to the amount of start-up cost. Rather than adopting a Western-style accounting system, for example, a Russian firm can adopt a system of double or even triple bookkeeping, in which the first is for the top management to understand what really happens at the enterprise, the other is the standard system, and the third is for tax authorities and other outside observers. Enterprises also have the option of computerizing the existing system to improve the flow of information. The option with the smallest start-up cost is the one that is usually chosen.

The Problems of Innovations and Incremental Learning

Industrial adjustment in an economy with rudimentary market institutions and in a planned economy is similar. Effective adjustment in both cases requires either the creation of institutions that are missing or the ability to learn fast incrementally. Thus, many Soviet managers appeared to be quite entrepreneurial and adaptive in the post-1992 economic reality. However, lessons from the recent past show that innovation is not without its drawbacks.

Consider the example (Kuznetsov 1989, 1993) of the persistent problem experienced by Soviet railroads of unloading agricultural fertilizers which used to freeze solid in winter during rail shipping. In order to unload a car containing such fertilizers, it had to be kept in a heated space for a number of days and even then it could only be unloaded manually. The technical solution to the problem was well known: producers needed to switch to granulated fertilizers that do not freeze. The mechanisms to induce producers to make such a switch were far less obvious. Usually such coordination problems were solved by the interested party using an informal network of individuals that included representatives of the agency which could solve the problem (the producer), an agency interested in seeing a solution (the consumer), and planning officials who would help to negotiate the solution and later legitimate it through a decree of the Council of Ministers and/or the Communist Party. To induce the producer to undertake the relevant change, the consumer had to offer some incentives, usually in the form of a transfer of its share of investment, construction capacities, or other scarce resources or services for the producer. To solve the problem, then, one had to establish an institution: an informal network capable of negotiating an exchange. Although this activity is by no means illegal, it had to be based on personal trust in order to make the agreement last. No wonder that it took such time and talent to establish such a network and that coordination between various branches of Soviet industry usually failed.

In the case of the freezing fertilizers, no one was willing to bear the cost of establishing such a network, and the solution was found along the lines of incremental learning. The railway agency asked the defense sector to invent a technical device that allowed thawing and unloading of those fertilizers that freeze solid during shipping. A cheap, effective device was invented and produced by one of the defense plants. This example shows that it is erroneous to assert that a Soviet-type system was innovation-averse. Rather, it innovated efficiently in response to economic imbalance. If this innovation had not appeared, it is likely that mounting pressure from both consumers and transporters of frozen fertilizers would eventually have induced the producer to switch to more efficient granulated fertilizers. After the technical innovation appeared, these pressures abated and a learning trap emerged which looked at more efficient outcomes. There are many facets of allocative inefficiency in Soviet-type economies, and technical change as a handicap rather than a promoter of socially efficient development is the most striking.[12]

Another current problem with incremental learning is the diversity created in attempts to avoid the start-up cost of investment in the creation of institutions. One enterprise in the Voronezh region institutionalized a pattern of part-time industrial employment by buying a failing agricultural cooperative near the city limits. Now the labor is employed (in facilities of the enterprise) only in winter, early spring, and late fall; the remainder of the time the labor is employed in dachas (summer plots) and in the newly acquired agricultural unit. The purchase of processing equipment is now being contemplated. The other enterprise in the same city, unable to sell its own food-processing equipment, utilized it by venturing into the food processing business. Given the extreme inefficiencies of Russian agriculture these ventures may be competitive, though the question is for how long and at what cost. The problem is not the diversification per se but rather the way it is accomplished. In both plants, workers and particularly engineers spend a portion of their time assembling and devising radio and electronic equipment, and part of the time engaging in agriculture-related activities. The management needs to decide upon a new organizational configuration for the plants but has postponed doing so not only because in conditions of pervasive uncertainty it is wise to delay an irreversible decision, but mainly because of the inability to come up with adjustment trajectories for the plant in question.

Still another result of incremental adjustment is regional segmentation of national markets. Unlike in the past, enterprises will no longer diversify before establishing the market for the output. They will send letters with the description of a prospective output to all enterprises in the region, make phone calls to potential customers, and bring samples to the most important ones through personal visits, although these are necessarily limited in distance because of transportation costs. East of the Urals, in particular, the market for new output seems to be limited by a distance of approximately one thousand miles. Because many suppliers became increasingly unreliable, or inputs are manufactured too far away, many enterprises choose to manufacture these inputs themselves rather than look for new suppliers. In addition, the advantage of "making" rather than "buying" is not only the supply guarantee but also provision of employment for an enterprise's own workers. The result is hardly cost-effective, but cost issues are irrelevant for enterprises which produce for the government or maintain a regional monopoly. As a result, defense enterprises notorious for their "roundaboutness" before the beginning of the adjustment now tend to produce rather than buy their inputs.

Types of Uncertainty in the Russian Defense Sector

There are several empirically significant sources of uncertainty which do not boil down to the uncertainty caused by high inflation.

Uncertainly specific to the military sector

Uncertainty specific to the military sector is caused by:

(1) Volatility in defense procurement. One-third of the enterprises in the 1992–94 sample experienced a 50–300 percent change in state defense orders. The remaining two-thirds expects significant variation in the future. When one cannot reasonably predict the utilization of assets, long-term adjustment becomes questionable.

(2) Volatility of payment for defense procurement. This is the most significant factor of uncertainty which makes enterprises chronically illiquid. In 1994 for enterprises in the sample the average time between the date of the payment due and actual payment was no less than four months. In 1994 the government procurement debt ranged from 38 percent to 50 percent of accounts receivable and payables of the defense sector.

Uncertainty specific to transition economies

Uncertainty related to privatization should be emphasized. Before the strategic owner of the enterprise emerges, one should not expect any commitment either from incumbent managers (who tend to spend most of their time maximizing their ownership stake in the enterprise) or from outside investors. In a number of R&D institutes in Moscow and St. Petersburg the competition for effective control has blocked not only adjustment but day-to-day operations as well. It should not come as a surprise, then, that half of the enterprises perceived as active and successfully restructuring were state-owned enterprises that chose to delay privatization. In all cases, the decision to postpone was assessed as the rational one. At these companies the management took full control (since there was no threat of dismissal), and adopted a long-term attitude. Delay in privatization also promises a more vibrant stock market and thus more sizeable revenues from the sale of stock. Note, however, that these were companies with attractive assets and thus with a considerable latitude for restructuring.

Exchange rate instability

The exchange rate is a source of uncertainty due to its profound impact not just on export-oriented producers but on the company's decision to undertake export activities or add new export-oriented lines of production. For instance, the export-oriented machine-building plant in Tomsk has obtained a long-term credit abroad and the guarantee of consumers in the West and Southeast Asia. Given the risk of lending to Russia, loan conditions are quite attractive. The company, however, did not take advantage of the loan because managers believe that if ruble appreciation accelerates the loan will become unprofitable. In a number of other cases it is clear that the switch to export has not been made because companies are reluctant to incur sunk costs that will emerge should exporting become unprofitable.

Uncertainty because of trade liberalization

The output of consumer durables in the defense sector increased in 1992 and 1993, but fell in 1994 by about 30 percent. The major reason was import competition. Enterprise managers, particularly those in Moscow and St. Petersburg, responded by abstaining from risky diversification into new manufacturing and focusing instead on financial and real estate transactions. Thus, somewhat contrary to theoretical expectations, exposure to competition resulted in shortened planning horizons.

Solving the Puzzles: The Verbal Model of Enterprise Adjustment

There are three steps in the decision chain of the manager planning to undertake adjustment.

Step 1. Selection of a planning horizon in the context of maximizing the potential[13] *revenue of the company.*

Having chosen the planning horizon, the manager undertakes a search of feasible restructuring options by combining a range of adjustment alternatives. Since initial restructuring depends upon the manager's ability to raise capital, all restructuring options may be described in terms of organizational innovations. Like any other innovation, these options differ by the amount of fixed cost and their payoff. Certain innovations such as the creation of customer networks abroad require significant initial learning, while others, like perfection of rent-seeking, require no fixed cost at all. Thus, restructuring is a process of learning-by-doing in which the company's intangible capital grows because of the particular investment in organizational innovations.

Step 2. Given his planning horizon, assets, and the initial value of intangible capital, the manager chooses a restructuring option.

To turn the company around and remain in control, the management needs a certain threshold level of ability (intangible capital). The first decision the manager must make is whether the company will be able to reach this threshold level. If the answer is negative, then the planning horizon of the manager will be short because he suspects that he is unlikely to survive the transition. If the answer is positive, the manager will be anticipating a large revenue stream after the turnaround and his planning horizon becomes long. The initial estimate of intangible capital is usually made on the basis of the manager's ability to respond to technological and organizational challenges before the shock and immediately following the shock or market transition.

Step 3. There is a recurrent step in the manager's calculation: reevaluation of the parameters of the problem (value of intangible capital, planning horizon, etc.) on the basis of the previous outcomes; i.e., on the basis of the result of learning as a change in performance.

If the initial potential for organizational innovation is insignificant (i.e., the management is incapable), asset productivity is small, or the planning horizon is short, then it does not pay to adopt organizational innovation with fixed cost. In other words, it does not pay to resort to dramatic restructuring involving the creation of new institutions.

At the same time, the process of evaluation is also a learned skill which requires that estimates or expected values be readjusted constantly on the basis of company performance. Correction of estimates such as productivity of company assets and accumulated value of intangible capital is another facet of learning—learning in the incentive sense. How well can I (the manager) perform a task? Should I start doing it? How capable am I? How competitive is the company? The interaction of two major facets of learning—learning as performance improvement and learning in the incentive sense—explains the puzzles outlined above with the following four features:

Feature 1: Growing discretionary differences between firms grow wider.

Growing discretionary differences means that negligibly small differences in asset endowment generate diverging trajectories of asset dynamics. Suppose the manager makes an optimistic estimate about the competencies of the company. The manager is then likely to adopt a long-term attitude, in which the fixed cost technology will pay off. By definition, adoption of fixed cost technology implies productivity improvement. Progress in learning confirms the optimistic expectations. When she/he reestimates them and readopts a long-term planning horizon, and invests a lot in intangible capital which continues to grow quickly, this is a virtuous circle. Suppose, however, that initially the manager experiences a failure in learning. He/she then readjusts his expectations downward and is likely to adopt a short-term planning horizon. In that case, serious learning (investment in fixed-cost innovations) is unlikely to pay off. Hence there is no productivity increase and intangible capital does not grow. This is a vicious circle. If, in the one case, interaction between our two facets of learning results in a virtuous circle, a vicious circle is perpetuated.

Feature 2: Variation of performance over time.
This variation is also explained by the possibility of virtuous and vicious circles. More specifically, in the aftermath of the shock of 1992 the majority of defense managers were overly optimistic. They considered the shock to be transitional and expected to remain in control after the transition. Therefore, it was rational to adopt a long-term planning horizon, macroeconomic instability notwithstanding. However, their range of competencies of how to operate in the market economy was small. As a result, productivity deteriorated with the resulting readjustment of managerial expectations downward. As a result of repeated failures in adjustment, many realized that in the long run they were inviable and adopted a short-term attitude. Hence private and traditional rent-seeking flourished.

Feature 3: If the manager is capable, it pays for him to be optimistic; if the manager is unsure about the enterprise endowments he should be more cautious and invest to discover them.
If initial endowments are large, then optimistic expectations provoke a virtuous circle. If, however, they are inadequate, optimism about the future is likely to result in a vicious circle of frustrated expectations, shortened planning horizons, and stalled learning. This sheds light on why even the most entrepreneurial managers were cautious to start dramatic adjustment: during the initial "waiting" phase they were focusing on learning in the incentive sense, and deciding what their capabilities were. Lessons that managers learn in the process of adjustment vary widely and are often unexpected. The manager of a mechanical plant in Voronezh visited German firms for training courses[14] on marketing and returned with the resolute belief that the company's intentions to start exporting were unwise; i.e., its expectations for exporting should be adjusted downward. Three-and-a-half years after the beginning of the adjustment, the management now has a more sober assessment of themselves, their companies, and economic future. When contemplating diversification three years ago, the managers focused on high-tech output, while today the emphasis is on more mundane products that meet market demand. There is a growing awareness among management and labor that without outside investment all attempts to turn the company around are futile, and outside investors are unlikely to step up unless the incumbent managers leave. In one case in the sample, the manager voluntarily stepped down to give way to a strategic investor.

Table 10.4 summarizes the major components of real adjustment used by enterprises, and progress in mastering them both in the sense of performance improvement (learning-by-doing) and learning in the incentive sense.

Feature 4: There is an optimal intensity of the shock that maximizes restructuring.

Table 10.4: Learning Dynamics 1992–1995

Strategies toward fixed assets	Learning in the incentive sense: how well can I (the manager) perform a task? Should I start doing it?	Learning by doing
Export orientation	Growing awareness of sunk costs related to entry into export market. More pessimistic attitude toward export promotion.	Accumulation of expertise in preparing export contracts and making an enterprise more transparent for foreign partners. Foreign consultant firms are used extensively.
Diversification to meet internal demand	Growing attention to "mundane" output including services and diversification into agriculture.	Ability to carve up viable part of enterprise and create wage differentials to induce the unwanted labor separations.
Downsizing with the preservation of the main production lines	Awareness that without an outside investor such a strategy is often doomed. Readiness to step down from top management to clear the way for outside investors.	Marginal learning related to cooperation with banks and search for inputs from new suppliers.
Downsizing on the way to closure; enterprise is a social protection unit (the most widespread)	Result of repeated failures in the past. No matter what I do, I am going to fail because of the unfavorable economic environment.	Marginal learning related to private rent-seeking (asset stripping) and traditional rent-seeking (lobbying the government).
Splitting of the enterprise into different parts	More permissive attitude toward split up because of the presumed ability to retain some control over spin-offs.	Learning to create new organizational forms such as business groups and other networks of firms.

Too strong a shock provides a mismatch between initial endowments and the magnitude of the restructuring. As a result, managers are more likely to adopt a short-term planning horizon, which provokes a vicious circle of stalled learning. A manager cannot find a restructuring option which pays off given his planning horizon. Consequently, expectations become self-fulfilling: managers who believe their enterprises are going to fail do not invest in learning, and thus the enterprises do fail.

Our case studies show that often more learning-intensive options are preferred. Once chosen, each adjustment is often being perfected and improved, and becomes self-reinforcing. As the formation of rent-seeking networks advances, the switch to restructuring options becomes unattractive: rent-seeking crowds out restructuring. On the other hand, once restructuring options are mastered, rent-seeking becomes unattractive: learning-intensive restructuring crowds out rent-seeking. Because of the cumulative nature of learning, early choices determine long-term outcomes.[15]

Enlisting New and Old Social Networks

In addition to learning to operate in the market environment and learning in the incentive sense, there is another, often overlooked facet of learning: learning to deal with inherited personal and social networks.[16] Personal networks of industrial managers were of vital importance in times of extreme uncertainty following the shock of 1992. Network capital—one of the components of intangible capital—still facilitates input/output decisions as well as contract enforcement. There is a growing realization, however, that inherited personal and social networks are increasingly turning into an impediment to adjustment and becoming unreliable. For instance, the general director of a large radio-electronic plant in St. Petersburg was skeptical that he could dramatically restructure the management structure of his company: "Because of implicit obligations to my deputies and to other staff, it is difficult for me. Someone from the outside must do it." The same manager allowed certain units of his enterprise to split up while using the parent company's infrastructure and R&D. The start-up company had attractive assets, so the parent company management obtained a loan to acquire a controlling share of its stock during the share auction. There was an agreement—based on personal trust—that the start-up company would not redistribute shares without consulting the parent enterprise. A spin-off company broke the agreement, and the management of the parent company lost control as well as a share of its investment. In retrospect, it is probably an efficient outcome, but had the managers of the parent company predicted it, they would have never allowed it to happen in the first place. One of the functions of a network is the provision of information and diffusion of the learning experience. The failure of the "engineered spin-off" became known to other enterprises and formed their attitudes toward similar actions.

The strong, charismatic leaders of some enterprises now choose to alienate themselves from any networks in order to be free from relevant implicit and explicit obligations. The prevailing attitude, however, is to carve out new networks that combine viable elements of the old ones, and to seek closer association with banks, trading companies, and other agents of the nascent private sector. Associations of graduates of elite Moscow colleges such as Moscow Physics-Technical

Institute, University Imeni Baumana, and the Aviation Institute play an active role in the process. Once the major source of human capital for the defense industry, they have now become major suppliers of skilled labor for the banking and trade sphere. Graduate associations, some of which are quite active, provide a cross-fertilization of expertise between reform-oriented directors and the new banking elite. New networks are being formed to provide and distribute information. Through such networks, for instance, banks ascertain which assets are potentially competitive and thus worth including in emerging business groups. At the same time, half of the enterprises that were subject to classification fell into the category of the fragile adjustment driven by the top manager who does not seek the benefits and obligations of network participation. This observation aversion is probably due to the mistrust of institutionalized, formal, or informal inter-enterprise obligations which were typical of Soviet-type planning. Secondly, the process of transformation of old networks has just started. In the future, reform-oriented networks are likely to become bank-led business groups, while rent-seeking networks will turn into corporate sectoral associations and lobbying groups.

The formation of rent-seeking networks, a process which is particularly pronounced at the regional level, parallels the carving-out of a restructuring-oriented network. Management of defense enterprises, many of which are single employers in their respective communities, have always been considered a "shadow" local government with an authority exceeding that of the actual one. Currently, when these enterprises are collapsing, local governments first resist any attempt to declare a bankruptcy for the ailing plants, and second, seek subsidies for them (which are actually subsidies to the plant's social infrastructure) through their own channels.[17] The manager of a large tank plant in Siberia was not receiving any defense orders, and had begged the federal government to close the plant.[18] He obtained some personal rents from its assets and wanted a safe retirement. It was the local government that threatened that if the manager stopped "fulfilling his social obligations" (i.e., maintaining the social infrastructure) it would engineer a comprehensive audit of the plant with the objective of revealing abuses of managerial authority. This is an example of the forced managerial entrenchment which illustrates the rapidly forming rent-seeking alliance between local government and unsuccessful enterprise management.

Two policy implications follow from this analysis. First, until substantial progress in learning-intensive restructuring is made,[19] the government should abstain from discretionary and in particular sectoral[20] policies that are particularly prone to rent-seeking. This would encourage an early choice of restructuring options (which is likely to perpetuate itself) from a portfolio of adjustment responses. Second, the federal government should make more specific and transparent subsidies to the social infrastructure, thereby discouraging alliances between anti-reform enterprise managers and local authorities.

For the majority of Russian defense plants, the prospects of becoming competitive in the high-tech area are bleak. An abrupt decrease in military and civilian R&D and investment in equipment and a massive brain drain of skilled labor to the service sphere has been accompanied by high-tech manufacturers' increasing inability to procure high-quality inputs at any cost. Given the dramatically decreased demand, it is no longer profitable for downstream producers to supply a negligible amount of inputs. This is particularly true for high-quality materials from ferrous and non-ferrous metallurgy and composite materials.

In addition, the capital goods manufacturing in which the defense sector is the most competitive has experienced a negative demand shock. The investment slump and the corresponding loss of competitive advantage may persist even when the shock itself is gone.[21]

A framework is proposed that will enable an answer to this question in the future rather than provide an answer itself (Table 10.5). Emerging adjustment patterns suggest three prospective sources of competitive advantage for Russia's defense sector: R&D and high-technology; skills and human capital; and low-cost manual labor. Combined with the adjustment cost required, six patterns of development result. Some additional caveats on the basis of available evidence are possible. First, penetration of the Western market is out of the question without some form of partnership with foreign companies. On this road, high-tech enterprises will discover the unpleasant surprise that their inclusion in the division of labor is not unlike that of Mexican *maquiladoras*: it is dictated by the logic of division of labor within the Western corporation rather than by the intention to enhance competitive advantages of the Russian enterprise. Thus it will generate little backward and forward linkages to the Russian economy. Secondly, many defense enterprises will cease to exist in their current organizational configuration: they are likely to be swamped by the expansion of domestic business groups originating outside the defense sector,[22] to form alliances with foreign partners or, in the case of export-oriented producers, to expand themselves via the assimilation of other enterprises.[23]

Conclusion

Since learning should imply a consistent improvement in performance, is there any learning in the Russian defense sector in which output and productivity are still declining? The answer is yes, because results of learning are affected not only by the intensity of the effort but also by the economic environment. Because of macroeconomic instability and a string of other uncertainties, the adjustment of the Russian industry has been almost devoid of investment in fixed assets. No absolute winner is likely to emerge in these conditions. Starting from a waiting phase that immediately followed the shock, the adjustment proceeded to an experimentation phase, the defining feature of which is very active learning in the

Table 10.5: Likely Patterns of Growth in Various Segments of the Russian Defense Sector

Competitive advantage Adjustment cost	1) High technology with export potential	2) Human capital: educated labor and engineering staff	3) Low-cost manual labor
High adjustment cost: substantial investment is needed	a) Large hysteresis effect (internal friction) b) Small hysteresis effect *Selective growth* Development of viable segments of companies via emergence of start-up firms —joint projects with foreign high-tech firms.	*"Dependent (export-oriented)" development* Focus on technologically undemanding tasks within a multinational company. Likely example: *Permskye motory*	*Start-up growth* Enterprise bankruptcy; scrapping of fixed assets; labor is utilized in activities totally unrelated to the original company.
Relatively low adjustment cost	*Unstable growth* Transfer of learning experience of the relevant enterprise via its diversification and formation of a business group. Maintenance of technological capabilities through reliance on foreign inputs and expertise. Example: Moscow Aviation Production Association (MAPO)	*Stable growth (largely for domestic demand)* Turnaround and subsequent diversification as a result of the acquisition of relevant assets by the Russian business groups originated outside the military-industrial complex (in the financial sphere; gas and banking sectors). Example: Radio plant (Berdsk), "Sputnik" enterprise (Voronezh)	*Assembly operations* Assembly operations as a result of turn-around investment performed by Russian or foreign capital. Change of incumbent management

incentive sense (growing awareness by the directors corps of what they are capable of), and now has just entered into a shake-up stage defined by the change of strategic owners and incumbent management. Ironically, the process of privatization has proved to be a hindrance to restructuring in some cases. State-owned firm managers were more secure in their positions and could choose a long-term planning horizon.

After the emergence of strategic owners and given a certain degree of macroeconomic stability, one can expect investment-driven restructuring. In the course of this restructuring stage the Russian defense sector is likely to change more profoundly than any other sector, both technologically and organizationally. As Stiglitz (1987) argued, the ability to learn must itself be learned. Up until now, and in particular during the experimentation phase, restructuring in the defense sector could be defined as learning to learn in preparation for the "real thing"—investment-driven restructuring. One of the positive outcomes of learning to learn is the weakening of managerial entrenchment: having learned that they are incapable of turning the company around, managers are stepping down to clear the way for outside investors. At the same time, there are many more cases of managerial entrenchment as a result of a growing alliance between incapable managers and local authorities.

Notes

1. I define institutions broadly as any long-term explicit or implicit agreements about patterns of social behavior—formal or informal social contracts.

2. The ubiquity and magnitude of such start-up costs are documented in De Soto (1989).

3. Because of the prevailing share of civilian goods in its output, the Russian defense sector is a purely administrative entity having little in common with defense industries of industrialized countries. For that matter most of the conclusions of this paper are likely to be valid for the Russian industry and enterprises in transition in general.

4. Such an activity takes three forms. First, there are shell firms created by the management which would perform auto repair, certain construction services, etc. Second, there is unrecorded activity which is related to the observation that in 1994 up to 60 percent of payment flows were on a cash-only basis. Third, for socialist enterprises, there is a traditional informal activity whereby the labor devotes a share of its time producing goods (motorcycles would be a typical example) for its consumption.

5. In 1993 the average amount of conversion credit per enterprise was close to $40,000.

6. As the manager puts it, "Although I was repeatedly warning the director of the neighboring plant that he should abide by more equal distribution of income, he didn't listen; as a result the office building was set on fire, which of course came as no surprise to me." Apparently, Hirschman's (1970) tolerance for income inequality, which is an important constraint for adjustment, is quite low in Russia.

7. This was the strategy pursued by the first of the two enterprises that were compared earlier.

8. An illuminating example is the case where a bank in Voronezh forced a loan to enterprise management (which was given a bribe to receive the loan) in the expectation that when bankruptcies began, it would be a first claimant on the plant property. Note that as long as the enterprise holds an account at the bank, the bank's risk is minimal since it can always freeze the enterprise's funds in the account.

9. Use of funds for investment purposes is not necessarily the best possible situation. For instance, the Novosibirsk Aviation Production Association in 1993 received one billion rubles (about a million dollars) to facilitate conversion. These funds were allocated to finalize development and test preparations of a new commuter passenger plane. Had they been deposited into a bank account, the enterprise funds would have at least earned interest (which began to turn positive in the beginning of 1994), since this is a highly risky project where the major market is either disappearing (commuter air traffic in the major part of Russia), going bankrupt (Northern Russia), or is becoming demanding and difficult to enter (foreign markets).

10. Kuznetsov (1995) shows that such a network implies settlement of business disputes on the basis of personal reputation and the threat of expulsion in case of non-compliance. There are designated individuals within networks—modern equivalents of law merchants—who prepare settlements and make recommendations to top management of enterprises of the network.

11. In consumer durables manufacturing one can clearly observe the phenomenon of waves of excess competition emerging as a result of the lack of information about market demand. For instance, in 1993 the market for refrigerators became profitable, and at least ten defense-sector companies, unaware that other firms were doing the same, entered the market. The database was created with the modest objective of avoiding such situations.

12. Note a peculiar structural rigidity that emerges as a consequence of this type of technical change. Should the producer of these devices go out of business, the supply of them would be costly to redevelop: importing them is out of the question (the problem could not possibly have emerged anywhere else in the world), and the production of them is costly to reproduce domestically (usually no documentation exists for this type of output and consequently the know-how is tacit know-how embodied in human capital).

13. One may consider the real interest rate to be a discount factor.

14. It is noteworthy that the management of all plants in the sample have attended business courses in the West ranging from one week to two months. Managers value this experience mainly because it helps them to evaluate the advantages and shortcomings of their companies.

15. This is a familiar argument of the lock-in in non-linear dynamics pioneered by B. Arthur and P. David.

16. Network is viewed here as an informal institution—a club—where the shared good is mutual trust and mutually shared information (See Kuznetsov, 1995). Formal organizations like industry associations are outside the purview of this analysis.

17. Regional administrations tend to lobby the presidential apparatus, while enterprise managers focus more on the government and its agencies such as the Committee for Defense Industry.

18. The only government agency which actively supported the director's proposal was the Ministry of Defense. The ministry's pro-restructuring attitude is not surprising: it wants a small but viable (rather than large and collapsing) defense industry.

19. One criterion of such maturity is the presence of competitive enterprises which became viable without government subsidies.

20. The next section indicates that the line between sectoral and horizontal industrial policies may prove to be unclear.

21. The irreversible loss hysteresis effect is very similar to the one perceived as a cause of the high European unemployment rate. Long-term unemployment is different from transitional unemployment because it perpetuates itself: once a worker long out of a job loses skills, he is unlikely to be hired irrespective of his wage and the amount of time spent searching for a job.

22. The Russian business groups, unlike those in semi-industrialized economies, do not lend themselves easily to outside observation. For instance, Rossiiski Kredit has created a business group consisting of 26 enterprises from the ferrous and non-ferrous metallurgy sector and defense industry with total employment exceeding 300,000 people. In all enterprises, the bank holds the controlling share of the stock. Since this is forbidden by Russian legislation, the relevant stock is held by the bank's daughter companies. Likewise, the group is not officially registered as is popular in the Russian industrial policy debate ("financial-industrial group") since by law its employment cannot exceed 100,000 people; moreover, the group does not exist as a legal entity at all.

23. An example of this tendency is the state-owned Moscow Aviation Production Association (MAPO), which recently assimilated a failing aviation design bureau.

Bibliography

Amsden, A. "The division of labor is limited by the type of market: the case of the Taiwaneese machine tool industry." *World Development* 5, no. 3 (1977): 217–233.

Amsden, A., and T. Hikino. "Project Execution Capability, Organizational Know-how and Conglomerate Corporate Growth in Late Industrialization." *Industrial and Corporate Change* 3, no. 1 (1994): 149–172.

Arthur, B. "Self-reinforcing Mechanisms in Economics." In *The Economy As an Evolving Complex System*, edited by Philip W. Anderson, Kenneth J. Arrow, and David Pines. Reading, MA: Addison-Wesley, 1988.

Chandler, A. *Scale and Scope: The Dynamics of Industrial Capitalism.* Cambridge, MA: Harvard University Press, 1990.

David, P. "Clio and the Economics of QWERTY." *American Economic Review* 75 (1985): 332–37.

Hirschman, A. "The Changing Tolerance for Income Inequality in the Course of Economic Development." *Quarterly Journal of Economics* 87 (1973): 544-65.

Klimenko, A., B. Blokhina, and A. Shmarov. "Money which chooses us." *Kommersant-Weekly* (28 February 1995): 17–28.

Krueger, A. "The Political Economy of the Rent-seeking Society." *American Economic Review* (June 1974): 291–303.

Kuznetsov, Ye. "Fantasies for Central Authorities." *Socialisticheskya Industria* (18 June 1989).

———"How Can an Economy Adjust to Simultaneous Market and Government Failures? Lessons from the Former Soviet Union, Contemporary Russia and Countries of Late-Late Industrialization." *Communist Economies: Economics of Transformation* 5, no. 4 (November 1993): 473–498.

———"Adjustment of the Russian defence-related enterprises in 1992–94 :macroeconomic implications." *Communist Economies: Economics of Transformation* 6, no. 4 (November 1994): 203–243.

———"Grafting Market Institutions in the Transition Economies. The case of Russia." Brookings Institution mimeo, 1995.

Stiglitz, J. "Learning to learn, localised learning and technological progress." In *Economic Policy and Technological Performance*, edited by P. Dasgupta and P. Stoneman. Cambridge: Cambridge University Press, 1987.

Soto, H. *The Other Path.* New York: Harper and Row, 1990.

Teece, D., R. Rumelt, G. Dosi, and S. Winter. "Understanding Corporate Coherence: Theory and Evidence." *Journal of Economic Behavior and Organization* 23, no. 1 (January 1994): 1–30.

Unger, Kurt. *Ajuste Estructural y Estrategias Emresariales en Mexico. Las Industrias petroquimica y de maquinas herramientas* (Centro de Investigacion y Docencia Economicas: Mexico, D.F., 1994).

11

Defense Enterprise Adaptation in St. Petersburg

Andrew J. Aldrin

Introduction

Few steps in the process of economic and political transformation in Russia today are more vital than resolving the crisis of reform and stagnation in Russian defense industry. Witness the contrast between the Arsenal factory and the Vavilov Optical Institute in St. Petersburg. Arsenal, a stereotypical Soviet-era defense industrial dinosaur, built reconnaissance spacecraft for the navy until 1992, when the defense order was abruptly canceled. Since that time, the factory has been separated into six autonomous enterprises, with the central holding company possessing only 20 percent of the shares of these enterprises. At least two of the enterprises are reportedly producing commercial goods at a profit, and the leadership of the holding company professes a willingness to allow these enterprises to succeed or fail based on their own merits. The plant has entered into several ventures; and the reorganized factory appears to be on the road to profitability.[1] In stark contrast, however, is the experience of the Vavilov Institute, which develops world-class optics. By any technical standard Vavilov possesses a staff more capable of producing a wide array of goods for a broader worldwide market. In spite of this, however, Vavilov has been unable to adapt to the market. The organizational structure of the institute remains unchanged from the Soviet era; attempts at developing market goods have been rejected by the staff; and, in the ultimate rebuff of reform, the director was recently fired by the workforce for being too interested in pursuing marketization rather than pure research.[2] Why is Arsenal reforming while Vavilov remains stagnant? The following discussion will attempt to cast at least some light on this question.

Given the high degree of uncertainty endemic to any research in Russia today and the limited number of broad-based examinations of defense enterprise reform, this study can only be regarded as a preliminary attempt to tease out some of the key variables involved in reform. While it might be politically convenient and analytically parsimonious, we ought not expect a single factor to emerge as the primary cause of enterprise reform. Instead we would be well advised to heed the words of James Q. Wilson on the subject of innovation that "trying to find one theory to explain them all [innovations] is like trying to find one medical theory to explain all diseases."[3] Therefore several factors will be examined in this paper to determine if they have a causal relationship to enterprise reform. Included are the relationships between enterprise reform and: (1) enterprise function (i.e., research, design, or production); (2) industrial sector; (3) state vs. private ownership; (4) changes in leadership; (5) level of state support; and (6) involvement with Western firms. For each area, propositions are formulated and tested against a database of 41 defense enterprises in St. Petersburg. Explanations are enriched through examinations of data gathered through a series of interviews conducted with defense enterprise managers in Russia and Ukraine from 1992 to 1994. While these variables are carefully chosen on an individual basis, the author makes no claim of a single overarching theory connecting them. Nevertheless, the ultimate goal of this paper is to begin to develop an understanding of the evolution of reform within individual enterprises and the diffusion of reform across enterprises.

Amid the confusion of Russian economic transformation, at least this much is clear: Russian defense enterprises cannot survive on subsidies and meager defense orders from the Russian government. The capacity of the Russian government to support the Soviet defense industrial base through defense orders, direct or indirect subsidies, and arms sales promotion will be sharply constrained by economic realities. The approved 1993 budgetary levels devoted to defense industry were incapable of supporting more than 15 percent of the pre-1991 total Russian defense industrial base.[4] To make matters worse, the Ministry of Finance has been slow to allocate funds, and has refused to allocate as much as 50 percent of appropriated expenditures in some sectors.[5] Clearly, therefore, Russian defense enterprises will have to look outside the Russian government to survive. They will have to survive in the marketplace.

There are three marketplaces in which Russian defense enterprises will have to compete: the civilian market, the domestic military market, and the foreign military market. All three are growing increasingly competitive, and only the most efficient firms will be able to survive. The civilian market in Russia has been invaded and occupied by Western firms. To compete effectively, Russian firms will have to offer significant price advantages while establishing something at least close to comparable quality. Similarly, the international arms market has recently become inundated with high-technology Western weapons being offered at increasingly competitive prices. Many Russian defense enterprise managers see this market as their salvation since it does not require any significant retooling of production lines. Even more importantly, the international arms market offers the best opportunity for earning hard currency revenues.[6] However, this market is highly competitive, and Russian defense enterprises will have to become much more efficient to compete effectively. Finally, the Russian domestic arms market is more competitive. The Ministry of Defense has now acquired the right to spend its money as it chooses and it often has many suppliers to choose from for production of a new weapon system. Therefore, no matter which market they turn to, Russian defense enterprises will face intense market competition. To survive they must become efficiently organized producers.

Russian defense enterprises are extraordinarily ill-equipped to do this. They suffer from two seemingly contradictory organizational flaws: over-centralization and compartmentalization. The Soviet economic system was intentionally developed to reduce the decision-making burden placed upon individual firms.[7] A great many of the decisions that are made at the enterprise level in market economies were made by administrative organizations in the Soviet economy.[8] Enterprise-level decision-making focused almost exclusively on internal issues ranging from meeting production quotas to social conditions for the workforce. Consequently, defense enterprises developed a structure to fit this environment. Decision-making was almost totally centralized in the office of the plant director. The director's office contained a small staff, often consisting of a few deputy directors, their secretaries, and a small accounting group. The director himself made the vast majority of decisions without consultation with the rest of the management team. This decision-making structure was remarkably consistent across different enterprise types.[9]

If individual enterprises were over-centralized, industrial sectors suffered from the opposite problem of over-compartmentalization of research, design, and production. In the classic Soviet scheme of industrial organization, applied technological research was performed in scientific research institutes, design work in design bureaus, and production in factories. There was little interaction between enterprises performing their respective functions. This problem was addressed, in part, by the industrial association reforms of the late 1970s, which attempted to combine production organizations with design bureaus. The reform was largely ineffective, and the barriers between the various functional entities remain in place to this date.

While such a decision-making structure may have worked well enough in a planned economy, it is totally unsuitable for survival in a market economy. Instead of a small handful of ministerial officials, enterprise directors now must deal with literally hundreds of customers and suppliers, as well as government officials. It is impossible for enterprise directors alone to meet the decision-making demand which now faces them. They must decentralize decision-making within their enterprises[10] by moving toward something akin to a multi-divisional structure.[11] At the same time, design bureaus and scientific research institutes are not structured to produce goods for sale in the marketplace. Their primary products are research. These organizations will have to either convert their own facilities toward production or integrate their facilities with factories capable of producing the final products of their research.[12]

Russian defense enterprise managers have only recently come to understand the importance of reorganization. An extensive 1994 survey of defense enterprise directors touched upon the issue of reorganization. Figure 11.1 shows that in 1994 more than 50 percent of the respondents placed reorganization at the top of their list of priorities for development of the firm. This almost doubled the percentage from the previous year. Simultaneously, directors de-emphasized the need to shift to greater civilian production. It would appear, therefore, that in 1994 many enterprise directors came to realize that before they could engage in a change in product line, first they would need to reorganize their enterprises.

Figure 11.1: Basic Plans for Development of Production Enterprises According to the Opinions of Their Directors[13]

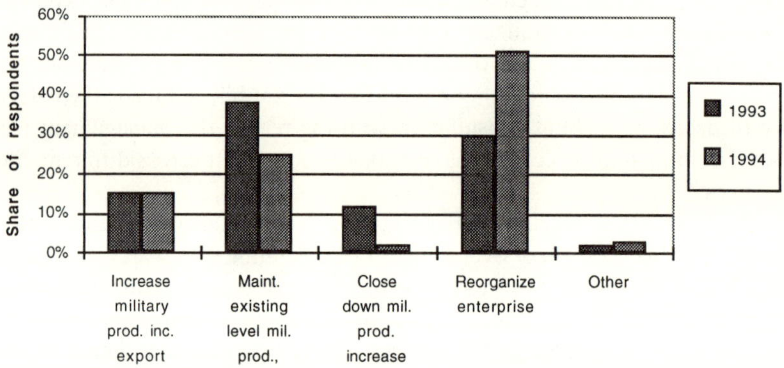

Of course, there is an important difference between intention and action. This study therefore aims to measure and explain the reforms that are actually taking place. There are many reforms that defense enterprises must undertake to survive in a market economy. But perhaps the first observable reform that must be undertaken is the reorganization of decision-making and production. Reorganization is also a useful indicator because it is a reform that an enterprise can undertake without any capital infusion. Thus it provides an indicator of the reformist tendencies of management. It should also be recognized that changes in organizational structure are not an end point, but rather a leading indicator that reform is taking place at an enterprise. Therefore, observation of reorganization provides a useful tool in building theories of diffusion of enterprise reform.

Data and Methods

The basic source of information for this study is data gathered by the St. Petersburg defense conversion committee on 41 defense industrial enterprises. Questionnaires were completed by the central office of the respective enterprises, and, to the extent possible, data were independently verified by the committee and by the author's personal interviews. Additional information was provided by the Department of Commerce database.[14] Combined, these data included information on organization, employment, and product lines. Data on military product lines, ministerial affiliation, and 1991 employment were provided from a database maintained by the Central Committee Department for Defense Industry. These data sources were compiled into the basic database used for this study.

St. Petersburg provides several early indicators of the unfolding reform process, making it a useful region for study. While the city has a very high concentration of defense industry, there is still a large and developing commercial sector which can both absorb defense industrial workers and provide a source of innovative management. St. Petersburg also has a higher concentration of high-technology defense firms than most Russian cities. Between 15 and 16 percent of all Russian defense R&D is located in this city.[15] Many believe that it is Russia's relatively well-endowed technological base that will be the key to economic growth over the long term. Additionally, the city's close proximity to the West has made it a magnet for Western businesses looking for a foothold in Russia. For these reasons, we would expect that reforms would begin to appear in St. Petersburg in advance of other cities, with the exception of Moscow.

Additional data were provided from two sources. Over the course of the past three years the author has conducted interviews focusing primarily on managerial and organizational issues at 17 defense industrial enterprises. These included five enterprises and scientific research institutes in St. Petersburg, at each of which several interviews were conducted over the past 18 months. These interviews provided sources of insight for developing the basic approach for this study as well as verification of the findings from the above data set.

Information was also provided from a 1994 survey of the opinions of managers of more than 150 defense enterprises across Russia. The enterprises in this survey are representative of all regions of the Russian Federation, as well as the range of enterprise size and industrial sector.[16]

Data Limitations and Qualifications

By the standards of Western research on Russian defense industry, the database used for this study is fairly ambitious. However, by the standards of most Western social science, the data set is rather limited. Western social scientists are accustomed to seeing studies incorporating hundreds if not thousands of records. Given the political realities of the environment for scholars of Russian defense industry, such databases are likely to remain difficult to obtain. Moreover, despite its comparatively small size, the data set does represent a significantly large portion of defense industry in St. Petersburg to support defensible conclusions within the context of that region.[17]

The second potential problem lies in accuracy of reporting. Central offices of defense enterprises may either deliberately have misinformed interviewers or made honest mistakes. Where possible data were verified by independent sources. Independent interviews were conducted with some firms which indicated that the data represented here are accurate. There were no reports from field interviews that contradicted information gathered through the questionnaires.

Operationalization and Observation

Observing and operationalizing many of the characteristics and dynamics of enterprise reform presents great difficulties. How does one observe that an enterprise leadership has developed an understanding of true costs? Well thought-out strategic plans are similarly difficult to investigate. One positive, observable, and objective indicator of reform is reorganization of decision-making structures.

There are, of course, varying degrees of reorganization. This study categorizes the levels of reorganization as follows:

Level 1: No reorganization. Creation of small marketing departments of fewer than ten people.

Level 2: Minor reorganization. Creation of significant marketing and/or legal departments with more than ten people.

Level 3: Reorganization. Creation of marketing departments and substantial decentralization of decision-making for individual production lines or enterprises.

Level 4: Substantial reorganization. Creation of marketing department; complete decentralization of decision-making to enterprise; and/or creation of holding company; fundamental shift in output from research and design to production.

Level 5: Complete reorganization. Separation of enterprises into completely separate entities.

Such a categorization scheme provides a reasonably good indicator for discussing the level of reform taking place at an enterprise. At level 1, we can be reasonably sure that there is very little reform taking place at an enterprise. Similarly, at levels 3, 4, and 5, reform is clearly taking place. At level 2, there is very little reform, but it is useful to distinguish between a total lack of reform and some effort. Slightly less than one-half of the 41 enterprises in this study exhibited no signs of organizational reform. One-fourth of the enterprises exhibited either substantial or complete reorganization, and the remaining one-fourth were at levels 2 and 3 of reorganization.

Table 11.1: Defense Enterprise Reorganization in St. Petersburg

Level of reorganization	Number of enterprises in study
1—No reorganization	20
2—Minor reorganization	5
3—Reorganization	5
4—Substantial reorganization	10
5—Complete reorganization	1
Total	41

The first observation to be made, therefore, is that about one-half of the enterprises are undertaking efforts at reform. Many defense enterprises are not, as the popular conception holds, dinosaurs refusing to change with the times. On the other side of the coin, one-half of the enterprises have not undertaken any efforts at reform. The obvious question arises: Why are some enterprises undertaking reform while others remain stagnant?

Data were also collected on voluntary versus involuntary staff reductions that supported the relationship between reorganization and reform. In the face of declining revenues, enterprise directors are faced with two basic choices. They can reduce salaries across the board and wait for staff members to depart voluntarily, or they can operate proactively by undertaking forced layoffs of a large portion of the workforce. The latter provides an unambiguous indicator that an enterprise is pursuing a reformist course. Unfortunately, data on involuntary workforce reductions could not be reliably collected for a large number of enterprises. Nevertheless, while 40 percent of the enterprises declined to provide data on this question, those that did respond provided confirmation of the validity of reorganization as a measure of reform. Seventy-five percent of the firms demonstrating higher levels of reorganization[18] reported that more than 50 percent of their workforce reductions came as the result of layoffs. Only 31 percent of the enterprises at lower levels of reorganization reported that more than 50 percent of their staff reductions had come from layoffs. These data therefore provided a strong indicator that enterprises that have undertaken reorganization are also likely to undertake other, more difficult reforms.

The remainder of this paper will seek to answer this question by positing various hypotheses and examining the data that either confirm or reject them. The limited size of the database precludes statistically significant hypothesis testing. Nevertheless, observing the distributions of the degree of reorganization across the organizational characteristics of firms, structural categories, and the extent of exogenous influences permits us to assess analytically a number of propositions in a manner that is sufficiently robust so that interesting conclusions can be made.

Factors Leading to Enterprise Reform

The following section examines a series of factors which could be expected to have a relationship to enterprise reform. Structural factors such as enterprise function and industrial sector may have an effect on the propensity of the firm to reform. Privatization and leadership change might prove to be catalysts for change. Also, exogenous influences, such state subsidies or foreign investment, may affect the speed of reorganization. The first section will consider the characteristics of defense enterprises; the second examines exogenous factors which may affect the reform process.

Enterprise Characteristics

There are significant differences in the basic structure and functions of Russian defense enterprises. Furthermore, there are reasons to expect that these differences will have important effects on the propensity of enterprises to reform. The ensuing section will examine several factors as they relate to observed reorganizations. As part of their Soviet-era legacy, Russian defense enterprises have different well-defined functions. Some enterprises perform only basic and applied research while others are involved only with production. This characteristic is termed enterprise function. The Soviet legacy also left firms in separate columnized ministries which operated in relative isolation from each other. This legacy is examined in the "industrial sector" subsection. Thirdly, much has been made of the importance of privatizing enterprises for creating the conditions of corporate governance which will lead to greater restructuring. This factor will also be examined. Finally, changes in leadership by themselves may be another factor which will lead to larger organizational changes.

Enterprise Function

There are three basic organizational types from Soviet-era defense industry. With one exception, these enterprise types perform discrete functions. Scientific research institutes (NIIs) perform purely scientific work. Design bureaus (KBs) design products, and production associations (POs) produce products. Scientific-production associations (NPOs) were created to combine the activities of KBs and POs. These functions, when linked together, may be thought of as comprising a complete product cycle, in which a product is brought from its initial conception in R&D to the design and prototyping phase, and ultimately into mass production. With the breakdown of these linkages in the post-Soviet period, each enterprise has had to fend for itself. The extent to which an enterprise is reorganizing, and the degree to which it will ultimately be viable in a market environment, may be highly related to the enterprise's function.

There are competing influences, however. On the one hand, it could be argued that the limited organic production capabilities of scientific research institutes and design bureaus would push these enterprise types toward greater reorganization. On the other hand, scientific research institutes and design bureaus face greater institutional barriers to reform, and therefore they will be less likely to reform. This section will examine the relationship between these competing hypotheses.

Scientific research institutes that can continue to perform pure research only may not need to change their flat organizational structures a great deal. They may need to reduce staffs, perhaps dramatically. Research enterprises may also seek to integrate themselves with other production-oriented organizations, or go into limited-scale production themselves. But given the general lack of available capital, downsizing is the more likely option. The reality is that there is very little they can do organizationally to make themselves better equipped to deal with markets. With basic and applied research highly dependent on the state for funding, many Russian research institutes will ultimately not survive.

Design bureaus were closer to production in the product cycle, but they typically do not engage in large-scale series production of goods for the market. Market economies have a limited ability to support these types of organizations. They are caught in the middle between research and production. To survive, design bureaus have three basic options. The first is to do nothing, and hope that there are sufficient orders from the government to support the firm and reduce their staffs accordingly. The second is to diversify their product line, develop dedicated design teams for new products, and expand production capabilities. The third option is to integrate the design bureau with a production organization. Given the inability of the state to support these organizations with defense orders, we would expect to see the second and third options regularly exercised.

Scientific production associations, which combine the design assets of a KB with the production capabilities of the POs, are clearly the most complex organizational entities. These will require a significant reorganization of decision-making. In their current organizational form the design bureau remains a separate entity from the production group. The majority of NPOs are directed by either the chief or general designer rather than the general director, who is relegated to largely administrative duties.[19] In a market economy, decision-making cannot be performed solely on the basis of technological criteria which dominate the mindset of the chief designers. At an absolute minimum NPOs will need to integrate market analyses into their decision-making. In order to survive, however, most of these organizations will need to decentralize decision-making and adopt divisionalized structures.

The fourth basic organizational type is a production association which, in its purest form, is only involved with production. To the extent that these associations have design groups, they are involved primarily with solving production problems. These firms are closer to the market than the design bureaus or scientific research institutes. Therefore, they do not suffer from the major structural problems faced by the other firm types. In many cases, production associations would be well advised to incorporate design groups within their organizational boundaries, but for the most part, they simply need to orient their organizations more toward meeting market demands. Decision-making will need to be decentralized, separate production lines may be split off as separate divisions, and marketing departments need to be created and vested with decision-making authority.

From a requirements standpoint, NIIs are most in need of reorganization in order to survive, followed by KBs, NPOs, and factories.

Alternatively, it could be argued that financial and institutional barriers will dictate the progress of reform. Research institutes and design bureaus do not have series production lines. The financial barriers to creating these will act as a powerful deterrent to reorganizations involving the shift to production from research and design. The financial barriers are greatest for scientific research institutes, which, for the most part, possess no productive capacity. They must go to the state or to capital markets for the investment required to build production lines. Since they possess at least prototype production facilities, design bureaus face lower financial barriers to reorganization. If the reorganization instead involves amalgamation of research or design organizations with factories, the financial burden would shift to the factories, which would then be responsible for maintaining the acquired research facilities. Neither of these scenarios applies to NPOs, since they already possess both productive capacity and organic research departments. The financial barriers argument therefore suggests that NPOs would be on the forefront of reorganization, with factories close behind, design bureaus third, and finally scientific research institutes.

A third argument relates to institutional resistance within enterprises to change. With the exception of factories, all Russian defense enterprises are under the direction of the head of research. Even in NPOs the lead organization is the design bureau rather than the factory. In these organizations there will be a more powerful incentive to maintain the core technology of the firm than would be the case in factories.[20] In the Soviet system factories were forced to be more flexible than other organizations. They might be producing a transport aircraft one year, and fighters the next. Along these lines we would expect that factories would be most likely to engage in necessary reorganization since they have the weakest attachment to a core technology.

A fourth factor involves proximity of enterprises to the market. Scientific research institutes are greatly insulated from market forces. It is many years before their end products (i.e. new technologies) are exposed to markets. As such, they are not as subject to market influence as NPOs and factories which must sell their products directly. Design bureaus lie somewhere in the middle. While their prototype designs (primarily for weapons systems) may be subject to at least internal market scrutiny as soon as they are completed, most of the external market forces will not act until the system is in full production. Consequently, this argument would suggest that the pattern of reorganization would have factories and NPOs as the most reformist, with design bureaus not far behind and scientific research institutes lagging behind.

The available data set includes seven scientific research institutes, fourteen design bureaus, eleven scientific production associations, and nine production associations. Their categorizations were determined on the basis of independent descriptions of their activities in 1991. Those which were described by industrial administrators as engaging in research were classified as scientific research institutes, design as design bureaus, etc.[21]

Figure 11.2: Enterprise Function and Reorganization

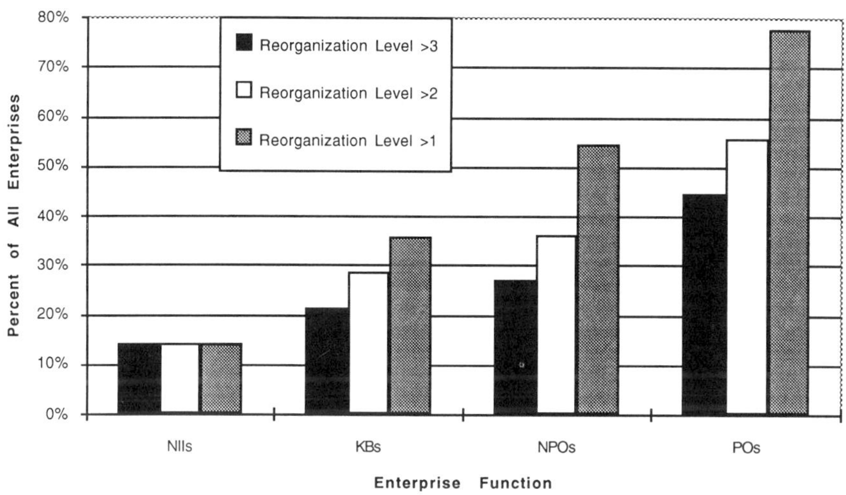

The data in Figure 11.2 show a strong relationship between enterprise function and reorganization. The relationship most closely supports the argument that organizations headed by designers or scientists would be more resistant to organizational change. Less than 15 percent of scientific research institutes and only 36 percent of design bureaus exhibited any signs of reorganization. On the other hand, 78 percent of factories and 55 percent of scientific production associations showed signs of reorganization. Perhaps more significantly, 44 percent of factories and 27 percent of scientific production associations were judged above level 3. This would provide these organizations with the organizational tools to survive in the marketplace. The difference in the levels of reform between NPOs and factories suggests that market proximity and financial barriers were less significant than institutional attachment to the core technology of the firm.

Interview data provide more details on the above relationship. Two scientific research institutes, TsNIIMash and Vavilov, have made only halfhearted attempts at developing and producing marketable products. Both have pinned their hopes on receiving sufficient state support to survive. Another scientific research institute, TsAGI, has adopted a mixed strategy of downsizing while creating several spin-off organizations producing unrelated consumer goods, but protecting the core technology of the institute without any major reorganization. According to Victor Tyurin, the commercial director of TsAGI, the spin-offs too will depend upon state orders and support for their survival.[22]

Interviews at design bureaus indicate that they are moving in the direction of production. The Klimov design bureau in St. Petersburg is entering what it calls a completely new phase in its history in which it will move from design only to design and production of engines. It is severing its relationships with the factories which produced its designs until recently and it has entered into a joint venture with Pratt & Whitney, which will provide much of the needed capital to develop new production facilities. OKB Mars, a designer of software and landing control systems, is similarly moving, albeit chaotically, in the direction of production. The Sukhoi design bureau is becoming increasingly reliant on the production of small aerobatic airplanes to support its design organization. This does not represent a significant reorganization, however. There have been no efforts at integration among the design bureaus interviewed. Sukhoi, in fact, was quite bitter regarding its relations with Russian production organizations in general and was seeking partnerships with foreign production organizations rather than working with Russian producers.

There were notable developments at three scientific production associations interviewed in 1992. Both the ELAS NPO and the Lavochkin NPO leaders had legally separated the activities of the design bureau from the factory. Moreover, the leaders went with the design bureaus rather than the factories. These moves were undertaken in the interests of saving the design bureau possibly at the expense of the factory. It is a purely defensive strategy and is again consistent with the notion of protecting the core technology of the firm.

Factories exhibited more economically rational strategies. Both LOMO and Arsenal in St. Petersburg reduced their staffs dramatically and decentralized decision-making to a significant extent. The Saratov Aviation Plant similarly decentralized decision-making.

It seems clear that the institutional barriers to the necessary reorganization of scientific research institutes are much higher than those for other types of organizations that are more engaged in production. Series production of goods is furthest away from the core technology of research institutes. All indications are that they are working very hard to maintain their core technologies.[23] It is perhaps more surprising that we do not see any research institutes integrating with design and production organizations. Instead, we see research institutes apparently being cast off by larger production organizations. Research is a costly and high-risk activity, with payoffs coming only in the long run, and in tough times it is rational for organizations to cut back research.

Design bureaus have existing prototype production lines. Their core technology involves not only design of products, but design of production lines as well. Therefore, they have lower barriers to engaging in series production. Even so, the level of reorganization at design bureaus is not substantially higher than at research institutes. They have not shown signs of splitting off independent design groups working on new product lines. Instead, most appear to be doing what they were doing before 1991, only a lot less of it. Again, it seems that protection of core technology might be standing in the way of rational change.

In the case of both design bureaus and research institutes, another impediment to reorganization is a general lack of investment capital for creating new production lines. Since neither NIIs nor KBs have operating serial production facilities, this is a significant obstacle. Western investors are very reluctant to invest in projects or facilities that do not have production lines already in operation.[24]

Science production associations and factories showed the greatest tendency to reorganize. The organizational imperative for these organizational types is approximately the same. However, there was a significant difference in the level of reorganization between these types of enterprises. A 50 percent greater share of POs than NPOs engaged in reorganization at all levels. This difference becomes more interesting when we consider that NPOs have a significantly greater attachment to their core technologies, by virtue of the fact that most are headed by the design office and general designer rather than the factory manager. In this study, nine of the eleven science and production associations appear to be controlled by the design groups. Since these NPOs have a greater institutional attachment to their core technologies, the difference in the level of reorganization of factories and NPOs could be attributed largely to the attachment to core technology.

What emerges from this discussion is the finding that technologists may be acting as a brake rather than an engine for the reform process, at least within the organizational confines of the institute. This is particularly significant in Russia, where the defense industrial base is dominated by designers and research heads. Approximately three-fourths of the institutes in this study fall under this grouping.[25] Given their recent track record, we should perhaps not be surprised that Russian policymakers are operating under exactly the opposite assumption.[26]

Industrial Sector

It is natural to assume that various industries will respond differentially to the crisis of Russian defense industry. Russian defense industrial enterprises cover a broad spectrum of activities from production of electronic components to ships. It can be argued that firms producing large, complex systems, such as ships, tanks, and aircraft, need to retain a higher level of integration. Therefore, we would not expect these industries to engage in a level of reorganization as high as other industries producing smaller, less integrated goods.[27] It is certainly necessary for these firms to create marketing departments, and to decentralize decision-making. However, they are probably more likely to succeed as a single integrated entity than as a holding company or as completely independent entities. Therefore, we could expect that enterprises that produce less complex goods in the electronics industries would be more likely to reorganize than those producing larger integrated products such as ships.

Asset specificity differs widely across defense industries. Electronics firms, for instance, could produce a wide range of consumer goods without much modification to either staffs or capital equipment. However, submarine, rocket, and tank-building factories would be hard pressed to produce anything other than those which closely resemble products they currently produce. Furthermore, the number of customers that will buy goods from these heavy industrial producers is much more limited than for the electronics industries. Less asset-specific firms with a greater number of product lines and customers face a more complex decision-making environment which requires greater decentralization. Therefore, due to their high degree of asset specificity, we could expect that shipbuilding, tank building, and aerospace industries would find it more difficult to adapt to a more marketized enterprise structure.[28]

Under the Soviet defense organizational system, every enterprise was assigned to a particular defense industrial ministry according to the output type. This provides us with a convenient means of operationalization and observation of the independent variable. The firms in this study were divided into three groups according to former defense industrial ministry. Firms from the Ministry of Shipbuilding Industry (MSP) and the Ministry of Defense Industries (MOP, which produced equipment for the ground forces) were put into one heavy industrial group. The aerospace industrial group consisted of enterprises from the former Ministry of General Machinebuilding (MOM), and the Ministry of Aviation Industry (MAP). Finally, firms from the former Ministries of Electronics Industry (MEP), Means of Communication Production (MPSS), and the Radar Production (MRP) were grouped together into the electronic industries sector. Enterprises from these sectors were then observed for their levels of reorganization. The observations are plotted in Figure 11.3. The distribution of enterprises was heavily skewed toward the heavy industries, with twenty-five of the cases falling into this category. Five cases were in the aerospace industry and the remaining eleven in the electronics industry.

Figure 11.3: Industrial Sectors and Reorganization

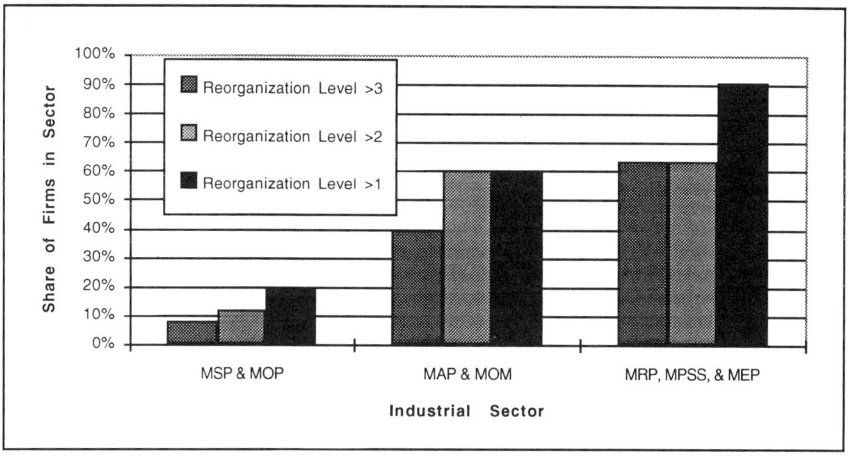

Figure 11.3 shows an unambiguous relationship between industrial sectors and reorganization which is consistent with the proposition that industries which are less complex and asset specific will be more likely to reorganize. More than 90 percent of electronics firms have engaged in some level of reorganization, and 64 percent have completely decentralized decision-making (level 4 or 5). Also as predicted, the heavy industrial group showed very little tendency to reorganize, particularly at high levels. Eighty percent of this group did not engage in any reorganization. Aerospace industries fell in between. Sixty percent of the firms in this group did engage in reorganization above level 2. This level is probably appropriate for this group.

Interview data provide mixed support for the above proposition. Two firms from the former Ministry of Defense Industries (Vavilov and LOMO) were interviewed in 1993 and again in 1994. The leadership of Vavilov felt that its staff and equipment were only suited to performing research on very high quality optical equipment. The leaders did not believe that their equipment and staff, which were tailored to research, could easily be converted to production. Likewise, while LOMO has been decentralizing its decision-making processes and giving each of its production divisions considerable authority over product design and marketing, its director, Klebanov, has made clear that reorganization will go no further. Both enterprises, then, lend support to the argument that industrial enterprises with high asset specificity are unlikely to undergo extensive reorganization.

Interviews in the aerospace industry are similarly consistent with the proposition. The Arsenal Production Association has undergone significant decentralization, converting itself into a holding company with several divisions. More than half of the staff, however, remains in the spacecraft design and production division, which remains a more or less unified company. This is the complex, asset-specific side of Arsenal. Other divisions producing a range of civilian goods from large compressors to consumer appliances and furniture are also separated, but are smaller and more independent.[29] The Saratov Aviation Plant provides an example of appropriate reorganization for an aviation factory. At SAP, the director has given the individual shops a substantial degree of freedom over daily operations, but has preserved strategic decision-making for the enterprise in the central office.

The MinEcon survey provides other illuminating data which are depicted in Figures 11.4, 11.5, and 11.6.

Figure 11.4: Enterprise Directors' Expectations of Financial Situation in 1994

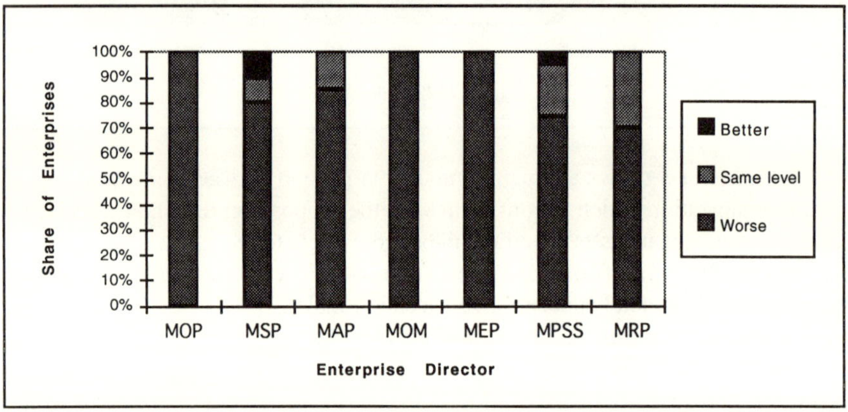

Figure 11.4 shows that the expected financial condition of enterprises in the electronics sector (MEP, MPSS, MRP) is somewhat better than in other sectors. Conditions are only slightly better in the heavy industrial sector (MOP and MSS) than in the aerospace sector (MAP and MOP). These data suggest that enterprises in this sector would have more organizational slack necessary for organizational reform.[30]

Figure 11.5: Enterprise Directors' Expectations of Future Military Production

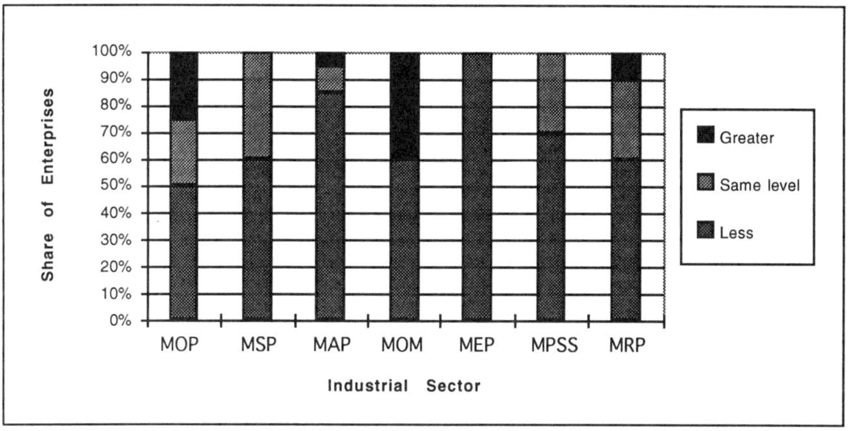

Figure 11.5 shows that the expectations for future military orders are higher in the heavy industries, somewhat lower in the aerospace industries, and lowest in the electronics industries. This is symptomatic of high asset specificity. Furthermore, high expectations of military orders acts as an impediment to reform. Enterprises that believe that they will be able to continue their current line of production are less likely to undertake organizational reform.

Figure 11.6: Enterprise Directors' Expectations of Future Civilian Production

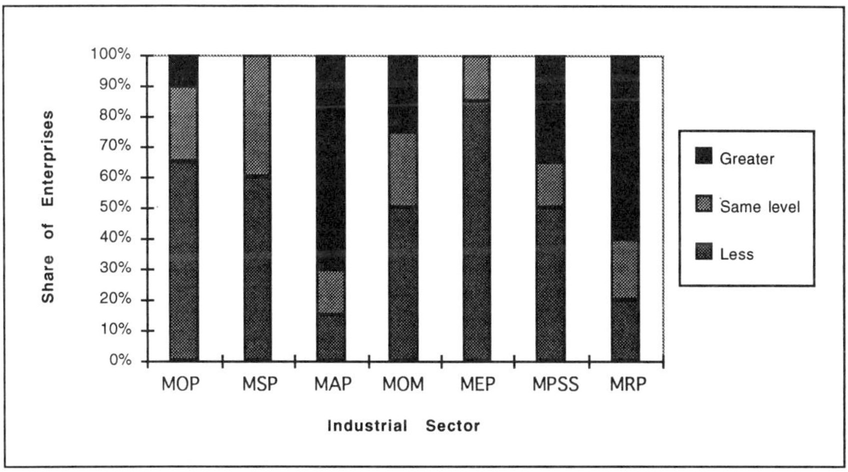

Figurc 11.6 shows that enterprise directors in both the aerospace and electronics industries have fairly high expectations for increased civilian sales. These should

act as an incentive to reform. However, aerospace enterprises expect to sell civilian versions of primarily military aircraft, launch vehicles, and spacecraft. Their requirement for organizational reform is lower. The heavy industries have relatively low expectations for civilian sales.

The poorly perceived conditions within the Ministry of Electronics Industry are interesting and contradict the propositions regarding asset specificity and complexity. With respect to asset specificity and complexity, MEP should be in better shape than any other ministry. This ministry was primarily concerned with production of electronic subcomponents, such as IC chips and circuit boards. Perhaps the best explanation for their poor performance is that their products are simply too far below world standards to be competitive on the Russian market. Also, their products are readily substituted with higher quality Western components. The same survey also shows that firms from MEP have by far the most pessimistic outlook for future state and non-state investment, which suggests that they have been abandoned by the government as well as the market. Of the two enterprises from MEP in the St. Petersburg study, one exhibited the highest degree of organizational reform of all the enterprises in the study, while the other showed only modest, level 2 reform.

The above data provide clear indication of a close relationship between industrial sector and organizational reform. Electronics industries, with the notable exception of enterprises coming from the former ministry of the electronics industry, are most likely to engage in organizational reform while the heavy industries are least likely. Aerospace industries lie in between. The support for this proposition from interview sources is also strong.

The causal connection between asset specificity and reform also seems clear. The evidence from the MinEcon survey suggests that the heavy industries are highly asset specific, with the enterprise directors expressing a high dependence upon military orders from a single customer. In the aerospace industries, the producers do not have a great deal of flexibility regarding the basic type of product they produce, but they may produce it for a variety of military and civilian customers. Electronics industries have a great deal of flexibility in both product line and market.

Privatization

Privatization has been the cornerstone of Russian economic reform. Western and Russian economists and policymakers have operated on the basic assumption that enterprises will perform more efficiently under private ownership than under state ownership. The state has an interest in factors, such as maintenance of employment, which are often antithetical to efficient economic behavior. Michael McFaul was explicit in his connection of privatization to reorganization, noting that "privatization also should rationalize the size of enterprise, meaning that large conglomerates will be broken into several entities while each newly privatized enterprise will likely be scaled down dramatically."[31] It could therefore be expected that enterprises which have undertaken privatization will be more likely to engage in reorganization.

Operationalization and observation are similarly straightforward. The percentage of privatized and non-privatized (state-owned) enterprises which have engaged in specific levels of reorganization are depicted below in Figure 11.6. Seventy-eight percent (32 of 41) of the defense enterprises in this study remain under majority state ownership.

Figure 11.7: Share of Privatized Enterprises Engaged in Reorganization

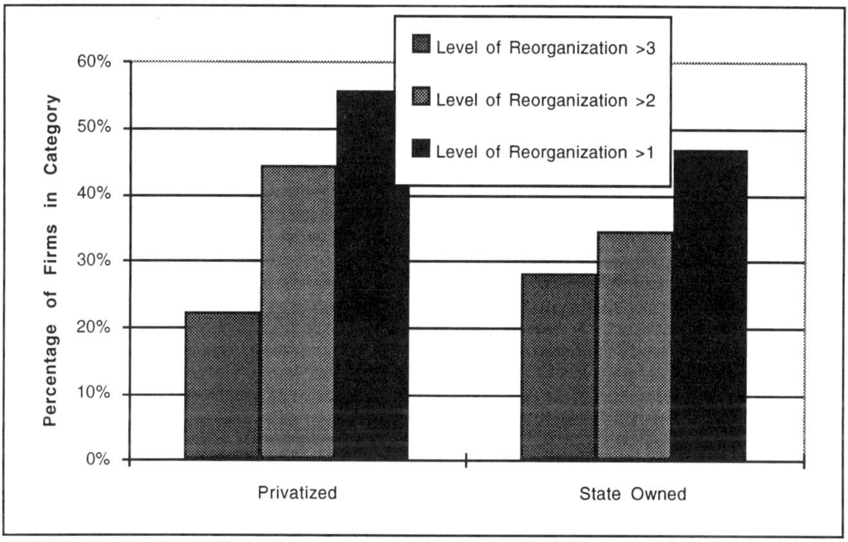

While the data in Figure 11.7 show that a somewhat larger proportion of privatized enterprises have engaged in lower levels of reform, at higher levels of reform (above level 3) a greater proportion of state-owned enterprises have actually engaged in reform than those that are privately held. Clearly, then, the data from St. Petersburg do not support this proposition.

Interviews with enterprise leaders support the proposition but suggest a somewhat different relationship. The first deputy director of the Arsenal Production Association, Vitaliy Sychev, argued that it was absolutely necessary for the facility, particularly the consumer products division, to completely privatize because the amount of time the state leadership needed to approve proposals made by the enterprise (usually on the order of three weeks to two months) was simply too long for the enterprise to be able to operate successfully in a market environment.[32] Similar arguments were advanced by Ilia Klebanov, the director of the LOMO enterprise.[33] Interestingly, the first deputy director of the Klimov NPO made essentially the same argument, but in reverse, arguing that since it had no requirement for short turnarounds on decisions, there was no need to privatize. By remaining state owned, the leadership felt they had a significantly better chance of acquiring state orders. A very different argument was advanced by the director of the Saratov Aviation Plant, who believed that employee ownership would lead the workers to be more productive.[34]

The MinEcon survey indicated that directors of privately held enterprises were somewhat more interested in reorganization than directors of state-owned enterprises. Fifty-eight percent of directors of privatized enterprises listed reorganization as their primary means of developing the enterprise, while the figure was approximately 48 percent for state-owned enterprises. These figures are roughly comparable to those gathered in St. Petersburg. More than twice as many privatized firms as state-owned firms (approximately 17 percent to 7 percent) responded that their preferred form of reorganization was decentralization of decision-making.

The available data paint two very different pictures. On one hand, the St. Petersburg data clearly indicate that there is no connection between privatization and reform. The interviews, however, suggest exactly the opposite. There are two potential reasons for this discrepancy. First, it could be argued that since many enterprises have only recently privatized, there is a certain lag time between privatization and reform. This explanation would of course be consistent with the general economic theory behind privatization, since it would be expected that there would be delay between the creation of private ownership and the initiation of new organizational forms by the new owners. However, the fact that the share of privatized enterprises expressing an interest in reorganization (from the MinEcon study) does not differ greatly from the share of enterprises that have actually engaged in reorganization does not lend support this argument. Rather, it suggests that most of the privatized firms that are interested in reform have already begun the process.

A second explanation is that privatization has done nothing to wrest control of the enterprises from the hands of the directors. McFaul argues that "actual restructuring and pressures toward increased efficiency will begin only when hard budget constraints are introduced. If the Russian state has the capacity to enforce hard budget constraints, then even inside owners will be forced both to make the difficult changes of enterprise reorganization and to accept the risks of allowing greater outsider ownership."[35] The evidence cited here is consistent with this argument, but it does not provide unambiguous support. For the present, we can only observe that there is no clear connection between privatization and enterprise reform.

Leadership Changes

The Soviet defense industrial system invested a great deal of authority in the managers of defense enterprises. Within almost all defense enterprises, the director or general designer had complete authority. Almost all decisions must cross his desk before they may be implemented.[36] The authority of plant managers extended not only downward through the facility, but upward to the government as well. Directors and general designers exerted a significant degree of influence over their government managers. The introduction of new weapons systems was more often the result of managerial lobbying than the result of military requirements being enforced upon defense enterprises.[37] As a result, it could be expected that changes in enterprise leadership would have a significant effect on the organization of a firm.

Leadership change may bring new talents to the helm of the enterprise, with important consequences for the organization.[38] Scholars of Western institutions have observed that there is a strong relationship between organizational behavior and leadership background.[39] Soviet-era managers almost uniformly come from purely technical backgrounds and therefore were trained only to think in terms of technical production problems. As a result, they were poorly prepared for dealing with problems of marketing, sales, and law. Leadership turnover would therefore have the positive effect of bringing in leaders who are better trained for a market economy. This leads one to expect that enterprises which have recently appointed new leaders are more likely to engage in organizational restructuring.

This proposition was tested by observing the percentage of enterprises undertaking organizational restructuring which have appointed new leaders against the percentage of those which have not. These observations are provided in Figure 11.8. Data on leadership changes were gathered for 40 enterprises. Seventeen enterprises had changes in leadership since 1991 and the remaining 23 had the same director.

Figure 11.8: Leadership Turnover and Enterprise Reorganization

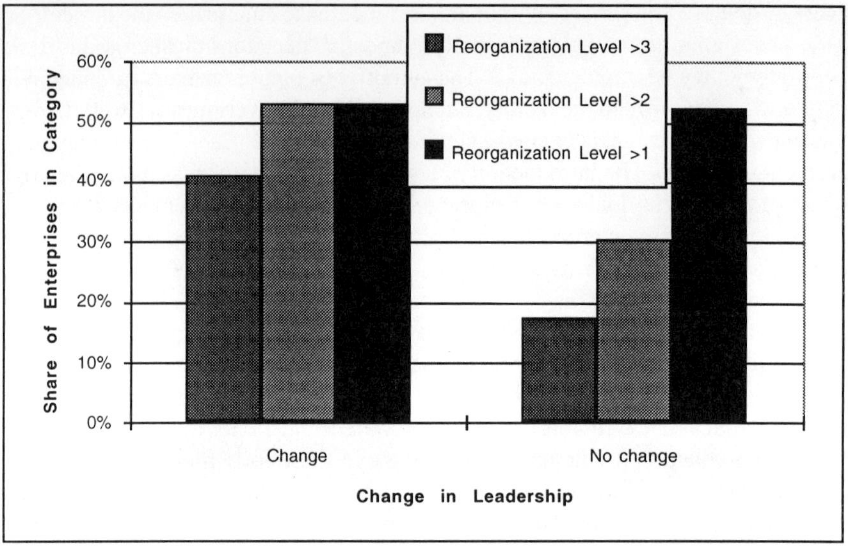

The data collected here provide clear support to the hypothesis. Enterprises that have appointed new leaders since 1991 are more than twice as likely to have reorganized at levels 4 or 5 than enterprises that have not changed leadership. The difference is somewhat smaller above level 2 and there is no difference at the lowest level of reorganization. Therefore, these data indicate that leadership change appears to be particularly important for achieving higher levels of reorganization. On the other hand, old enterprise directors are every bit as likely as new directors to pursue change, but they appear to grow noticeably more timid when it comes to more substantial reforms.

One should note that the educational and training systems in Russia have not yet had sufficient time to train a new cadre of managers with marketing, law, and business backgrounds. Consequently, the backgrounds of the new leaders remain very similar to that of the leaders who preceded them. It will take many years before new managers can be trained. When this occurs, management changes may assume even greater significance.

Moreover, exceptional leadership, even if old, can effect enormous organizational change, while ineffectual new leadership can produce meager results. The depth and breadth of reforms undertaken at the Leningrad Optical Mechanical Association, the Saratov Aviation Plant, and the Arsenal Production Plant would have been impossible without the leadership of Ilia Klebanov, Aleksandr Ermishin, and Vladimir Petrovski, respectively. By contrast, the Vavilov State Optical Institute has floundered under three different leaders.

Unfortunately, as social scientists we are hard pressed to operationalize the process of leadership change as a useful theory.[40] While it is clear that new leaders are more likely to reform, there is little more which can be added to this notion. The appearance of new leadership appears to be a random occurrence. The available data fails to identify any significant connections between leadership change and enterprise size, sector, or product cycle. Further research into the causes of leadership change is warranted.

Exogenous Factors

There are many external factors acting upon Russian defense enterprises. Many of these factors relating to economic and political turmoil affect all firms with relative equality. The following section will consider two factors that vary across firms. State support is being handed out based upon many factors, and it seems likely that such promises will have an effect upon firms' behavior either positively or negatively. Additionally, interaction with foreign firms seems likely to have an effect on the behavior of Russian defense enterprises. The relationship between these two exogenous factors and enterprise reorganization will be examined in the following section.

State Support

Over the course of the past three years, the Russian state has formulated several programs intended to lessen the shock to defense enterprises of the drastic reduction in defense orders from the Ministry of Defense. While these programs may not have been able to provide substantial assistance, many argue that the mere possibility that these programs may provide assistance has led enterprises to resist market forces in the hope that the state would bail them out. This argument lends itself to the assertion that enterprises which expect to receive state support and protection are less likely to engage in reorganization.

Operationalization of this proposition is through reference to a presidential decree containing a list of enterprises to remain under state protection which was circulated through the Russian government in August 1993. The list contained 479 enterprises, with fifty-three located in St. Petersburg. Enterprise directors lobbied heavily to be included on the list. Therefore, it provides a good indicator of enterprises which expected at least some level of state support. Figure 11.9 illustrates the percentage of enterprises which appear on this list that are engaged in certain levels of reorganization. The enterprises were fairly evenly divided between the groups, with twenty-three enterprises appearing on the list and eighteen which did not appear.

Figure 11.9: State Support and Enterprise Reorganization

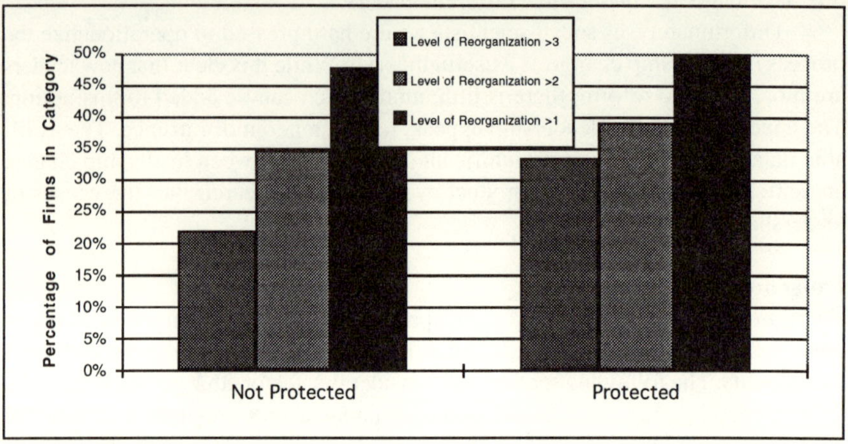

The data depicted in Figure 11.9 not only fail to support the proposition, they indicate the opposite. That is, enterprises which expect to receive some state protection are more likely to engage in reorganization than those which do not expect to receive such support. This difference at the higher level is particularly significant, with 22 percent of enterprises not expecting state support engaging in greater than level 3 reorganization, compared with 33 percent of those which appeared on the list.

Interviews with enterprise directors paint a different picture from the above data and lend support to the proposition. Directors at three firms which did not appear on the list commented that they did not believe that the state was capable of providing any support which would make a difference, and therefore they did not attempt to be included. All three enterprises (LOMO, NPO Klimov, and PO Arsenal) engaged in significant levels of reorganization (level 3 or above). The Vavilov State Optical Institute, which did appear on the list, has pinned its future on the existence of state support.

Interview data notwithstanding, the larger body of data provides a clear enough picture of the relationship between promised state protection and organizational reform. Defense enterprises which are promised state protection do not appear inclined to wait for support to come. Instead, they are somewhat more likely to engage in serious organizational reform. A probable reason for this is that the most effective enterprise directors have simultaneously lobbied for inclusion on the state protection list while simultaneously pursuing market reorganization. There is nothing preventing them from doing both. It is only natural to expect that effective leaders would hedge their bets.

These cases also say something about the credibility of state policies. Enterprise leaders simply do not believe that the state can or will deliver on its promises.

Foreign Involvement

Western partnerships can be a powerful force in driving the adoption of organization reform. Organizational theorists have long recognized that an appeal to outside expertise is a catalyst for reform in at least two ways. Due to their expertise, outside consultants can see problems and develop solutions which the leadership of the organizations often miss.[41] Perhaps more importantly, though, outside experts are often used by organizational leaders to lend legitimacy to difficult decisions, such as downsizing, which the enterprise leaders knew were inevitable before they invited the experts to assist them. Thus, dissatisfaction is directed at outside experts rather than at organizational leaders.[42]

Given the city's proximity, relatively pleasant surroundings, and the opportunities presented by a large number of high-technology firms staffed with well-educated workers with a relatively high technological level of capital assets, it should come as no surprise that St. Petersburg defense enterprises have been well exposed to Western firms. Western firms have established a variety of arrangements with Russian firms ranging from management consulting to fully operational partnerships. It is only natural to expect that these interactions would have some effect on the level of reorganization.

Mechanisms for reorganization exist on at least two levels. Most Western firms would not be comfortable entering into partnerships with Russian firms as they are currently structured. Along these lines, Russian firms would learn how to reorganize from their Western partners. Some Western management firms have been sent to St. Petersburg with support of their respective governments specifically for the purpose of helping Russian firms to reorganize. We can therefore posit the proposition that firms which have had considerable interaction with Western firms will be more likely to reorganize than those which have not.

Operationalization of this proposition is straightforward. One may simply observe which firms have engaged in working relations with Western firms and compare this with observation of levels of reorganization. Unfortunately, observation is problematic. In the first place, Russian firms were reluctant to disclose their partnership's potential or otherwise with Western firms for fear of revealing commercial secrets at a time of sensitive negotiation, and also for fear that the emerging joint venture would be taxed out of existence. Therefore, observation was through data compiled by the Department of Commerce office in St. Petersburg on joint ventures. These data probably represent conservative figures. The Commerce Department data only reported eight joint ventures among the 41 enterprises in this study. There are probably other joint ventures which for the reasons noted above have gone unreported. The direction of causality proves challenging in examining the relationship between foreign involvement and reorganization.

Figure 11.10: Western Partnerships and Reorganization

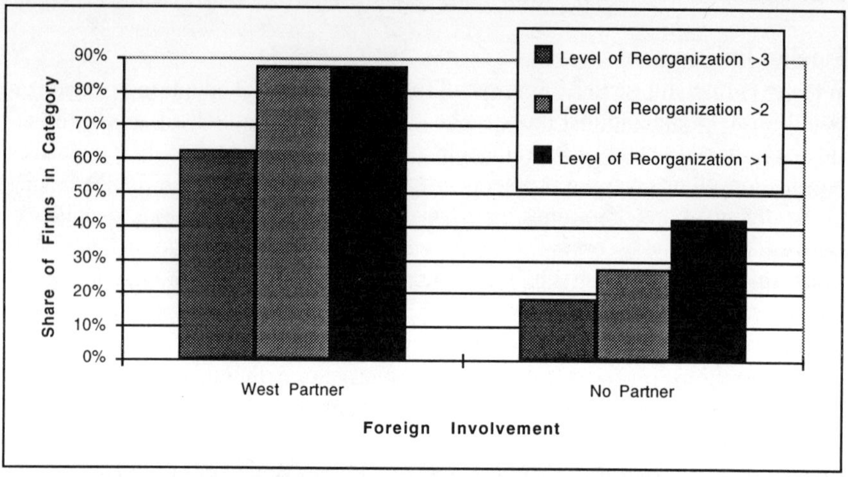

The data in Figure 11.10 show a very strong relationship between partnership and reorganization which is consistent with the proposition. Eighty-eight percent of the firms with Western partners have engaged in level 2 or above reorganization. This contrasts with 27 percent of the firms without Western partners having above level 2 reorganization.

Before drawing conclusions, there are questions of causality which become immediately apparent. Is reorganization caused by partnership with Western firms, or do Western firms tend to partner with Russian firms which have already reorganized? The above data are not capable of addressing this issue. For more information on this, we must turn to the richer data provided by interviews.

Several interview sources provide a fairly clear indication that causality lies in the direction from partnerships to reorganization. The Klimov design bureau began serious consideration of a transition to a production-oriented outfit only after being approached by Pratt & Whitney regarding the possibility of producing Pratt & Whitney engines in Russia. LOMO began its reorganization scheme only on the recommendation of the U.S. consulting firm McKinsey. The Khrunichev production association only absorbed the Saliut design bureau after the separation created problems in its emerging partnership with Lockheed. The Saratov Aviation Plant developed its privatization and reorganization plan on the recommendation of Science Applications International Corporation and scholars from Stanford University.

However, these cases have their equivocations. Both Saratov and LOMO exhibited a strong interest in reorganization prior to interaction with Western firms, and in fact sought out the assistance of these firms. Nevertheless, the fact remains that reorganization took place only with the assistance of Western firms.

A survey of 150 defense enterprises performed for the Ministry of Economics suggests that there are many enterprise directors with an interest in joint ventures. In the survey, 27 percent of enterprise directors said that their main direction of reorganization would be to seek out and form partnerships with foreign firms. This response was second only to the 29 percent of directors who felt that establishment of financial-industrial groups was the best direction.

The above data provide clear support to the proposition that cooperation with Western firms has a strong effect on the tendency of Russian firms to engage in organizational reform. Western firms bring with them a variety of management expertise which is sorely lacking in Russian defense enterprises. Through this process of interaction, Russian firms gain a realistic appreciation for markets and costs. More importantly, they learn from their partners how to perform these activities themselves. As partners in joint ventures Russian managers further learn how marketing and cost assessments figure into decision-making.

All indications are that Russian managers are quick studies—perhaps too quick. In the case of one firm receiving assistance from a Western management consulting firm, the enterprise leadership became so adept at producing business plans and pursuing financing that virtually all of their efforts became focused on these two activities to the virtual exclusion of more important management decisions. This suggests that the type of reforms undertaken by a Russian firm may tend to be a reflection of the Western partner.

Given the data problems noted above, these conclusions can only be regarded as tentative. It is entirely possible, and in fact likely, that there are a number of partnerships that are not recorded in these data. Moreover, this analysis counts each partnership to be the same. Not only are there differences between consulting arrangements and productive partnerships, but there are significant differences in the level of cooperation of different partnerships. If nothing else, though, these data indicate that cooperation between Western firms and Russian defense enterprises is an important area that demands further attention.

Relationships among Propositions

No fewer than four variables emerged from the St. Petersburg cases which appear to provide explanations for the reformist (or conservative) tendencies of enterprises. This presents some difficulties for parsimonious explanations. Before advancing conclusions as to the strength of these explanations, the interrelationships require some examination.

There are apparent problems with covariance between the variables of industrial sector and enterprise function. The analysis showed that the electronics industries showed a very strong propensity to reform. Similarly, there was a strong tendency to reform among production-oriented factories. Of the eleven enterprises coming from the electronics industries, four are factories and another four are NPOs. However, all the three KBs and NIIs in the electronics industry exhibited level 4 or above reorganization. If anything, this suggests that the relationship between sector and positive reorganization is more powerful than the initial analysis would suggest. On the other side of the coin, eighteen of the twenty-five enterprises from heavy industries are either KBs or NIIs. Only three of the heavy industrial NIIs and KBs exhibited any signs of reform, but three of the seven NPOs and factories exhibited at least level 3 reorganization. Furthermore, twelve of fourteen NIIs and four of seven KBs in the study were from the shipbuilding industry. Therefore, we would have to conclude that sector does not provide a clear negative explanation of reform. That is, it does not tell us why enterprises do not reform.

These problems stem from two sources. First, there is a heavy concentration of naval industry in St. Petersburg. Approximately 35 percent of defense industry in St. Petersburg is concentrated in the former Ministry of Shipbuilding (MSS). Sixty-five percent of MSS enterprises are either KBs or NIIs. More importantly, there are structural differences between the former ministries. In the electronics industry there is a heavy concentration on production-oriented organizations. Approximately 81 percent of the enterprises in these industries in St. Petersburg are either NPOs or factories. In the heavy industries, the figure is 37 percent.[43] This is due almost entirely to a proliferation of NIIs and KBs in the shipbuilding industry.

Many of the structural differences across ministries have historical origins. The heavy industrial ministries were created with the advent of Soviet planning. The Ministry of Aviation Production was also created before World War II. These ministries share the separation of research from design, and design from production. MOM and MRP were created in the years following the war. In these ministries there is much greater integration of design with production. MPSS and MEP were created in the 1970s and their structure tends toward even further integration of research, design, and production. Therefore, what we are observing may be the byproduct of the evolution of the Soviet system rather than a clear connection with asset specificity.

A stronger case can be made for the significance of enterprise function. In the first place the difference across categories is greater on the positive side for enterprise function than for industrial sector. There was less than a 5 percent difference between the aerospace and electronics sectors above level 2 reorganization (from 60 percent to 65 percent). The difference between factories and NPOs at this level of reorganization was much more dramatic (36 percent to 56 percent above level 2 and 27 percent to 44 percent above level 3). On the negative side there was only one factory which failed to engage in any level of reorganization. Therefore, it appears that enterprise function is a more powerful explanation than industrial sector.

Western involvement also appears to be confounded with industrial sector and enterprise function. Of the eight cases of Western involvement, four are from the electronics industry. The relationship between enterprise function and Western involvement is much stronger. Ten of the eleven cases of Western involvement are either NPOs or factories. Furthermore, in the single case of Western involvement with a design bureau, the Western firm Pratt & Whitney came in with the proposal to convert the firm from a design bureau into a factory producing Pratt & Whitney designs. The logic behind this relationship was noted above. Western investors are not interested in investing in technologies; instead, they wish to invest in actual products that are capable of being produced.

As a variable, change in leadership does not appear to suffer from problems of covariance. There are no clear connections between leadership change and enterprise function, industrial sector, or foreign involvement. In fact, from all indications leadership changes are entirely random events.

There are qualitative differences between these variables which should also be considered. Industrial sector and enterprise function are relatively stable conditions. This study suggests that among these conditions, enterprise function plays the greatest role in providing the underpinnings for reform. However, these factors alone may not be enough. It may be necessary for either leadership or foreign involvement to serve as a catalyst. In more than three-fourths of the enterprises exhibiting level 2 reorganization, there was either a change in leadership or a partnership with a Western firm.

Conclusions

In the end, this paper has not revealed any single causal explanation for reform in Russian defense enterprises. While it appears that reform is taking place, the evidence gathered here indicates that there may be several factors which encourage reform. We should be neither surprised nor distressed that we have not come out of this with a single explanatory variable. After all, the causes of innovation have long vexed theorists searching for monocausal explanations.[44] Even though the research presented here is exploratory in nature, it has at least pointed to some clear directions for future research. It also provided us with a somewhat better foundation for conceptualizing the problem.

This study suggests that there may be a discernible evolutionary pattern to reform and stagnation in Russian defense enterprises. Three factors that emerged as significant explanations were changes in leadership, primary involvement in production, and foreign partnerships. Combined, these factors exerted a powerful influence. All three of the enterprises that exhibited these three factors had reorganized above level 2 and only three enterprises (7 percent) engaged in above level 2 reorganization did not possess at least one of these qualities (i.e., was a non-factory, with no foreign partner, and the same leadership).

There are interesting temporal dimensions to these factors. Enterprise function is an inherent characteristic of a firm. It represents a condition which provides more fertile ground for reform. In this sense, we can say that it is easier for factories to reform, and very difficult for scientific research institutes. In some cases, (e.g., PO Arsenal) enterprise function alone may have been enough. More often, though, some additional push was needed. Leadership change and foreign involvement provided the necessary catalysts for reform for many enterprises. For the majority of enterprises engaging in higher levels of reorganization, either a new leader was appointed, bringing with him a reform plan, or a foreign partner brought with it the reform package. In this respect, new leaders were helpful, but foreign partners were decisive. It should be kept in mind, however, that foreign companies have the luxury of choice, while a prospective leader's choice is constrained to either the enterprise in which he currently works or exit to a new profession. In the end, the true test will be whether enterprises engage in reform without the catalysts of new leadership or the intervention of foreign partners.

New leadership or a foreign partner can initiate reform and serve as the driving force behind it, but both provide only a limited effect across the entire industry. Ultimately, reform will have to diffuse without reliance upon the chance appearance of a new, dynamic leader or a foreign partner. While the appearance of new leaders appears to be almost random, it is likely that there will be a much higher concentration of foreign partners in Moscow and St. Petersburg. Therefore, reform is likely to proceed at a slower pace in other defense industrial regions for this reason alone.

This paper suggests that the seeds of reform are beginning to take root in Russian defense industry. The relative significance of leadership changes and foreign involvement indicates that Russian defense industry is still in the initial stages of what may become a transformation of the entire defense industrial base. To an extent far greater than many had supposed, though, healthy organizational adaptation appears to be taking place in St. Petersburg defense industrial enterprises. Roughly one-half of the enterprises in this study are undertaking at least some reforms. As management personnel become better trained, and if privatization takes hold, we can expect to see another wave of leadership changes over the coming years. With stabilization of the Russian polity and economy, we can also expect to see greater investment from domestic capital sources and increased Western involvement. With both will come greater reform. As some factories move to profitability, we might begin to see integration of scientific research institutes and design bureaus into their structures. However, reform itself will not be enough to ensure survival; it will only give enterprises the organizational foundations for building a management structure which can effectively make decisions in a market economy. Enterprises will have to produce goods and services of sufficiently high quality and at a sufficiently low price to compete in the emerging Russian market, if not the world marketplace. Reorganization will at least provide Russia with a better set of tools to accomplish this.

By the same token, this chapter has also indicated that one-half of the enterprises are not reforming. Without reform these enterprises are unlikely to survive. Tragically, perhaps, the first to go will be the most technically qualified segment of Russian defense industry—the scientific research institutes. While the past three years have demonstrated that these institutes are capable of maintaining some semblance of existence on the meagerest of government subsidies, government support is clearly declining, and we must wonder just how long these enterprises can hold out until the government is forced to declare them bankrupt.

This chapter also suggests future paths for research. In the first place, we must develop a better understanding of our dependent variable, reform, and more precise measures for observing it. Reorganization is an important early indicator, but we will have to go much further to understand the type of reforms which are most appropriate for specific enterprises and developing a means of accurately observing those reforms.

Second, we need to gain a better understanding of the interaction between Western companies and Russian defense enterprises, which appears to be one of the most important influences on reorganization. While there is a wealth of anecdotal information available from Western businesspeople, this information needs to be collected, structured, and analyzed with academic rigor. We also need to establish a baseline to describe the reform process in a city that is not as well penetrated by Western companies as St. Petersburg, although activity may depend on cooperation with the Russian government in order to gain access to firms in these regions. In order to gain a greater understanding of the impact of the Western marketplace on Russian companies, case studies need to be done on Russian firms that are attempting to enter international markets. In the context of international collaborations, Russian firms' need to reorganize for productive efficiency may turn out to be different from their need to reorganize to develop new lines of production. Finally, we need to perform case studies of the different types of cooperation between Western and Russian firms. There may be important differences in the effects on enterprise reform between consulting arrangements, production partnerships, and marketing partnerships, to name only three types of cooperation. Although this may well prove to be a productive line of research, one cannot help but wonder if this realization is coming too late. Relations between Russia and the West are clearly cooling. Both Russian and Western businesspeople have grown frustrated over the slow progress over the last three years, and have aired mutual recriminations. The Russian government is reconsidering its policies of relatively open disclosure of the activities of defense enterprises. All this comes at a time when, as this chapter suggests, interactions between Western firms and Russian defense enterprises may be the most important conduit for reform.

Notes

1. The leadership is quick to acknowledge that the state is unlikely to allow the core design bureau to go bankrupt. See interview with Vitaliy Sychev, St. Petersburg, March 15, 1994. For discussions in the open literature of the conditions at PO Arsenal, see *Segodnya*, November 1993, p. 9.; and U.S. Department of Commerce, *Russian Defense Business Directory* (Washington, D.C.: United States Government Printing Office, 1995), p. 190.

2. Interviews with Vavilov Institute leadership, March and December 1994.

3. James Q. Wilson, *Bureaucracy* (New York: Basic Books, 1989), 227.

4. While precise calculations are impossible, a rough estimate can be generated by assuming that the average salary in 1994 was 70,000 rubles a month. In 1991 defense industrial employment was reportedly 6.9 million. The 1993 budget called for a combined 1,574,977 million rubles to be spent on defense industry. Using an overhead rate of 100 percent, the 1993 budget would support 937,000 workers, or 14 percent of the 1991 defense industrial base.

5. See *Konversiia,* April 1994.

6. In 1994 Russian hard currency arms sales were reportedly $1.7 billion. See *Delovoi Mir*, July 11, 1995, p. 7. Western experts estimate that Russian arms sales could go as high as $4–5 billion in the near future. See Randall Forsberg, "Competitive Arming or Cooperative Arms Control," in *The Arms Production Dilemma: Contraction and Restraint in the World Combat Aircraft Industry*, edited by Randall Forsberg (Cambridge: Harvard University Press, 1994).

7. See Joseph Berliner, *Soviet Industry from Stalin to Gorbachev* (Ithaca, NY: Cornell University Press, 1988), esp. Chapter 5, "Planning and Management," pp. 97–143.

8. For an excellent description of the socialist economic system see Janos Kornai, *The Socialist System: The Political Economy of Communism* (Princeton: Princeton University Press, 1992).

9. The only major exception to this would be the creation of scientific-technical councils in some research institutes which had to ratify the initiation of major new research projects. In design bureaus and scientific production associations, the director is supplanted by the chief or general designer, but the structure remains the same.

10. On the necessity of localized knowledge for adaptation see Fredrich Hayek, "The Use of Knowledge in Society," *American Economic Review* 35 (1945): 519–530. The need to find a balance between centralization and decentralization is also dealt with in Oliver Williamson, "Economic Institutions: Spontaneous and Intentional Governance," *Journal of Law, Economics and Organization* 7, special edition (1991): 159–187. Burns and Stalker deal with the importance of a decentralized structure in an environment of high uncertainty; see Tom Burns and T. M. Stalker, *The Management of Innovation* (London: Tavistock, 1961).

11. See Neil Fligstein, "The Spread of the Multidivisional Form among Large Firms, 1919–1979," *American Sociological Review* 50, no. 3 (1985): 377–391; and Oliver Williamson, *Markets and Hierarchies* (New York: Free Press, 1975).

12. On issues of integration of research with production see Gary Pisano, "The Governance of Innovation: Vertical Integration and Collaborative Arrangements in the Biotechnology Industry," *Research Policy* 20 (1991): 237–249; Paul S. Adler, "Interdepartmental Interdependence and Coordination: The Case of the Design/Manufacturing Interface," *Or-*

ganization Science 6, no. 2: 147–167; and James D. Thompson, *Organizations in Action* (New York: McGraw-Hill, 1967). For a historical perspective on the subject see David Mowery, "The Relationship between Intrafirm and Contractual Forms of Industrial Research in American Manufacturing, 1900–1940," *Explorations in Economic History* 20 (1983): 351–374.

13. Source: Interdepartmental Analytic Center (MATs). *Vyiavlenie osnovnykh tendentsii rasvitiia promyshelennogo kompleksa (na primere vpk) na osnove rezul'tatov oprosa rykovoditelei predpriiatii,* Report No. 94/2-II, Moscow, 1994.

14. Department of Commerce, Defense Business Directory, U.S.-Russia Defense Conversion Subcommittee, September 1993.

15. Leonid Blaukhman, "Razvitie oboronogo kompleksa Sankta Peterburga," mimeo, Leningrad State University, 1994.

16. One summary of this survey was published as "Konversiia: shto den' griadyshchii nam gotovit?" *EKhO* (October 1994): 24–37. More complete information is provided in A. K. Ponomarev, *"Vyiavlenie osnovnykh tendentsii razvitiia promyshlennogo kompleksa (na primere vpk) na osnove rezul'tatov oprosa ukovidiltelei predpriatii,"* Mezhvedomstvennogo analitichekogo tsentra, Moscow, June 1994.

17. The Central Committee database of the most significant defense enterprises in the USSR counted 89 defense enterprises in St. Petersburg. Local officials have referred to numbers as high as 150. In either case, the 41 enterprises examined here represent a significant fraction of defense industry in St. Petersburg.

18. The measurements for reorganization are discussed below. This figure refers to levels 3 and higher.

19. Interview with the director of the Institute for the Improvement of the Qualifications of Engineers of the Ministry of General Machinebuilding, Moscow, September 1990. See also Peter Almquist, *Red Forge: Soviet Military Industry Since 1965* (New York: Columbia University Press, 1991).

20. Thompson makes the basic argument that organizations will make every effort to protect what they perceive to be their core technology. See James D. Thompson, *Organizations in Action* (New York: McGraw Hill, 1967).

21. While some organizations had functions at variance with their official titles, the Central Committee database on Soviet defense industrial enterprises was clear in its explanations. The database contained more than 900 enterprises across the former Soviet Union.

22. Presentation by Victor Tyurin at OECD conference on regional conversion, Zhukovskii, Russia, December 10, 1994.

23. The importance of core technologies to organizations was raised first in James D. Thompson, op. cit. See also Bruno Latour, *Science in Action* (Cambridge: Harvard University Press, 1987) for a discussion of the adherence of scientists to their chosen technologies.

24. Presentation by Michael Lehner of the Defense Enterprise Fund. Pepperdine University, March 1995.

25. The levels for other regions could be expected to be somewhat lower.

26. See Viktor Glukhikh, op. cit.

27. This argument was advanced by Michael McFaul in "Privatization at Four Enterprises," in *Defense Industry Restructuring in Russia: Case Studies and Analysis*, edited by David Bernstein. (Stanford, CA: Stanford University Center for International Security and Arms Control, December 1994): 111.

28. See Oliver Williamson, "Transaction Cost Economics: The Governance of Contractual Relations," *Journal of Law and Economics* 22, no. 3.

29. Interview with Vitaliy Sychev, deputy director of PO Arsenal. St. Petersburg, March 1994.

30. On the issue of the importance of organizational slack for reform, see Jeffrey Pfeffer, *Power in Organizations* (New York: Ballinger, 1981).

31. As quoted from Michael McFaul, "Agency Problems in the Privatization of Large Enterprises in Russia," in *Privatization, Conversion and Enterprise Reform in Russia*, edited by Michael McFaul and Tova Perlmutter (Boulder, CO: Westview Press, 1995): 40.

32. Interview with Sychev, op. cit.

33. Interview with Ilia Klebanov (director of LOMO), St. Petersburg, December 1993, op. cit.

34. Interview with Aleksandr Ermishin (general director of Saratov Aviation Works), Saratov, March 31, 1992.

35. See McFaul, op. cit. p. 50.

36. The only systematic exception to this rule appears to be major scientific research institutes, in which some major decisions, such as opening a new line of research, must be approved by the Scientific Council.

37. For a more detailed discussion of this relationship see Peter Almquist, "Soviet Defense Industry: From a Seller's Market to a Buyer's?" in *Soviet Military Power in a Changing World*, edited by Susan Clark (Boulder, CO: Westview Press, 1991).

38. There is a growing literature on the importance of leadership for organizational transformation. See in particular Warren Bennis and Burt Nanus, *Leaders: The Strategies for Taking Charge* (New York: Harper and Row, 1985); and Noel M. Tichy and Mary Ann Devanna, *The Transformational Leader* (New York: Wiley, 1986).

39. See Neil Fligstein, "The Spread of the Multidivisional Form among Large Firms, 1919–1979," *American Sociological Review* 50, no. 3 (1985): 377–391.

40. This point was made by James Q. Wilson, *Bureaucracy* (New York: Basic Books, 1989).

41. See John Bessant and Howard Rush, "Building Bridges for Innovation: The Role of Consultants in Technology Transfer," *Research Policy* 24 (1995): 97–114.

42. See Thompson, op. cit., and Pfeffer, op. cit.

43. These figures were derived from the Central Committee database on defense industrial facilities, which included 89 enterprises in St. Petersburg.

44. See in particular George W. Downs Jr. and Lawrence B. Mohr, "Conceptual Issues in the Study of Innovation," *Administrative Sciences Quarterly* 21 (December 1976); and Stephen Peter Rosen, *Winning the Next War: Innovation and the Modern Military* (Ithaca, NY: Cornell University Press, 1991), especially pp. 1–53.

12

The Role of Third-Party Facilitators in Public-Private R&D Collaborations in the United States

Stephen L. Gomes

Introduction

Science and technology are fundamental to economic growth and international competitiveness. U.S. preeminence in research and innovation traditionally has been a major source of economic strength. In recent years, however, other nations have begun to threaten to surpass the United States in significant technology areas that are important to U.S. competitiveness. Such U.S. technologies as those related to large civil aircraft and civil helicopter production, semiconductor lithography machines, robotics, and construction automation are under increasing pressure from foreign producers. This deterioration in competitive position has made policymakers begin to question whether there are aspects of U.S. science and technology policy that must undergo substantial change.

In particular, U.S. policymakers and industry leaders are beginning to focus on the processes of technology transfer, especially via the mechanism of public-private partnerships. Some analysts have concluded that the traditional mechanism of "spinning off" technologies has been ineffective, that investments have failed to be directed toward certain strategically important areas, and that diffusion of scientific ideas and technological innovations has been insufficient. (Alic et al., 1992; Branscomb, 1993; Alic, 1995) In particular, the countries that are experiencing the most rapid economic growth and generating the most competitive industries seem to be those where there are highly developed partnerships between the public and private sectors that pool resources and expertise in collaborative research and development (R&D) ventures.

In the American context, public-private partnerships offer an opportunity to improve economic competitiveness without significant federal budgetary outlays. Given concern about the federal deficit, risk and reward sharing on a market basis, through public-private joint ventures, is preferable to direct government subsidization. Moreover, the market-oriented model promises to be more effective than more bureaucratic alternatives in commercializing technology. Introducing market forces through the involvement of industrial partners virtually requires that the development of a technology be more relevant commercially than an R&D process that relies solely on the decisions of government agencies insulated from, and with little knowledge of, the commercial marketplace. In addition, the potential rewards of public-private collaboration are enormous. Of the approximately $140 billion annual U.S. R&D expenditure, an estimated 10 percent of total federal civilian R&D has a 3–5 year technology development time frame and therefore is a promising candidate for rapid commercialization. Based on these assumptions, federal agencies are developing innovative ways of working with industry and academia on R&D in order to transfer and commercialize technology.

In addition, some federal laboratories are experimenting with the inclusion of third-party value-added facilitators in public-private partnerships. These non-profit facilitators play an active role in establishing partnerships, resolving conflicts, and enhancing the ability of the partners to sustain productive collaboration. This paper will discuss the case of the American Technology Initiative (AmTech), which plays the role of a value-added facilitator for NASA in the space agency's efforts to commercialize technology. AmTech's experience suggests that third-party facilitators can provide a variety of services that supplement the competencies brought to the table by the public and private partners, and can help to establish and sustain profitable partnerships.

Rationale for Public-Private Partnerships and the Involvement of Third-Party Facilitators

Theorists and policymakers both have concluded that there are two conceptually distinct modes of technology transfer. Alic et al. (1992) define the distinction as "spin-offs" versus "dual use." Spin-offs are alternative commercial uses for technology developed by national labs that emerge without programmatic effort. Dual use implies integration of R&D from the government research and industrial sectors. Spin-offs imply a pipeline model of technology development, flowing from an idea developed at a lab to a product developed by a company. Dual use describes a process with a feedback loop, wherein commercial demands shape the development of a technology from an early stage and in an iterative process.

Policymakers at NASA have made a similar distinction, recognizing the deficiencies of the previous mode of technology transfer, and drawing conclusions about what mechanisms should be instituted in order to gain the advantages of a transfer process more in accordance with dual use. NASA chartered a committee, led by Dr. Jerry Creedon of NASA Langley, to analyze the agency's programs for transferring and commercializing its aeronautics and space technology. The Creedon Committee's report in December 1992 identified two categories of technology transfer activities: non-targeted and targeted technology transfer. The Creedon Committee defined non-targeted technology transfer as "technology which is transferred via 'acquisition and dissemination' of information and is used as is, or is extended by the user without further NASA assistance." Most of NASA's 30-year program to transfer technology falls into this category. While noting that these mechanisms to acquire and disseminate technical information are important, the committee asserted that relying solely on this process would not enable the Agency to fulfill its potential to contribute to the nation's economic competitiveness. Instead, the team recommended that NASA become active in its efforts to transfer its technology to the private sector. The Creedon Committee called this type of process "targeted technology transfer."

Targeted technology transfer involves NASA's conscious involvement to collaborate with industry to commercialize its technology. The team broke down this category into two sub-categories: primary and secondary targeted technology transfer. Primary targeted technology transfer occurs when the technology is part of NASA's primary mission and is developed from the outset with the purpose in mind of transferring it to an identified aerospace user. NASA's entire aeronautics program represents this category. Newer programs, like the Centers for the Commercial Development of Space, are examples from the Agency's space program. Secondary targeted technology transfer refers to when technology originally developed for a NASA mission is extended by NASA to meet the identified needs of a specific user for a non-aerospace application. The committee noted that NASA dedicates very little effort or resources to this category, although this area offers greater opportunity for transfer of NASA technology. This is the only area where the Creedon Committee recommended that NASA increase its budget.

Bringing together government, industry, and research institutions into joint collaboration involves marrying different interests in a mutually advantageous union. Government benefits from public-private partnerships because R&D expenditures can be leveraged with private sector resources in order to meet additional mission-directed objectives. Moreover, government has an interest in ensuring that critical technologies are transferred from government and universities to the private sector. Also, there is a clear interest in gaining government access to manpower and state-of-the-art technologies residing in the private and nonprofit sectors. Finally, in a period of shrinking budgets and downsizing, collaborations are useful because they generate royalties for government agencies and inventors through technology commercialization.

For industry, collaboration can reduce the cost of product development by sharing R&D expenditures with the government. Commercial firms in the collaboration can obtain exclusive or non-exclusive commercial rights to technology developed through the research partnership. Also, industry benefits by gaining access to R&D undertaken at government and nonprofit research centers, and to specialized government equipment and facilities.

Universities and research institutions, for their part, benefit by gaining support for cutting-edge research in commercially suitable technologies. Researchers, including students, gain the opportunity to contribute to important new discoveries leading to commercial products and processes. Collaboration provides academic researchers and students with access to state-of-the-art facilities in government and industry. Finally, universities and research institutions may derive royalty income based on their intellectual property rights in technologies that are commercialized successfully.

Despite these widespread benefits from public-private partnerships, many government laboratory employees have personal reservations. The case of NASA is illustrative. It is difficult for people brought up in the NASA tradition of large science missions to accept that increasingly, NASA's most pressing mission is to serve the commercial sector. Many civil servants are resistant to the idea that someone will make a profit based, in part, upon taxpayer-funded R&D, although this is fundamental to the process of technology transfer and commercialization. Government researchers, for their part, are often disinclined to devote attention to nonresearch considerations important to the commercial sector, such as cost and development time, manufacturability, reliability, maintenance, and market appeal. NASA personnel policy does not provide incentive for researchers to focus on technology transfer and commercialization activities in addition to their assigned government research projects.

At the same time, a number of unresolved policy debates inhibit the government's ability to effectively enter into close partnerships with the commercial sector. Government concerns over unfairness of opportunity and conflicts of interest, as well as other questions, interfere with its ability to select capable partners and establish meaningful partnerships. The private sector, on the other hand, finds standard government procedures and protocols cumbersome, in the context of the strict time frames of the commercial marketplace.

Nevertheless, government laboratories and their industrial partners are persevering in their collaborations, because of the important mutual benefits from joint-sponsored research. All partners are able to leverage resources, share R&D risks, and produce more research. Identifying commercially relevant R&D in advance speeds technology transfer and commercialization, allowing all sides to realize the benefits of dual-use technologies. Intellectual property rights are negotiated up front, thereby providing NASA rights for government use and the private sector partner with clear commercial benefit. Finally, there are intangible benefits from the transfer of knowledge between NASA and industry technologists in the course of their close collaboration. Dialogue with their scientific counterparts in the commercial sector stimulates the creativity of otherwise isolated NASA scientists, allows them insight into how their technologies fit into the larger commercial world, and opens opportunities for further collaboration in research.

For any technology to be commercialized successfully, four elements must somehow come together: a useful technology, a dedicated champion with business acumen, market demand for the technology, and available working capital to bring the technology to market. First, technologies must be innovative and potentially useful commercially. In general, the vast wealth of human talent and R&D resources at national laboratories guarantees that useful technologies can be found. Second, a champion with business acumen is necessary to be a motivating force driving a collaboration, marshal the resources necessary to establish and sustain it, and provide the long-run commitment necessary to overcome various obstacles that inevitably arise. A champion is necessary not only to overcome bureaucratic hurdles within the government bureaucracy, but also to attract the interest of industrial partners, who are often hesitant to enter partnerships with government and are generally averse to risk in any unproven technology—unless the possible rewards can be effectively communicated. Typically laboratories offer commercialization ideas on a "technology push" model; that is, they target opportunities that are technologically interesting. Effective commercialization, however, also requires the active involvement of a private sector champion intimately acquainted with market demand. Industrial partners can make technology development more responsive to market forces. Finally, working capital is necessary to finance the development, commercialization, and marketing of a new technology. At a minimum, this would involve leveraging government funds with industry investment. In some arrangements, venture capital firms and other financial institutions might contribute capital.

In the case of technologies developed in NASA laboratories, and the same is true at other government laboratories for the most part, only one of these elements always exists—the technology. The other elements do not commonly reside within the field center or laboratory. More importantly, few NASA technologists embody all these qualifications, and it is at the technologist's level that the process of technology transfer begins. Partnerships with industry and universities provide many of the other elements necessary for commercialization.

Nevertheless, forming and sustaining these partnerships is difficult, and there may still be organizational deficiencies. Third-party business facilitators can help to deal with these problems. David Gibson (1994) describes an active role in public-private partnerships for "boundary spanners," individuals and organizations who "facilitate cross-organizational collaboration." Facilitators expand social networks, provide logistical and organizational support, and offer the steady commitment to the collaboration necessary to overcome obstacles and sustain cooperation. Gibson's analysis of the case of Austin and the Microelectronics and Computer Technology Corporation (MCC) suggests that these boundary spanners can have a tremendous impact, although the benefits are often subtle and seen only in the long run.

In the case of NASA, AmTech provides boundary-spanning activity as a value-added facilitator. This is the case to which we now turn.

Case Study: AmTech

American Technology Initiative, Inc. (AmTech) is the first and only nonprofit corporation entirely dedicated to a nonprofit facilitation approach to the proactive formation of R&D partnerships between the public and private sectors using the joint venture mechanism. Within NASA, AmTech has pioneered the development of this mechanism, which enables the pooling of public and private resources from the inception of the joint R&D venture to ensure the transfer and commercialization of technology. AmTech directly assists in the negotiation of R&D plans, finances, in-kind resource contributions, intellectual property rights, commercialization commitments, and other terms and conditions and maintains a "stay-in-the-deal approach" as an in-house mediator, coordinator, and monitor. AmTech is directed by a nationally recognized board of trustees with experience in business and technology issues across government, academia, and industry. In accordance with AmTech's charter, all surplus income will be invested to support new technology transfer and commercialization mechanisms, fellowships, and research in areas of technology critical to national needs.

AmTech's role as a value-added facilitator is connected with the development of NASA's Joint Sponsored Research Program. The Joint Sponsored Research Program is a NASA-industry partnership to promote dual-use technology development, accelerate technology transfer and commercialization, and leverage R&D resources. The unique aspect of this program is the formation of formal collaborative R&D agreements between NASA and a company (and if required, a university or other nonprofit research institution) under the authority of the National Aeronautics and Space Act of 1958, as amended, otherwise known as the Space Act. The Space Act authorizes the NASA administrator, at his or her discretion, to enter into "other agreements" to fulfill the Agency's mission. This is referred to by federal lawyers as the "other transactions authority." In March 1992 the NASA administrator delegated this authority for the specific purpose of entering into Joint Sponsored Research Agreements (JSRA).[1]

While the authorization exists to enter into these new agreements, a tiny fraction of NASA civil servants have the facilitation know-how to craft a JSRA. To fill this gap, NASA entered into a cooperative agreement with AmTech, a California nonprofit corporation, to conduct research on how to form these agreements, and when appropriate, facilitate the formulation of a JSRA.

AmTech's efforts have resulted in a detailed understanding of the elements necessary for crafting a JSRA and the process that must be followed. AmTech identifies common areas of interest between NASA research and industry research where both parties can agree to cosponsor this research at a university or nonprofit research institution. In return NASA receives technology rights for government use and the company receives exclusive rights for commercial use of the technology and may retain its ownership of the intellectual property co-developed by the alliance or partnership.

AmTech is intimately involved with the creation of NASA's partnerships with industry and determining how they are structured. First, AmTech works with NASA to identify R&D projects that are candidates for commercialization. It is important to note that this is R&D that NASA is already pursuing to meet some Agency-approved mission. If a JSRA cannot be crafted, NASA will still pursue this work unilaterally or in some other form of collaboration. With candidate NASA R&D in hand, AmTech then seeks potential industry partners: the initial criteria is simply a company that is pursuing similar R&D for its own purposes. A match is made when NASA and a company agree to cosponsor a specific research project which may also include a university. At this point, AmTech facilitates the negotiation of the agreement between all parties including preparing the first draft of the legal partnering agreement. The end product is a JSRA that details the research to be conducted, and the money, people, and equipment that are to be utilized. NASA and the company will contribute to the effort, and the intellectual property rights and royalties (if any) are shared. AmTech has no implementation role in the agreement other than being available to facilitate future communications between the parties, monitor its progress, ensure commitments, and, in some cases act as a fiduciary for funds flowing between participants. AmTech sees its role as monitoring, administering, and facilitating the ongoing relationship between the participants throughout the life of the joint R&D project in order to maximize the chances of success. Moreover, AmTech seeks to create an organization dedicated to learning through research, experimentation, and feedback resulting from real work experience with joint R&D projects.

As a result of the JSRA, federally funded R&D is undertaken at a university, nonprofit institution, federal lab, private sector lab, or some combination of multiple locations. The scope of joint R&D and the rights to resulting technology are pre-negotiated consistent with the needs of the parties. Technology transfer and commercialization objectives are incorporated into the R&D process and are implemented from the beginning of the R&D project. Participants share the management of the specific technical and administrative responsibilities of the R&D project.

Since its inception, AmTech, in collaboration with NASA, has successfully implemented eight prototype joint ventures and has developed, negotiated, and drafted agreements for fourteen other types of NASA-industry collaborative R&D projects. One such partnership is the Advanced General Aviation Transport Experiments consortium (AGATE), the largest partnership in NASA's history. This partnership involves approximately 120 separate participants in the general aviation industry. The Environmental Research Aircraft and Sensor Technology

(ERAST) Alliance involves NASA, the Department of Energy, the Ballistic Missile Defense Organization, and several highly competitive companies, which previously refused to work together, in a collaboration related to developing a remotely piloted aircraft that could assist in gathering data on the impact of pollutants on the ozone layer.

The participants in the two prototype projects have appreciated a number of important benefits from this unique arrangement. First, the time to commercialization can be significantly reduced. In the first prototype JSR project, the technology is currently under license negotiation, less than twelve months after completion of R&D. (The norm for commercialization of federal technology in the past has ranged from six to ten years.) Second, U.S. economic competitiveness has already been improved through public-private collaboration. A software development effort undertaken through a consortium under the second prototype JSR project is leading to a U.S. standard for aircraft design software. Prior to this JSR project no one company had the incentive to pursue this research unilaterally.

Value-added facilitators for targeted technology transfer are employed in order to bring missing skills to the technology transfer process. In the case of AmTech, the facilitator is a third-party nonprofit expert who works directly with the NASA technologist, legal staff, and center management. AmTech supports NASA's collaborations in a number of key ways, which can be generalized to the role of value-added facilitators in public-private partnerships. Facilitators must provide critical expertise in a wide variety of fields, they must expand collaborators' networks, and they must provide strong advocacy on behalf of the partnership. In the course of these activities, facilitators must be careful to maintain active, adaptable involvement in the partnership, while preserving their neutrality.

AmTech has demonstrated over its five-year history that an effective facilitator must offer an in-house, interdisciplinary one-stop shop, with a team of experts in technology, policy, business, finance, and law (strategic partnering, government, and intellectual property) that can provide critical expertise on a custom basis as required by each individual technology development partnership or alliance. A dedicated nonprofit facilitator offers a unique mix of staff capabilities—business, engineering, public policy, law, finance—not found in any other public sector or private sector organization. Such an integrated team approach provides important value that is greatly appreciated by all participants. The interdisciplinary team assures that necessary expertise is readily available at any stage of project development or commercialization. The facilitator's experts coordinate their input to a project and its participants, eliminating the confusion that often surrounds conflicting advice provided by unrelated advisors. Third, the integrated expert team approach eliminates the usual professional and discipline-based compartmentalization that often delays deal structuring and negotiations. Fourth, the cost of the facilitator's dedicated expertise, offered through a public benefit nonprofit corporation, is far less than if obtained through normal professional channels (i.e., a law firm, engineering consultant, etc.).

Of the various functions that AmTech provides to NASA partnerships, facilitation know-how is most important, and the most difficult to quantify. Facilitation know-how is essentially a bridging function linking the sometimes conflicting cultures of government science and technology, which is driven by mission and schedule, and business, which is market driven. Industry technologists and government technologists are different, almost by definition, due to the process of self-selection that sends some people into government service while others enter the world of commerce. In short, there is a cultural gap between these groups. Bridging this gap is a requirement for successful targeted technology transfer, and facilitators can meet this need.

A broad range of core competencies is required for effective facilitation. The facilitator must be researcher, educator, advisor, catalyst, and project manager. Required competencies include

- program design and implementation
- policy development and advocacy
- marketing, i.e., identifying and recruiting prospective partners
- deal definition and structuring
- partner consensus building
- negotiation
- definition of legal terms and conditions
- resolution of policy, legal, business, and operational issues
- technology commercialization
- project administration
- dispute resolution
- project facilitation

Facilitators must also be able to expand participants' networks and access to new opportunities. The facilitator must offer an established network of leaders and experts on which it can draw in order to gather support for a project or policy, or resolve issues and breakdown barriers. It is particularly valuable for a facilitator to be able to tap its own network to *independently* initiate activities and provide access to leaders and decisionmakers on behalf of the partnering agency. For example, AmTech tapped its network in the multimedia field to proactively identify and initiate a collaboration opportunity with NASA.

Facilitators must provide strong advocacy on behalf of the collaboration. The importance of a champion to the successful transfer or commercialization of technology is widely acknowledged in the field of technology transfer. Champions are understood to be necessary to overcome typical barriers such as "not invented here" and the frequent aversion to new technical solutions and new ways of doing business. Further, champions are critical because they garner critical resources and support. In building public and private sector partnerships and consortia, the facilitator is frequently required to function as champion at multiple levels of the government and commercial organizations. The facilitator must have broad in-house expertise and an established network of influential leaders to permit it to successfully interface with and influence decisionmakers at all organizational levels—namely, technical, marketing, legal, managerial, and director levels. Broad access and targeted communication are often the keys to winning support for a project.

Facilitators must be intimately involved in the research and operational activities of the collaboration. The work of transferring and commercializing technology, and building R&D partnerships and consortia, must be continually responsive to the current management's policies and interests, as well as the particular requirements of a technology and the deal participants. Therefore, the facilitator must have research capabilities and be organized to perform rapid study and evaluation of policies, technologies, and commercial interests. Operational involvement in a partnership maintains a party's critical awareness of the partnership's direction, health, and progress. Where the facilitator is seen by project participants as actively involved in the project, empirical data show that participants readily turn to the facilitator to provide a balance between participants' interests.[2] If the facilitator is not operationally involved in the project, participants view the facilitator as distant and out of touch with the real issues, and are likely to try to resolve disputes by themselves, often in an adversarial manner, to the detriment of the partnership.

Likewise, facilitators are more effective when they are seen as neutral parties in the collaboration. Cooperation involves real risks to each partner. To create a partnership or R&D consortia, the trust of participants in each other must be gained. An independent neutral party has credibility with prospective participants in identifying to each the strengths of the others and the benefits of collaboration. Participants are more secure in revealing to an independent neutral facilitator their private assessments of the proposed transaction. With this information in trust, the facilitator can carefully address individual concerns, including sensitive issues regarding another party's capabilities, motives, funding, etc. In complex partnerships or delicate negotiations, the facilitator's neutrality is an even more essential tool in moving individual players into a team and converting perceived risks into acknowledged shared rewards.

Finally, facilitators must be able to adapt to rapidly changing conditions. New strategies will be required to maintain collaborative projects over a duration of years, as participants' requirements and expectations change, and their originally projected budgets expand or decrease. On a programmatic level, the facilitator must be adept at tailoring its operations strategy to the climate of the client organization. It must effectively develop strategies to build collaborations even in environments of limited management support, and must be capable of initiating new strategies to overcome barriers encountered in the current environment. One particular example is turnover in participants in the collaboration. As individuals change job assignments, leave the organization, or retire, turnover is inevitable. New participants may enter a collaboration with the requisite skills to carry on the work, but can seriously disrupt its momentum by being unaware of important understandings that have developed among the team members. The facilitator must act as the neutral "keeper" of the project and provide the collaboration with continuity as new players are integrated. The facilitator must be attentive to each player's understanding of others players' backgrounds and ongoing expectations in order to establish and maintain a working team relationship.

Conclusion

There is no set formula for building meaningful partnerships, alliances, or consortia, or for transferring and commercializing technology. Each technology presents its own challenges; each group of players present their unique expectations, terms, and conditions. The ability to build consensus and visualize broad frameworks for collaboration are essential skills for successful alliance facilitation. Facilitating transactions requires extensive expertise, built on experience. It requires personal credibility and neutrality, as trust is an essential ingredient in building—and maintaining—a team out of divergent organizations. The job requirements cannot be simply quantified; they are ever-shifting and involve qualitative considerations.

While the AmTech research experiment on public-private R&D collaboration is still in its early stages, a number of preliminary findings are suggested by its experience. First, public-private R&D collaborations seem to transfer and commercialize technology faster, more successfully, and in a more market-responsive manner than previous mechanisms. Traditional methods of transferring technology were passive, relied on government pushing already-made technology into the private sector, and were ultimately unsuccessful. Partnerships where collaboration begins at an early stage of research, and where industry is closely involved and takes a financial risk in the venture, are forced to respond to the pull of the market, and therefore to develop technology that is more likely to be commercialized successfully.

In addition, government and the private sector derive mutual benefits from pooling their resources, ideas, and capabilities. More R&D objectives can be met, at lower costs.

Neutral third-party facilitators are essential to the success of R&D collaboration. In most traditional contracting and assistance relationships with the private sector, the government directs the work and specifies the results. Collaborative research requires negotiation and a much closer co-venturing relationship, which is difficult given the cultural clash and the legacy of distrust between government and industry. Third-party facilitators can help to overcome this distrust, bridge cultural differences, and provide participants with a neutral forum for negotiation.

Several issues remain to be addressed. Policymakers must come to grips with the fact that the facilitation and formation of joint R&D projects is a labor-intensive process. With further experimentation and experience, the efficiency of this process can be increased. However, at best it is likely to be no more efficient than perhaps the venture capital investment decision-making process. Clearly, effectiveness in the transfer and commercialization of technology can only be increased with additional investment related to facilitation quality and efforts, but the trade-offs between costs and benefits must be weighed carefully. A second and related point is that a fair amount of further experimentation will be necessary before the concept of a market-driven R&D arrangement with broad applicability will emerge. At this stage, public-private R&D collaborations appear to be generating unique, specially customized relationships. Third, the knowledge, ideas, and know-how related to technology transfer that has been generated by the numerous national organizations involved with R&D collaboration and consortium building should be formalized and institutionalized as a research program, perhaps within business schools. In this regard IC^2 Institute is already leading the way. Finally, at the policy level, greater incentives are needed to foster and reward innovation leading to development, experimentation, and implementation of improved market-driven R&D programs.

The purpose of the AmTech experiment is to demonstrate that it is possible to compress the time for technology commercialization from a NASA field center through employment of value-added facilitators. The facilitator's unique expertise should accelerate the process of technology transfer and commercialization, promote dual-use technology development, and contribute to national and regional economic competitiveness. The metrics for success are enhanced economic development, technology transfer to existing companies, technology transfer to new firms, and the transfer of knowledge. If these experiments are successful, NASA will transform itself from its role as a civilian fixture of the Cold War to a national technological engine for economic growth through the accomplishments of its aeronautics and space missions. It may also provide a model for the transfer and commercialization of high technology from public research institutions that can be widely replicated.

Notes

1. One of NASA's most innovative attorneys, Rick Dunn, left NASA for DoD, where he succeeded in obtaining the "other transactions authority" used to create the TRP program.

2. These data are derived from AmTech's experience in business facilitation and from the official alliance notification files maintained by the U.S. Department of Justice.

References

Alic, John A., Lewis M. Branscomb, Harvey Brooks, Ashton B. Carter, and Gerald L. Epstein. *Beyond Spinoff.* Boston, MA: Harvard Business School Press, 1992.

Alic, John A. "Science, Technology, and Economic Competitiveness," this volume.

Branscomb, Lewis, ed. *Empowering Technology*. Cambridge, MA: MIT Press, 1993.

Gibson, David, and Everett Rogers, *R&D Consortia on Trial.* Boston, MA: Harvard Business School Press, 1994.

13

General Trends in International Technology Partnering: The Prospects for European Economies in Transition

John Hagedoorn and Bert Sadowski

Introduction

This paper explores some of the first steps necessary for a further assessment of the challenges, dangers, and possible benefits of strategic technology partnering between companies from the developed economies and the European Economies in Transition (EET), the former East European communist countries. We will first discuss some of the major characteristics of the current international distribution of scientific and technological capabilities among major international trading blocs. Against this general background, we examine international inter-firm technology cooperation between companies from the developed economies, the newly industrialized countries, the least developed countries, and in particular the EET.

The general overview of the international distribution of technological capabilities is followed by a brief discussion of data that provide some insight into the major characteristics of international strategic technology alliances. The next section discusses the distribution of alliances with EET firms in greater detail. Particular attention is paid to the sectoral distribution of these alliances, the changes in their patterns and rates of growth over time, and the concentration of strategic technology alliances in particular countries. The final section considers the implications of our empirical findings for the EET's prospects of "catching up" through technology partnering.

The International Distribution of Technological Capabilities

It is difficult to compare the technological capabilities of different countries because of the variety of methods of data collection in use on indicators such as R&D input and patents. For instance, the standard OECD definition of R&D developed in the 1960s is still not accepted or applied by most countries outside the OECD. It is only very recently that a number of EET countries have begun to change their official statistics according to OECD standards (European Commission, 1994). To some extent the same is true for the measurement of R&D output, as patent approval systems outside the OECD countries differ to a very large extent.

In spite of such statistical difficulties, however, the broad picture is abundantly clear: global R&D is highly concentrated in the rich OECD countries. In fact, more than two-thirds is probably within the "Triad" of the United States, Japan, and the European Union (EU) and the European Free Trade Association (EFTA). The Third World, in the most generous estimates, accounts for only about 6 percent, and without China its share of global R&D is probably less than 4 percent. The share of the erstwhile communist countries is much bigger, but about three-quarters of this was accounted for by Soviet military-space R&D, which is now in sharp decline. The only major exception in this rather lopsided international distribution is the group of four Dynamic East Asian Economies (DAEs). Their R&D expenditures have increased rapidly in the 1980s, and not only South Korea but also Taiwan, Hong Kong, and Singapore now have a level of R&D intensity comparable to that of most European countries (Freeman and Hagedoorn, 1995).

Figure 13.1 provides an overview of recent trends in the R&D intensity of major international trading blocs. Although the categorization of countries is somewhat different from the one used in this paper, it clearly illustrates the dominance of the OECD countries. Japan, the United States, and Europe have an R&D intensity above or close to 2 percent of GDP. The DAEs demonstrate a steady increase during the eighties from less than 1 percent to about 1.8 percent of GDP, thereby overtaking some of the less developed countries in southern Europe. The majority of developing countries have a very stable R&D intensity of well below 1 percent. As already indicated above, a striking trend is the rapid decrease of R&D intensity in the group of former communist countries in Europe: R&D intensity here over the past decade has tumbled from more than 2 percent to less than 1 percent of GDP.

These data and other sources such as NSF (1989) and OECD (1992) suggest the following major trends in global R&D over the past decades:

- R&D has rapidly increased in Japan, Germany, and a few other European countries.
- Slower growth patterns are found in the United States, the United Kingdom, and other OECD countries.
- Stagnation or decline is seen for much of Third World R&D, especially for the group of heavily indebted countries.

Figure 13.1: Trends in R&D Intensity (Total R&D as a Percentage of GDP)
Source: EC, 1994, p. 16

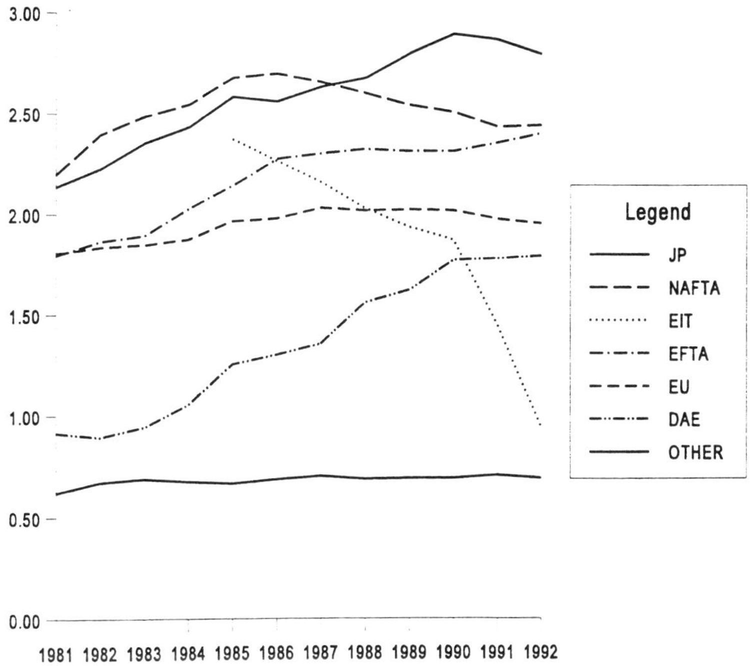

- Rapidly rising R&D expenditures and R&D intensities are found in the dynamic Southeast Asian countries. There is also some increase in China and the "second tier" Southeast Asian countries.
- In recent years there has been a sharp decline in former Soviet and East European R&D.

This last point is illustrated by the shift in the R&D intensity of countries such as Bulgaria, the Czech Republic, Hungary, Poland, Romania, the Russian Federation, and Ukraine since 1985 (see Figure 13.2). At the outset one has to remember that most former communist countries have not yet fully complied with the OECD definition of R&D. Therefore, the extremely high R&D intensity of countries such as Bulgaria and Czechoslovakia (higher than the R&D intensity of Japan for the larger part of the second half of the eighties) is probably inflated. While transitional, then, these figures do illustrate the strong decrease in R&D intensity in all of these countries since the end of the eighties with the exception of Poland, the R&D intensity of which has remained fairly stable. For some countries, i.e., Hungary, Russia, and Ukraine, this decline reduced the present R&D intensity to less than half of the R&D intensity during the mid-eighties.

Figure 13.2: Trends in R&D Intensity (Total R&D as a Percentage of GDP) in European Economies in Transition
Source: Calculated from EC, 1994, Statistical Annex Tables 1.2 and 1.7A

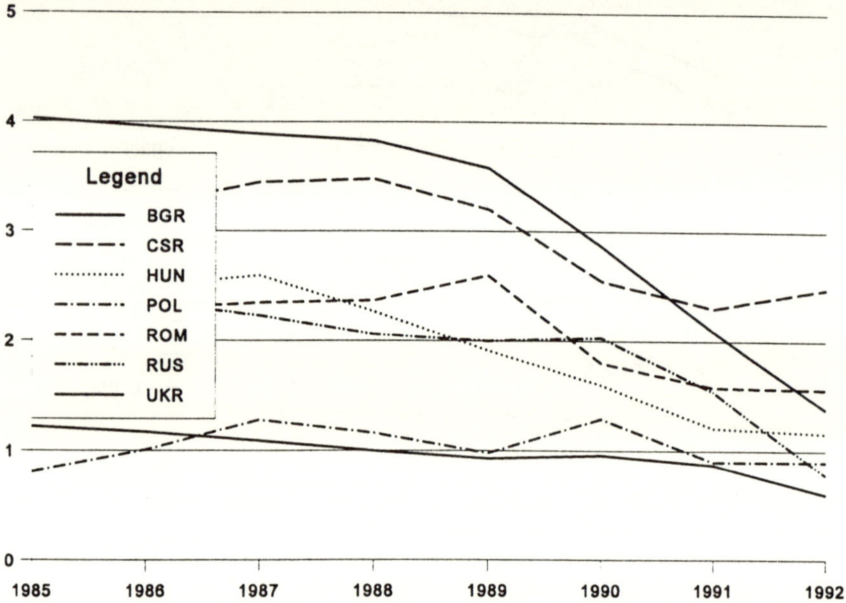

The world distribution of patents as an output indicator of technological capabilities is similar to patterns in the international distribution of R&D inputs. The United States and Europe, with similar patent systems, are the largest centers of both marketing and sophisticated inventive activities, and as such provide a yardstick for measuring inventive output. Other national patenting systems are too divergent to simply count patents in each separate country only. Freeman and Hagedoorn (1995) report that the vast majority of U.S. patents (taking into account the overrepresentation of the United States) granted during the eighties are held by the Triad. The combined share of Third World countries is probably between 1 and 3 percent of a truly global index of patents; i.e., even lower than their R&D share. The share of the former USSR and other Eastern European countries is also lower than for R&D. This is caused by their until recently extraordinarily heavy expenditures on military systems, which yielded hardly any foreign patents for both security and commercial reasons. Again the group of Dynamic East Asian Economies seems to have been able to catch up as their patenting activity also began to rise very rapidly in the late 1980s. Similar patterns for European Patent

Office patent applications during the eighties and early nineties are found in EC (1994).

Patents are of course different in nature from R&D activities per se. Patents are a measure of inventive output, not a direct measure of resource commitment. R&D activities are wider in scope than those inventive activities which yield patents. Nevertheless, at the country level there is a strong correlation between R&D statistics and patent statistics (Dosi, Pavitt, and Soete, 1990). The fact that both measures yield very similar results for the international distribution of technological activities is therefore a rather strong confirmation that the share of the EET (as well as that of the least developed countries) in activities generating new technologies is indeed low and, moreover, has been markedly decreasing in recent years.

For the developing countries, it is often suggested (World Bank, 1991 and OECD, 1992) that the import of technology is the most critical resource for development. Imports of technology are of course extremely important for all countries and not just for the former communist countries and Third World countries. But this does not mean that imported technology should be seen as an alternative to indigenous R&D and other scientific and technical activities. Internal R&D and science and technology are essential for the efficient import of technology. Otherwise the imported technologies can neither be understood, adapted to local conditions and resources, nor improved to keep pace with world competition. To a large extent, then, import and performance are complementary activities in technology, not alternatives to one another.

In this context we can relate the international distribution of technological capabilities and the performance of R&D to education and training. Pavitt (1990, 1992) has shown that the output of basic research is an important aspect of the creation of a capacity for the assimilation of new science and technology, for instrumentation and for problem-solving. The four Dynamic East Asian Economies have demonstrated perhaps most clearly that, as in Japan, successful industrialization depends on a combination of indigenous scientific and technological activities, with a major commitment to education and basic research and imports of technology. In that sense the "crisis" of the R&D infrastructure in the former communist countries poses a serious problem for both furthering the development of their technological capabilities and their search for opportunities in international inter-firm technological cooperation. Recent research on the effects of strategic technology partnering (Hagedoorn and Schakenraad, 1994) suggests that successful networking in technology is strongly correlated with R&D performance, not an alternative to it. The implication for the EET is that successful technology transfer through strategic technology alliances can only take place if this transfer is embedded in an existing indigenous technological strength that is reinforced by cooperation. Maintaining a certain level of basic research capabilities and developing a base of scientific and technological skills is a prerequisite for firms from the EET that wish to enter into the arena of international strategic technology partnering.

Patterns in International Strategic Technology Partnering

As shown in Figure 13.3, international inter-firm strategic technology partnering has increased substantially during the past decades. Particularly in the new core technologies—information technology, biotechnology, and new materials—international strategic partnering has escalated since the late seventies and early eighties. During the second half of the eighties, the growth of newly formed alliances in new materials and biotechnology decreased again somewhat, but the early nineties appear to have ushered in another period of growth in strategic technology partnering in these high-tech sectors.

Figure 13.3 also demonstrates that these core technologies together are responsible for a very large part of international strategic technology partnering. Since the early eighties the sum of combined core technology alliances has outnumbered those made in all other sectors of industry put together. Despite the growth of newly made strategic technology alliances in more "traditional" sectors,

Figure 13.3: Growth of Newly Made International Strategic Technology Alliances, 1970–1993
Source: MERIT-CATI

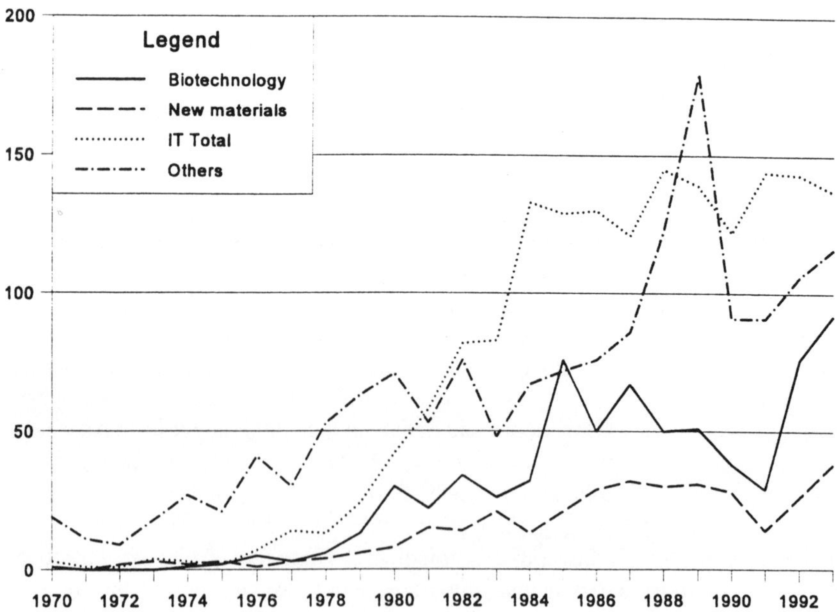

the vast majority of new international alliances over the past two and a half decades were made in the new core technologies.

These three new core technologies affect the economic and technological performance of companies and countries. In addition, these new core technologies are expected to shape the technological landscape for the coming decades. In an international comparison of the participation in strategic technology partnering, therefore, it makes sense to examine to what extent the international distribution differs with respect to the participation in core technology partnering. Companies that are active in partnering in core technologies are engaged in technology sharing at the cutting edge of technology. Table 13.1 reveals that of all international alliances made since 1980 more than 65 percent were made for new core technologies. Furthermore, the share of new core technologies has gradually increased during the past decade. Intra-Triad cooperation in particular is dominated by these core technologies, which have expanded their share from 70 percent in the early eighties to more than 75 percent during the early nineties. Cooperation between companies from the Triad and the DAEs fluctuated considerably during the past fifteen years, but in the most recent years nearly 90 percent of these international alliances have been made in the new core technologies. In contrast with these findings are the results for collaboration with firms from the EET. There we see that since the mid-eighties nearly 60 percent of the alliances have been made in fields other than information technology, biotechnology, and new materials.

Table 13.1: International distribution of international strategic technology alliances, shares of core technologies (information technology, biotechnology, new materials) in total, 1980–1993.

	1980–1984	1985–1989	1990–1993
Intra-Triad	70.7%	72.8	75.7
Triad-DAEs	62.2%	40.0	89.7
Triad-EETs	–	40.7	40.5
Total	66.1%	67.2	68.7

Source: MERIT-CATI

Looking at the international distribution of these alliances (see Figure 13.4), we notice that the increased growth of international strategic technology alliances is due largely to inter-firm partnering within the Triad. Beginning in the late seventies, the number of new alliances in the Triad grew rapidly. Some stagnation in this growth occurred in the late eighties, but growth picked up again in the early nineties. In total these intra-Triad alliances account for about 90 percent of all international technology partnerships. The dominance of the developed economies becomes even more apparent when one takes into account that the category of "other combinations" consists to a very large extent of inter-firm cooperation

Figure 13.4: Growth of Newly Made International Strategic Technology Alliances between Partners from Different International Regions.
Source: MERIT-CATI

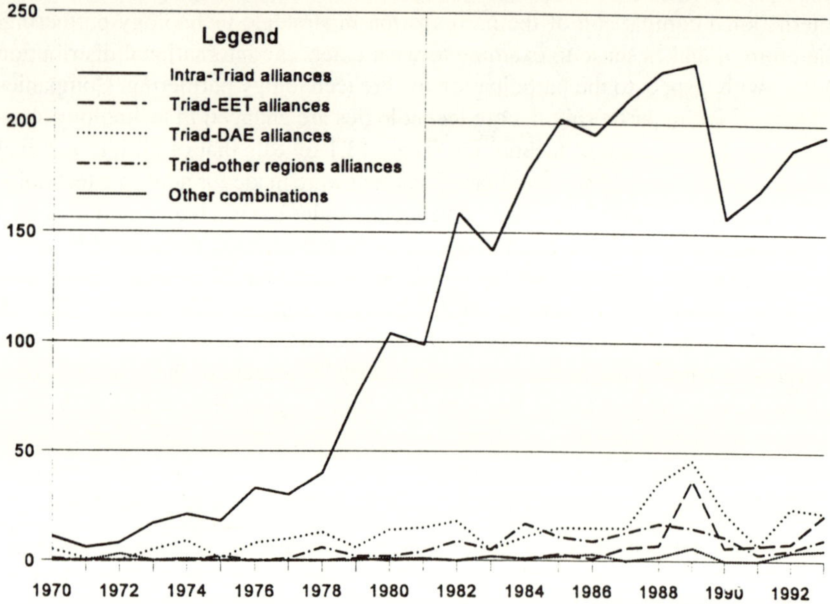

with firms from other countries in the OECD. The combination of Triad firms with companies from the DAEs was almost nonexistent during the seventies but has clearly grown since the eighties. Early comparisons with Triad-EET partnering are difficult to make. During the seventies and early eighties state-dominated international venturing excluded many agreements from our data because private companies hardly existed in the former East European countries. However, since the late eighties there has been growth in the number of alliances with firms from the EET.

Strategic Alliances with European Economies in Transition

The decline in R&D effort and performance in many EET countries during the early 1990s could have critical long-term effects on their competitive strength. This concerns those responsible for science and technology policy (Radosevic 1994). In addition to current problems with respect to privatization and market adjustment of firms, this decline can be traced back to an institutional structure based on central planning. The centralized science and technology system in the EET countries developed over the past forty years led to a decline of the R&D

system in the late eighties (Pavitt and Hanson 1987, Gomulka 1990, Kornai 1992). This decline is apparent if one notes the number of patents taken out in Europe and the United States or the share of scientific publications (EC 1994).

As suggested above, the transfer of technology can mitigate this decline in the EET. Strategic alliances between leading multinational firms and local enterprises in these countries provide a way for them to gain access to technology, and an opportunity for these "latecomer" enterprises—so characterized because they have been located outside the loci of world innovation and R&D for non-defense industries—to catch up with leading firms in the West. For civilian technologies in particular, enterprises in the EET do not face demanding buyers in their domestic markets on a large scale, and are cut off from leading industrial clusters and important marketing networks (Radosevic 1993, Hobday 1994). To overcome these "negative externalities," strategic alliances can be used to build successful inter-firm networks and to improve corporate performance. However, the problems remain of how to interest partners for technology transfer if there is little perceived return and little understanding of how to capitalize on the learning opportunities that exist in the transitional economies.

Strategic alliances with firms from capitalist countries were the exception in Eastern Europe prior to the opening of their economies to the West. Eastern Europe's share of international strategic alliances has been, in general, very small during the 1980s. A closer examination, however, shows that their total number increased rapidly in the late 1980s and early 1990s in contrast to the early 1980s. In 1989, the establishment of strategic alliances peaked: one-third of all alliances in the EET countries were formed in this year. The subsequent two years saw the number of alliance formations decrease sharply, with a record low in 1992. One year later the establishment of strategic alliances began to increase again (see Figure 13.5).

While the majority of alliances with firms from the EET have been in areas other than the new core technologies, it is interesting to note that the first peak in the new alliances, in 1989, was primarily due to alliance formation in the information technology sector. Most of these are state-regulated alliances created to improve the telecommunications infrastructure in the EET. The bulk of strategic technology alliances, however, has been in the aircraft industry, followed by the chemical and space industries. These sectors are in turn followed by heavy electrical equipment (HeaE), automotive (Auto), and engineering contracting (EngC) (see Figure 13.6).

The opening of the defense and aircraft sectors in Russia, Ukraine, and the Czech Republic played a major role in the second wave of alliance formation in 1992 and 1993. In these two years, the largest number of alliances was recorded in these countries for the aircraft and defense sectors.

By looking at the number of strategic alliances according to countries, see Figure 13.7, these findings receive further support. Two-thirds of all alliances were established in the former USSR, the majority of them after 1989. Hungarian firms

Figure 13.5: Growth of Newly Made International Strategic Technology Alliances with Companies from EET Countries, 1980–1993.
Source: MERIT-CATI

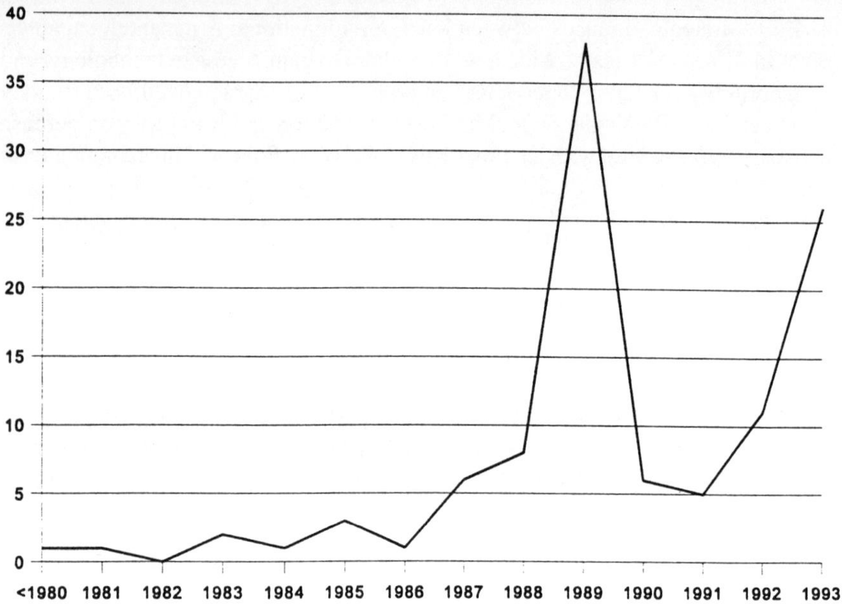

Figure 13.6: Sectoral Distribution of Newly Made International Strategic Technology Alliances with Companies from EET Countries, 1980–1993.
Source: MERIT-CATI

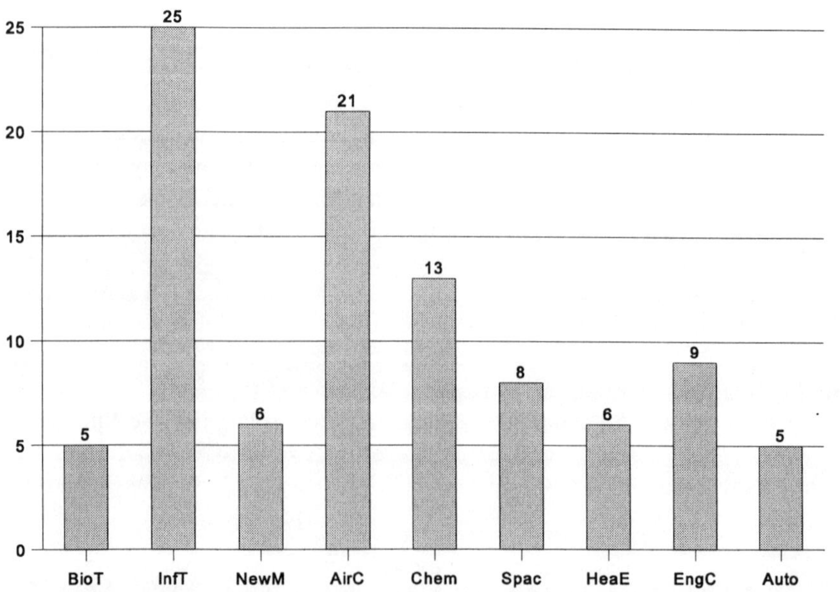

Figure 13.7: Distribution of Newly Made International Strategic Technology Alliances with Companies from EET Countries, 1980–1993.
Source: MERIT-CATI

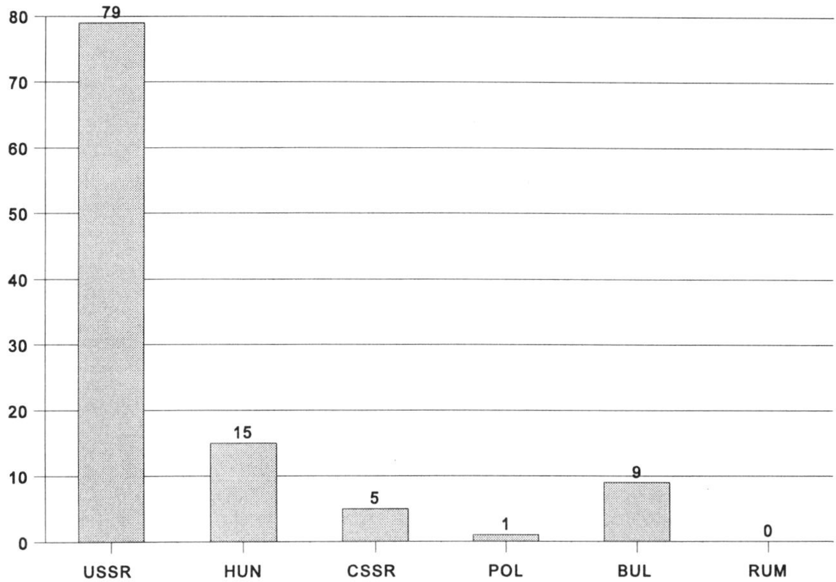

are second in terms of their attractiveness for strategic alliances, followed by Bulgarian firms and companies from the former Czechoslovakia. Firms in Poland and Romania are apparently lagging in their propensity to engage in strategic alliances.

During the 1980s and the 1990s, there have been only minor changes in the characteristics of the firms from EET countries that are involved in strategic technology alliances. Prior to 1989 these firms came exclusively from the state-controlled sector. After 1989, firms from the private sector gradually became involved. However, their number has been relatively small compared with the number of partnering firms from the state-controlled sector. In the former USSR this certainly has been due to the slow pace of privatization and market adjustment of the state-controlled sector. In other EET countries, the restructuring has not yet led to firms from the private sector that are able to pursue cooperative research and development with foreign firms.

It seems, however, that privatization and market adjustment have been beneficial for research institutes and laboratories in EET countries with respect to their propensity to engage in strategic technology alliances. Until 1989 these institutions were rarely independently involved in joint R&D efforts with firms from non-EET countries. They were part of state-controlled firms or were accountable to governmental ministries. In the early 1990s this began to change. After 1992 the number of research institutes and laboratories participating independently in strategic alliances began to grow. Between 1992 and 1993 it doubled and included more than half of all partnering firms of EET countries involved in strategic technology alliances (see Figure 13.8).

Figure 13.8: Characteristics of Firms in EET Countries That Are Part of an International Strategic Alliance, 1988–1993.
Source: MERIT-CATI

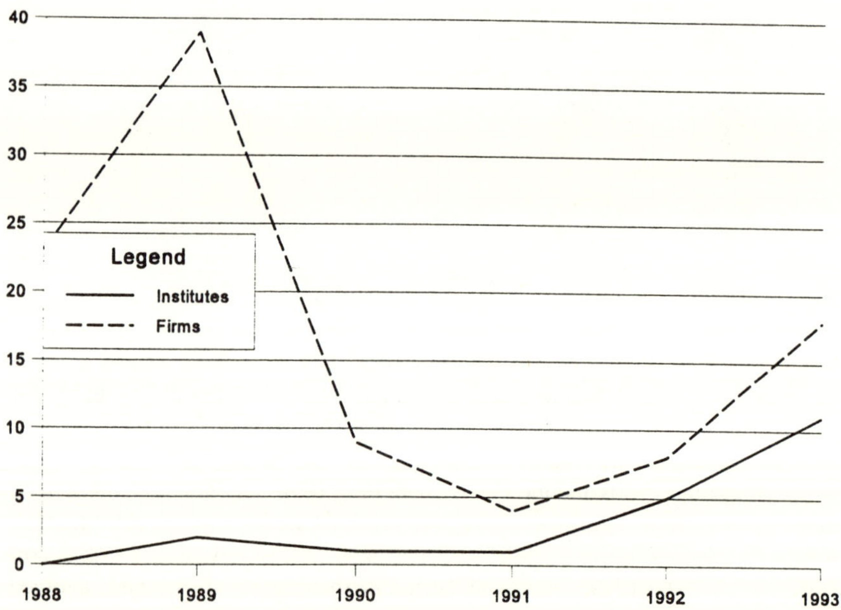

Conclusions

It is clear from an examination of the international distribution of technological capabilities and participation in inter-firm technology partnering that most countries outside the developed economies and some advanced newly industrialized countries will not quickly be able to catch up with their competitors. Strategic technology partnering and advanced technology transfer appear to be as heavily concentrated within these developed countries as other aspects of technological development such as R&D expenditures and patenting. Many companies that originate from developing countries and the EET are virtually locked out from inter-firm partnerships that concentrate on new core technologies such as information technology, new materials, and biotechnology. This implies that this pattern not only reflects the present distribution of worldwide technological capabilities, but it will also enforce the concentration of technological capabilities for new technologies in the foreseeable future. There are, of course, examples of high-tech inter-firm partnering of companies from the OECD countries with EET firms. However, many of these cases are actually examples of collaboration that focus less on joint R&D or advanced technology transfer than on market access or joint manufacturing.

The main exception to the general pattern of unequal development is the role played by companies from some East Asian countries. Although their situation should not be overstated, it is clear that these countries do play some role in strategic technology partnering. By and large this role appears to parallel their increased technological capabilities as indicated by their R&D efforts and their share in international patenting.

Companies that intend to participate in international technology collaboration must develop a level of indigenous technological capabilities that approaches worldwide technology standards. In order to even enter the game firms must possess enough assets to enable them to find capable partners. To successfully play the game firms must capitalize on the learning process that comes with inter-firm partnering. These learning capabilities have to be built up within companies. However, in the case of technological development, they partly depend on a broader technological infrastructure that involves national innovation systems with research institutes, universities, and technology spill-over effects from other extant companies.

Therefore, maintaining an advanced civilian technological infrastructure is a prerequisite for EET firms wishing to participate in inter-firm technology partnering at or close to the technological frontier. Divestments of parts of the existing infrastructure without parallel investments in new segments, brain drain to the more advanced economies, and low scientific and technological productivity are threatening the long-term positioning of firms from EET countries. The former German Democratic Republic is a case in point. Rapid deindustrialization in conjunction with the imposition of West German institutions on the East German R&D infrastructure has had a damaging impact on the R&D performance and technological capabilities of East Germany. By 1993 R&D expenditure in East Germany amounted to only 50 percent of the West German level. Moreover, R&D personnel was reduced to 30 percent between 1989 and 1993 (Meske, 1993 and 1994). This tendency was furthered by a migration of scientists and engineers to the well-developed R&D infrastructure in West Germany. This tendency has the potential to foster an East-West divide of R&D institutions in Germany in the long term. If these developments, including the migration of technical personnel and the closing down of large parts of the technological infrastructure, are reproduced on a much larger scale for the EET as a whole, long-term structural inequalities in technological capabilities affecting a substantial part of the world economy will result.

There are no easy solutions for the EET, and strategic technology partnering is no panacea. A comparison with the advanced DAEs and an analysis of the current situation in the EET both suggest that companies from the EET have to be supported by a well-developed technological infrastructure. In particular, firms that aim to compete beyond the world of the less-developed countries require an improved indigenous scientific and technological infrastructure before they can begin to gradually appropriate high-tech capabilities through inter-firm alliances.

References

Dosi, G., Pavitt, K., and L. Soete. *The Economics of Technical Change and International Trade*. New York: Harvester Wheatsheaf, 1990.

European Commission. *The European Report on Science and Technology Indicators 1994*. Luxembourg: European Commission, 1994.

Freeman, C., and J. Hagedoorn. *Convergence and Divergence in the Internationalization of Technology*. In *Technical Change and the World Economy*, edited by J. Hagedoorn, 34–57. Aldershot, England: Edward Elgar, 1995.

Gomulka, S. *The Theory of Technological Change and Economic Growth*. London: Routledge, 1990.

Hagedoorn, J., and J. Schakenraad. "The Effect of Strategic Technology Alliances on Company Performance." *Strategic Management Journal* 15, no. 4 (1994): 291–311.

Hobday, M. *Innovation in East Asia: The Challenge to Japan*. London: Edward Elgar, 1994.

Kornai, J. *The Socialist System: The Political Economy of Communism*. Oxford: Clarendon Press, 1992.

Meske, W. Die Umgestaltung des ostdeutschen Forschungssystems—Eine Zwischenbilanz. 93–401. Berlin: Wissenschaftszentrum Berlin für Sozialforschung, 1993.

———Forschung und Entwicklung in Deutschland—Institutionelle, regionale und soziale Disparitäten, in BISS public, Wissenschaftliche Mitteilungen aus dem Berliner Institut für Sozialwissenschaftliche Studien. 97–106. Berlin, 1994.

National Science Foundation. *Science and engineering indicators 1989*. Washington, D.C.: NSF, 1989.

OECD. *Technology and the Economy*. Paris: OECD, 1992.

Pavitt, K. *What Do Firms Learn from Basic Research?* In *Technology and the Wealth of Nations*, edited by D. Foray and C. Freeman. London: Pinter, 1992.

Pavitt, K., and P. Hanson. *The Comparative Economics of Research Development and Innovation in East and West: A Survey.* New York: Harwood Academic Publishers, 1987.

Radosevic, S. *International Technology-Transfer Policy Revisited: From Contract Bargaining to Sourcing.* Brighton: SPRU, University of Sussex, 1993.

———*Strategic Technology Policy for Eastern Europe.* Brighton: SPRU, University of Sussex, 1994.

World Bank. *The Challenge of Development.* Oxford: Oxford University Press, 1991.

Section Four

Inside the Commercializing Firm: Legal and Structural Issues

14

Spin-offs As a Restructuring Strategy for Post-Socialist Enterprises

David Ellerman

Introduction

Privatization is not enough. The enterprises of the post-socialist world are in need of massive restructuring. The spinning off of assets into new firms has emerged as a particularly appropriate restructuring strategy in the post-socialist economic environment. Therefore this paper will focus on the advantages and drawbacks of spin-offs, on the financial structures of spin-offs, and on government policies to promote spin-off restructuring.

Advantages of Spin-Offs

Unbundling Assets from the Socialist Era

In view of the highly distorted economic environment of the socialist era, it would be remarkable if the enterprises of the post-socialist world could emerge into a market economy without major restructuring. Assets had often been conglomerated into uneconomic masses that need to be unbundled and rebundled into more appropriate units. A spin-off provides the opportunity to assemble a synergistic set of assets which can then function as a new and separate business unit.

Greater Efficiency of Small and Medium-Sized Firms

Socialism usually suffered from the disease of gigantism. In economic terms, socialism passionately believed in the possibilities of reaping large returns to scale in most industries. This belief seemed rather misguided even in yesteryear, and it is certainly misguided in modern times. New technologies have usually allowed quite efficient units in smaller scales (e.g., the mini-mills in the steel industry). In human terms, small and medium-sized firms have always had the advantage. In general, organizations should strive to have greater proximity and visibility between the efforts expended by teams and the fruits of those efforts. This is clearly more possible in smaller organizations than in large ones where the results of a team's effort can easily be lost in the immensity of the operation.

A reduction in the scale of operations can in part be achieved by dividing an enterprise into profit centers. However, profit centers always seem vulnerable to accounting manipulation (e.g., the vagaries of transfer pricing) and "political" interference from top management. By spinning off the profit center as a separate business unit, the unit can achieve a smaller scale, a greater measure of autonomy, and a greater ability to capture the fruits of its efforts with less threat of expropriation by the parent enterprise.

Some care must be taken as to how an enterprise is broken into segments. Given a vertically integrated monopoly, it would be of little use to vertically disintegrate it into a series of smaller monopolies, each buying an input or selling an output to another. That would only reduce the scale of the monopoly and replace an internal transfer with a bilateral monopoly bargaining situation between the units. It would be better to break up a firm horizontally according to product lines and/or regions so that there are some contested markets at least around the edges.

New Management

The managerial dynamics of spin-offs are quite interesting. The older top management cannot provide the day-to-day management in a number of spin-offs even if they are maintained as subsidiaries, not to mention if they are spun off as separate firms. Thus this sort of a breakup will always dilute and disperse some of the direct control of the original top management. Ordinarily some of middle management would emerge as the management of the spun-off unit, although it is possible for an outsider to be recruited for the job. Top managers might drop down into a particularly promising subsidiary, but would then be giving up their previous position spanning the whole enterprise.

In view of the general need for new management in the post-socialist world and of the organizational inertia of the old structures, the break-up restructuring of spinning off various business units may prove a more effective means of getting new management in place than fighting to directly replace the top management of the parent company. In any case, methods need to be devised to motivate the parent, including compensating the parent for explicit release of rights to technology previously developed in the parent and to be used by the spin-off; giving the parent a right of first refusal for subsequent financing; preferential utilization of the parent's labor as the spin-off grows; or giving the parent the right of first refusal on licensed production.[1]

Advantages of Asset Deals

Spin-offs usually start as partially or wholly owned daughters of the parent enterprise. Assets are "dropped down" into a new corporate shell in return for some shares in the daughter, some cash put in by the other owners of the daughter, or the assumption of some explicit liabilities from the parent. The important aspect is that the spin-off is a new corporate entity taking assets from the old entity. It is not a continuation of an old corporate form with a transfer of old shares to new owners (e.g., as in mass privatizations or in some direct sale transactions). Workers need to leave the old legal entity and be rehired by the new one. This is very important in breaking up the set of psychological and institutional expectations (e.g., lifetime employment) that attached to the old enterprise. The old enterprise might also have some hidden liabilities (e.g., for past environmental damage) that could be left behind by organizing the spin-off. Hidden liabilities for the old firm that might emerge in the future would not fall on the spun-off firm.

Tapping the Vitality of the New Private Sector

One of the striking features of the early transitional economies is the phenomenal success of the de novo private sector as opposed to the privatized (or to be privatized) sector. Why the massive flow of entrepreneurial talent and resources to the new private sector? Plainly, because that is where the returns on the investment of entrepreneurial resources are the highest. Elementary economics shows that for the overall return to those resources to be the greatest, the rate of return must be equalized between the two sectors.

Thus we arrive at a plausible design goal of the restructuring program in a post-socialist economy: to sharply increase the rate of return to entrepreneurial activity in the to-be-privatized sector so that entrepreneurial resources will stay in or flow into that sector to carry out the needed restructuring and revitalization. How can this be accomplished?

In view of the amount of downsizing, rationalization, and rehabilitation that is needed in the old enterprises, entrepreneurs are understandably reluctant to invest their time, energy, and other resources in the old firms. But many of these problems are alleviated in spin-offs. One strategy is therefore to structure spin-off transactions that are so attractive to entrepreneurial managers from inside or outside the old firm that the spin-off option will compete with the alternative of investing those entrepreneurial resources in the new private sector.

Legitimizing Spontaneous Privatization

There is another way to view this problem of tapping the vitality of the new private sector. When managers in the to-be-privatized sector exhibit "premature entrepreneurship," this is known as "spontaneous privatization" in much of the Western privatization literature. Instead of trying to stop economic spontaneity in favor of a state-planned and state-administrated privatization program, a government might consider how to channel and legitimize spontaneous privatization so that privatization would be driven by a more natural economic engine.

There are a number of guidelines that, if observed, would help to legitimate an otherwise spontaneous process of privatization.

Valuation: The government needs to establish a standardized procedure for the valuation of the assets involved in asset drop-down and other transactions.

Reasonable credit terms: The government needs to establish acceptable credit terms for seller-supplied credits or lease buyout transactions.

Transparency: The transaction should be wholly "above the table" and open to the scrutiny of public authorities.

Fairness: The top managers should not be able to structure the transaction to their benefit only; a majority or supra-majority of the affected workforce should agree to the transaction.

Genuine payment: An installment payment scheme in which the payments were reloaned to the debtor or in which the debt was continually "rolled over" would not be a genuine payment scheme.

Settlement with creditors: The system of corporate law should establish regulations so that major assets may not be stripped out of an indebted entity without due involvement of the endangered creditors.

With such a system of regulations in force, spontaneous forms of privatization such as spin-offs could be a legitimate outlet for entrepreneurial forces within the to-be-privatized sector.[2]

Pure Plays for Outside Investors

Strategic (or trade-related) investors might only be interested in a relatively small and well-defined part of the old enterprise. They have no desire to take over the entire old firm and then close down or sell all the other parts just to get the desired segment. In any case, foreign investors would be ill-advised to consider such politically unpopular actions the "cost of doing business" in the to-be-privatized sector. Thus if the government does not allow the foreign investors to "cherry pick" the viable and relevant parts of the old firm (a "pure play") and to repackage them in a spin-off, then the foreign investor would probably be better advised to build his own pure play, e.g., by making a greenfield investment in the new private sector.

This point is also relevant to the hopes of many Western advisors and their post-socialist counterparts that strategic owners will emerge on the secondary stock market following a voucher privatization program. Leaving aside the high transaction costs of pulling together a clean majority ownership package on the secondary market, why would investors want strategic ownership of a huge misbundled set of assets and a corporate form laden with socialist expectations? Unless there were company-specific reasons to do otherwise (e.g., to get access to a distributional network), a rational investor would insist on a pure play on the relevant assets in a new corporate form, namely on a spin-off of the appropriate business unit. That transaction could either be an asset deal or a private placement of the daughter shares from the parent to the investor. The secondary stock market for the shares in the old firm (floated in the voucher privatization) would thus be largely irrelevant to the strategic investor. A corollary is that some post-socialist governments may need to reassess their expectation that the messy job of restructuring the old voucher-privatized firms will be undertaken by foreign strategic investors buying shares on the public secondary market.

Post-Voucherization Restructuring

In the early debates about privatization of the post-socialist economies, "private ownership" referred to real owners who would control and restructure the privatized enterprises. But with the widespread use of voucher-based mass privatization, private ownership was redefined to mean the private ownership of shares in quasi-public companies ("quasi-public" in the Berle and Means sense[3] of having widely dispersed passive shareholders). The privatization debate finally discovered the difference between privately owned shares and privately owned companies. Since the public "voucherized" companies had no effective owners, once again the question arose of fulfilling the original aims of privatization, namely restructuring the companies to function profitably in a market environment.

In the absence of foreign investors or domestic entrepreneurs able to invest in a company, the management-employee buyout (MEBO) is one of the few feasible methods of privatization that results in a set of owners able to act as a coherent unit (particularly in small and medium-sized firms) and bear the costs of their own actions or inaction. Thus a hard budget constraint can enforce real restructuring in a MEBO-privatized firm.

One restructuring option for a large voucherized firm is, in effect, to sponsor its own "second-stage" MEBO privatization program by facilitating the spinning off of its business units to their management and staff. Thus the process of going from state ownership to real private ownership could be seen as a two-stage process: the first stage voucher privatization program and then the second stage MEBO privatization program.

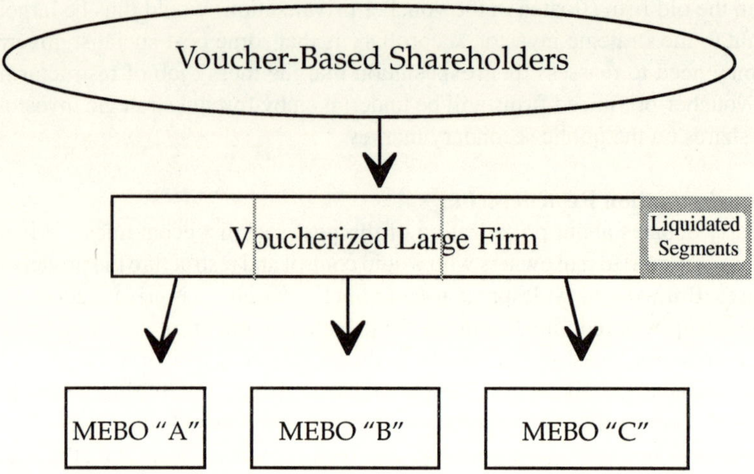

Second-Stage MEBOs

The original voucherized firm could survive as a mini-mutual fund channeling the payments from the spun-off MEBOs to the voucher-based shareholders.

Alternatively, the voucher-based shares could be redeemed for a bundle of preferred stock and other hybrid or debt instruments issued by the spun-off MEBOs (essentially an equity-for-hybrids or equity-for-debt swap). The original firm could then survive as a type of group headquarters or be liquidated.

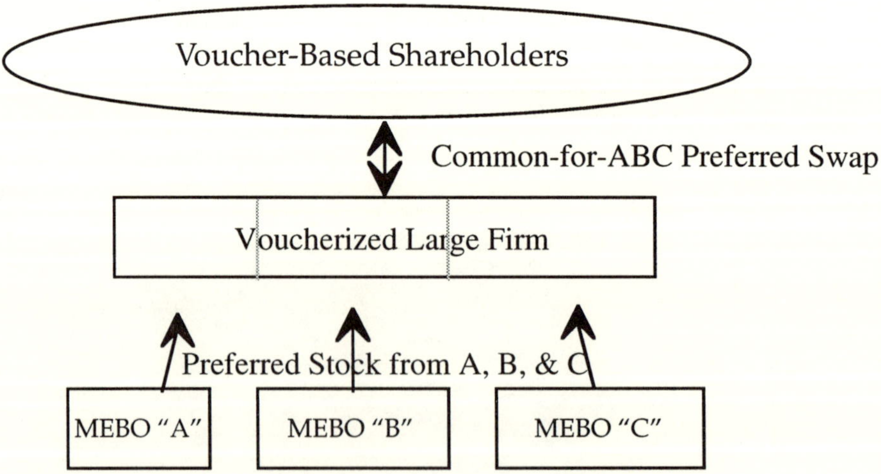

Redemption of Voucher-based Shares for Bundle of Preferred Stock from A, B, & C

Drawbacks of Spin-Offs

Less Financial Clout

In a market economy, the spun-off units would have greater creditworthiness since they have greater economic viability. But in the transitional economies, credit may in fact follow political directives, and the whole original enterprise would have more political clout and thus more financial clout than the separate parts. This might well be seen as a drawback of the spin-offs.

The spun-off units need not, however, stand alone. To use a political analogy, several segments of the former Yugoslavia "spun off" as separate countries which then sought to join, as soon as possible, the supra-national organization of the European Community. In a similar fashion, segments of a large enterprise might spin off but then become a part of a larger financial-industrial group or "post-socialist keiretsu." In that manner, they might hope to reap some of the benefits of a spin-off while not entirely giving up the financial and other advantages (e.g., access to social assets) of a larger grouping.

Bankruptability

It is increasingly unrealistic to expect post-socialist governments to let large voucher-privatized firms sink into bankruptcy. In a strong economic environment, the United States government still stepped in and bailed out large firms such as Chrysler and Lockheed when they were threatened with bankruptcy. In the middle of a transitional depression, the post-socialist governments can hardly stand by as numerous large firms go bankrupt. Without an alternative, the old soft budget constraint threatens to be continued in new forms.

Break-up (or bust-up) restructuring separates a large firm into a number of parts spun off from the parent enterprise. Some of the spun off units may prosper and absorb workers from the other parts that have to rationalize their workforce. Since the spun-off firms are much smaller than the original enterprise, each of them is more bankruptable in the sense that it could go bankrupt without creating as much sudden unemployment and social shock. From the viewpoint of the individual spun-off unit, that bankruptability may be a drawback of the spin-off strategy.

The top managers and the workers of the original enterprise understand instinctively that the ability to extract continuing subsidies and support from the local and national governments depends on the size and clout of the enterprise—which may be lost after bust-up restructuring. Thus bust-up restructuring presents a dilemma to the top managers and workers:

(1) to break up the enterprise by spinning off segments and gain the advantages outlined above, or

(2) to keep the old enterprise together and sacrifice the advantages of the spin-offs in favor of the greater political and financial clout of the whole enterprise.

A spin-off restructuring program would have to make the spin-offs attractive enough to overcome the attendant increase in bankruptability.

Financial Structure of Spin-off Transactions

The financial options are essentially the same as in a first-stage or state-sponsored MEBO privatization program except that the seller is the parent enterprise rather than the state. Bank credit or other third party credit is typically not available to finance MEBOs in transitional economies, so the transaction must use seller-supplied credit or be a leasing or hire-purchase transaction (which could be viewed as a seller-financed transaction with ownership only transferring at the end of the payments).

Seller-supplied credit can take a number of forms, the simplest being a term note to be paid off over a period of years. A more flexible arrangement is for the MEBO to issue redeemable preferred stock to the parent. Instead of payments on a term note, the preferred stock is divided into different issues that need to be redeemed at fixed time intervals. If a redemption payment is missed, that issue of preferred stock converts into common voting stock. Thus if the MEBO fails to keep up its payments, then it will become increasingly owned by the original parent enterprise and its attempt to break away from the parent will have failed.

In a typical example, the management and staff destined to work in a proposed MEBO spin-off will incorporate a new company using some amount of cash (possibly with personal borrowings). Then the parent drops down the necessary assets into the MEBO firm in return for some combination of common stock, preferred stock, or debt notes issued by the new company (or by assuming some debts of the parent company). If only a minority of the outstanding common voting stock went back to the parent, then the spin-off would be heavily leveraged but the insiders would have majority control (and thus a rather clear incentive and capacity to go their own way).

In designing a second-stage MEBO spin-off privatization program, consideration must also be given to the creditors of the parent firm. Ordinarily some consent of the creditors might be necessary before major assets were spun off to a daughter firm. It might also be possible to drop down liabilities (as well as assets) into the spun-off firm. Such a reassignment of debts would typically require the consent of the affected creditors. If the parent enterprise were distressed (as would often be the case), the creditors might prefer their chances of being paid by the spun-off MEBO firm.

The following is an example of a MEBO spin-off using an asset drop-down transaction.

Active	Passive
Assets 14,000	**9,000 Voucher-Based Shares**
	5,000 Liabilities

Parent Firm before Asset Drop-down Transaction

The staff of the to-be-spun-off business unit invests 1,500 in a new company. Assets worth 6,000 are dropped down to the MEBO firm in return for 4,000 worth of preferred stock and dropping 2,000 worth of debts.

Parent Firm

Active	Passive
Preferred Stock 4,000	**9,000 Voucher-Based Shares**
Other Assets 8,000	**3,000 Old Liabilities**
	(2,000 Debts Dropped to MEBO)

[Assets dropped down in return for preferred stock & taking over debts]

[Liabilities dropped to MEBO firm]

MEBO Firm

Active	Passive
Dropped Assets 6,000	**1,500 Insider Equity**
Cash 1,500	**4,000 Preferred Stock to Mother**
	2,000 Dropped Liabilities

Asset Drop-down Transaction

Government Policies to Promote Spin-off Restructuring

In the aftermath of the privatization or "voucherization" of state-owned enterprises, the government has lost the direct power of ownership over the enterprise. But many voucherized enterprises lack effective private owners (euphemistically called the "corporate governance" problem) and are slowly sliding into bankruptcy. As arrears pile up in distressed firms, the government gains new levers to promote restructuring. The court systems are quite unable to handle such a scale of impending bankruptcies. Widespread liquidation is not politically or socially feasible. Thus a government reorganization bankruptcy program may be the only alternative to de facto or de jure renationalization of the large bankrupt voucherized companies.

A government reorganization bankruptcy program would typically be implemented not by the courts but by some body that we will call the Enterprise Restructuring Agency (ERA). The ERA would have a form of trusteeship over the enterprises in its restructuring program. Enterprises could enter the program in various ways: involuntarily, after a certain level of arrears (at the instigation of creditors or by an automatic trigger); or voluntarily, as an alternative to liquidation. Government-sponsored restructuring programs are sometimes called "enterprise jails" or "enterprise isolation programs" because the sources of soft credit are removed to enforce adjustment and restructuring.

The ERA could use a centralized or decentralized approach to restructuring. In a centralized approach, a team of experts would "take over" the enterprise and devise a restructuring strategy. In a decentralized strategy, the burden of preparing the restructuring plans would fall on the managers or entrepreneurial teams (which might come from outside the firm). In view of the paucity of turnaround management expertise in the transitional economies, it is advisable to focus on decentralized strategies.

With different management or middle-management teams vying to spin off segments of the distressed business, the ERA will need to decide which restructuring plans are to be accepted in view of the interests of creditors and other stakeholders. Insofar as possible, the process should be "passively" regulated in the sense that the ERA officials do not need to exercise active judgment, but only need to check if the restructuring plans satisfy certain objective criteria.

To reduce transaction costs and bargaining stalemates, the ERA should enforce standard terms for asset drop-downs and other spin-off transactions.[4] A standardized valuation procedure would be needed to value the assets. Assets dropped down to a spun-off unit might be

(1) purchased in return for a note receivable or redeemable preferred stock with installment payments on standard terms;

(2) purchased in return for assuming (renegotiated) debt from the parent enterprise;

(3) transferred as injected capital into the daughter in return for shares; or

(4) leased by the daughter with an option to purchase after specified payments have been made.

A certain percentage (e.g., majority or 75 percent) of the workers originally involved in the segment being spun off would have to accept the spin-off proposal (e.g., by getting shares if not jobs in the initial spin-off unit). Within the standardized guidelines for each option and within a certain time deadline, the entrepreneurial team would have to devise a restructuring plan acceptable to the Enterprise Restructuring Agency. Otherwise liquidation procedures would commence to satisfy, as far as possible, the claims of the creditors.

Conclusion

A biological example may help to explain the systematic use of spin-offs as a restructuring strategy in the transitional economies. If organisms (such as mammals) have a fair amount of control over their environment and can protect their offspring, then the species can survive by having one or only a small number of offspring at a time. However, if organisms (such as fish or insects) have little control over the environment for their offspring, then they must have numerous offspring so that at least some will survive and can then grow to perpetuate the species.

It is clear which survival strategy is best adapted to the uncertain economic environment of most post-socialist economies. Bust-up restructuring with systematic spin-offs will maximize the chances that some of the offspring will survive and prosper—and even absorb some of the excess workers cast off by the less successful offspring.

Notes

1. David Bernstein, "Spin-offs and Start-ups in Russia: A Key Element of Industrial Restructuring," in *Privatization, Conversion, and Enterprise Reform in Russia*, edited by Michael McFaul and Tova Perlmutter (Boulder, CO: Westview Press, 1995), 205.

2. There is always the danger, however, that regulations will be driven more by the politics of resentment than by the rules of the marketplace. The political opposition often seems to operate on the "valuation theory" that any price a buyer is actually willing to pay in a privatization transaction is "obviously" too low a price. Western advisors often support that "theory" when they see the domestic buyers as *nomenklatura*, but are dismayed when the same logic is applied to foreign investors.

3. See A. Berle and Gardiner C. Means, *The Modern Corporation and Private Property*, revised edition (New York: Harcourt, Brace & World, 1968).

4. See the examples of Polish leasing or the draft law for internal privatization for Slovenia in *Management and Employee Buy-Outs As a Technique of Privatization*, ed. David Ellerman (Ljubljana: Central and Eastern European Privatization Network, 1993).

15

Commercialization without Privatization: Government Spin-Offs in China's High-Tech Sector

Corinna-Barbara Francis

Introduction

That China's approach to the transition to a capitalist market economy is unorthodox has been generally recognized. China's market reforms have been carried out under a strategy of liberalizing the economy without a systematic program of privatization. This sets China's reform experience apart from the general global trend toward privatization and from that of most of the former centrally planned economies of Eastern Europe and the former Soviet Union, which have made privatization central to their market reforms. It also raises questions about how China's market economy has flourished despite its unorthodox features.

In this paper I argue that a key to explaining this phenomenon is a type of market-oriented firm established by and linked to particular government and state entities.[1] Rather than systematically pursuing the privatization of the state sector, or even the development of a de novo private sector, Chinese market reforms have encouraged (and in some ways forced) individual state and government entities, including local rural governments, national-level institutions, national ministries, provincial government departments, municipal governments, universities, and state enterprises to go into business for themselves. The evidence suggests that the firms set up by these entities of the state, referred to here as government spin-offs, constitute an important intermediate sector between the traditional state-owned enterprises and the private sector, and are key to understanding China's unorthodox market system. While China's state sector remains largely sluggish, and its private sector remains small and constricted, this intermediate sector appears to be a key engine of economic growth.[2]

Government spin-offs help explain three particular features of China's unorthodox market system. First, they suggest reasons why China's economy has flourished despite the fact that it violates the conventional assumption regarding the need for clear boundaries between "the state" and "the market" for a well-functioning market system.[3] Neoclassical economics argues that markets operate best when states minimally interfere.[4] Institutional economics argues that when states exercise property rights this impedes them from acting as a third neutral agent enforcing the property rights regime which underpins the market.[5] China's market system, in which myriad state entities have become key players, and in which property rights are to an important extent lodged with state agents, fundamentally violates this norm.[6] Government spin-offs suggest why this might be so. They suggest in the first place that "the state," which is treated in a unitary and homogeneous fashion in the neoclassical economic literature, may operate in a differentiated way in which sub-units of the state act as market actors and exercise property rights while higher levels of the state perform more "traditional" state functions such as legislation, regulation, and taxation. Furthermore, they suggest that individual entities of the state may be effective owners of resources if they exercise the property rights associated with ownership, such as the rights to the returns from assets, the rights of utilization, the right to exclude, etc. This decentralized property rights structure, in which managers exercise de facto property rights, would be expected to increase their incentives to maximize the assets of their firms.[7]

Second, spin-offs help explain successes in China's economy despite the proliferation of fuzzy ownership types and the apparent lack of a well-enforced property rights regime. A second fundamental assumption of Western economic thinking is that clearly defined property rights and a well-enforced property rights regime are a precondition of a well-functioning market economy.[8] Secure property rights are considered to be vital for encouraging entrepreneurs to put their resources at risk and to invest their time and energy to develop resources to their fullest. Vaguely defined property rights are viewed as a formula for costly conflicts, inefficiencies, and the squandering of resources. The paradox here is that the economic sector where property rights appear to be the least well-defined—China's intermediate sector composed of mixed-type firms—appears to be one of the fastest growing and efficient ones in the economy. Spin-offs help explain this in two ways. First, they demonstrate that while property rights may be fuzzy in the sense of lacking legal definition, they may be clearly determined through informal norms and customs. This suggests, secondly, that property rights need not be enforced by the state, but can be enforced through informal norms, customs, and local sanctions. While the ownership of the capital assets of China's intermediate sector firms may not be clearly defined in a legal sense, property rights over these assets may nevertheless be mutually clear and determined through informal bargaining.

Thirdly, spin-offs shed light on how China has carried out its market transition without a systematic program of privatization and with a still-small private sector.[9] It is a nearly universal assumption in neoclassical economic literature and the current literature on market transitions in post-communist systems that the predominance of private enterprise is necessary to a competitive market system.[10] This view is grounded in the fundamental belief that if an agent cannot claim clear and exclusive ownership over an asset that he or she will not have the incentive to invest resources into the property, to monitor its use, and to not squander it.[11] This right, it is argued, enables owners to "capture the full value of their individual investments, thus encouraging everyone to put time and labor into the development of resources."[12] Public property is plagued by the "tragedy of the commons"—where individuals do not exercise exclusive right over assets they will not have the incentive to develop resources to their fullest. A predominance of private enterprise is also seen as necessary to enable competition to flourish, as public enterprise is generally associated with monopoly.[13] Both arguments are based on the prevalent dichotomy between public vs. private enterprise, and between "the state" and "the market" as mutually opposed organizing principles.

However, spin-offs suggest that the choice is not such a stark one and that intermediate ownership and enterprise types provide alternatives. As an enterprise type spin-offs are not strictly private, yet they capture some of the advantages associated with private enterprise. At the same time they avoid many of the problems encountered by privatized state-owned enterprises, e.g., the dispersed nature of ownership, the resistance to privatization by powerful state actors, the lack of new management, etc. They also avoid problems associated with de novo private firms in post-communist market systems, such as the lack of capital in private hands and the reluctance of skilled professionals to entirely sever their ties to their state-sector job and to move fully into the private sector. As such, spin-offs appear to offer an alternative to privatization as an effective strategy for the transition to a market economy.

Spin-offs are not only key to understanding China's unorthodox market system, they are also conceptually interesting. Issues of ownership and ownership reform in China have been a vexing issue for observers of China's market reforms as well as a topic of intense debate among observers of the Eastern European and Russian market reforms. Chinese spin-offs present a type of ownership arrangement that does not fit easily into conventional Western categories. They raise many questions. Who actually owns these firms? How does this ownership structure influence the incentive structure within which managers operate? Do they represent a viable long-term enterprise type? Or are they an organizational form effective for easing the *transition* to a market system but not good for the long-term health of the market system?

Despite the relevance of spin-offs we still lack an understanding regarding their formation, the nature of their ownership structure, and the dynamics of their relations to their parent units. One variety of spin-off, enterprises established and operated by local rural village and township governments (TVEs), has received more attention, but urban varieties have been relatively neglected. This paper seeks to address this deficiency through an examination of the computer and electronics spin-off firms in Haidian District, Beijing. It is based on a case study of eleven computer and electronic firms in Haidian district, in suburban Beijing.[14]

A Property Rights Approach

This article adopts a property rights approach to the study of China's government spin-offs. This approach has a number of advantages for understanding the governance structure of China's hybrid enterprises which the traditional approach to ownership has not facilitated. The latter sees ownership as "that sole and despotic dominion . . .over the external things of the world, in total exclusion of the right of any other individual in the universe."[15] This exclusive control is seen as providing the incentives for individuals to invest their time and labor into their property, as making it possible for owners to identify other owners, and as providing the basis for owners to exchange their property with others, all of which are seen as critical features of a capitalist market economy.[16] More conventional approaches to the question of firm ownership have tended to stress the individual owner-manager's exclusive control over assets and his or her power to exclude.

The property rights framework, in contrast, sees "ownership" as a bundle of rights which an agent or agents exercise over property or assets as determined either through law, social practice, norms, or custom.[17] "Ownership" is not taken to mean that the assets and resources are actually "owned," but rather the *rights* to use assets and resources in particular ways are exercised by the owner. Alchian and Demsetz explain that to "own" land gives the *right* to till or sell the land, but not to throw it at a passerby. Ownership confers socially recognized rights to action. This approach has also been associated with the "nexus of contracts" model, in which the firm is seen as a complex of contracts between a plurality of interests, none of which can be said to be "owners" in the conventional understanding of that term.[18] A range of terms have been offered to describe the ownership structure of China's new enterprise types and their political economies, including "vaguely defined cooperatives" and "local state corporatism."[19] The concern of this paper is less to coin a new phrase for spin-offs as an enterprise type as it is to describe empirically the structure of control over and the property rights characteristic of government spin-offs.

The property rights approach offers a number of advantages to the study of the ownership structure of mixed-type enterprises in the current Chinese context in particular. First, it suggests the possibility of disaggregating the rights that comprise the ownership bundle, including the right of utilization, the right to the products of the asset, and the right to transfer or alienate the asset through sale or exchange. Second, it sensitizes us to the plurality and variety of agents which may exercise these rights and to the fact that multiple agents may simultaneously lay claim to the same right. As Alchian and Demsetz put it, "the domain of demarcated uses of a resource can be partitioned among several people."[20] Conventionally, "owners" have been identified as those agents which exercise the critical controls over an asset. However, the property rights approach posits that the bundle of rights comprising "ownership" may not only be exercised jointly, but may also be distributed among multiple agents. These insights are useful in the context of China's mixed enterprises, where property rights are distributed in particularly complex ways among a variety of agents.

The property rights approach is further useful in highlighting the limited and arbitrary nature of the dividing line between the concepts of public and private property. Since property rights are divisible, the line between public and private ownership is ambiguous. Some rights may be exercised by "the state," others may be exercised by private individuals, and others still may be jointly exercised. Furthermore, this dichotomy overlooks the wide variety of other agents that may also exercise rights, such as corporate bodies and institutions, as well as sub-units of the state. The extent to which "the state" is not a coherent entity in China today, along with the presence of a wide variety of alternative agents exercising property rights in China, including individual sub-entities of the state, makes the distinction between public and private ownership particularly problematic.

Finally, a useful insight suggested by the property rights approach is that the structure and types of rights are not fixed in time but may be transformed—either because of legal changes or changes in customs—and that they may be determined by custom or informal norms. Rights need not necessarily be legally defined and may be contested. Because of the rapid transformation of China's state and economy today, there is a particularly wide range of resources and assets over which property rights are not legally clearly defined and which are informally determined. One of the goals of this paper is to clarify the different sets of agents who make property rights claims over the important category of government spin-offs.

China's Government Spin-offs: The Case of Haidian District's Computer and Electronics Firms

The term government spin-off is used here to refer to a new firm that is set up by or with substantial help from a government or state entity with the expectation that it will be a profit-making enterprise. These firms were generally referred to by my Chinese informants as *quanmin qiye*, a term which indicates public ownership, but which is distinct from the terms used to refer to traditional state-owned enterprises (which are referred to as *guoying* or *guoyou qiye*—state managed or state-owned enterprises).

China's market reforms have strengthened the incentives for and pressures on individuals and institutions to set up spin-offs. Government policies have allowed, and pressured through budget cuts, state entities to engage in commercial ventures in order to generate extra-budgetary revenue with which to maintain and improve the living and research conditions of their employees and their members.[21] Universities and research institutions in particular have been encouraged to get involved in the computer and high-tech business.[22] The revenues generated by these commercial ventures are expected to alleviate fiscal pressure on the government.[23] Spin-offs set up in this way not only pay taxes to the government, they also generate extra revenue for their parent unit and its employees which alleviates some of the state's fiscal responsibility. Within the computer and high-technology sector, universities and scientific research institutes have been very active in setting up commercial ventures. According to one report, at least 600 high-tech enterprises had been set up by colleges and universities in China by 1993.[24]

Haidian District, located on the northwest periphery of Beijing and the home of some of China's top universities—including Beijing University, People's University, and Qinghua University—became one of China's main centers of the computer and high-technology industry in the 1980s and 1990s. Prior to that Haidian had been a largely agricultural district, in addition to the universities and the retail businesses that served these institutions. In the early 1980s the Haidian district government initiated a plan to develop its local computer, electronics, and high-technology industries in an effort to utilize Haidian's comparative advantages in human resources and land availability. It began offering preferential policies to computer and electronics firms that located in its district in an effort to encourage local universities to establish computer and high-tech companies and to attract other companies to locate in Haidian. These policies included a three-year grace period on taxes for firms newly registered in Haidian, less restrictive import and export regulations, greater autonomy for firms to set their own prices, easier access to multiple entry and exit visas for their employees' business trips abroad (particularly Hong Kong), a streamlined process for obtaining building permits, and more.[25]

In the mid-1980s the district government formalized these preferential policies with the establishment of the Haidian Scientific and High Technology Experimental Zone, which also facilitated the regulation of the growing number of firms located in the district. The success of this strategy became evident in the large number of computer and electronics firms which have set up business in Haidian, whose main street is now informally referred to as "electronic alley" (*dianzi jie*). Of the computer and electronic firms operating in Haidian, roughly half are government spin-offs.[26] The institutions which have been most active in setting up computer firms include Beijing University (Beida), research institutes within the Chinese Academy of Science (CAS), and various institutes within a number of national ministries including the Ministry of Aviation and Aeronautics and the Ministry of Machinery and Electronics Industry.[27] Beijing University had set up at least 20 computer and electronics firms, and various institutes within CAS had similarly set up more than 20 such firms. The Ministry of Machinery and Electronics Industry alone operates 150 subsidiary firms, a large number of which are high-technology firms.[28]

Company Formation

The initial idea for launching a commercial venture often comes from individual professors, scientists, researchers or groups of colleagues within a university or research institute. The idea is typically to commercialize an already existing technology or scientific knowledge, often developed by these same individuals. This process, referred to by the Chinese as the "second phase of development" (*di'er ci kaifa*), involves product design and development, market research, distribution, and marketing strategies with the aim of adapting basic technology or knowledge into a marketable commodity capable of generating a profit through market sales. In some cases additional basic research may be needed to develop the necessary technology, but most often the basic research is complete. Commercial spin-offs thus often exploit technologies and scientific knowledge that had long been languishing on university shelves. Under the centrally planned economy neither individuals nor institutions had much incentive or freedom to commercialize technologies, since they could not directly engage in the marketing of these products and could not directly benefit from this effort. The old incentive structure encouraged individuals and institutions to compete for government-funded research projects, but provided little incentive for the commercialization of the technological and scientific discoveries that resulted from these.[29]

Reform-era policies have given individuals and institutions the incentive to commercialize technologies by allowing them to engage directly in commercial ventures and to retain the bulk of their profits. If the venture succeeds, revenue is generated which can be used to raise salaries and bonuses, to improve living and research conditions, to pay for recreational activities, better benefits, etc. Individuals working in a successful spin-off can expect higher salaries and bonuses than those remaining in the parent unit. And remuneration tends to be more closely tied to individual effort, skill, and productivity, giving hard-working and competent individuals the opportunity to work harder and make more money. Being involved in a commercial venture is also seen by many as more exciting and challenging than the regular academic and research positions. It taps the individual's creativity and energy, and gives him or her a greater sense of involvement in the reform process. Employment in a commercial venture opens up career prospects, giving regular researchers the opportunity to move into management positions much faster than if they stayed in their original unit.[30] However, it does not cut the individual off from the security of state employment since the individuals who transfer to work in a spin-off retain their formal employment with their state work unit, together with the employment benefits, job security, medical insurance, social security, and other benefits and perks associated with a public-sector job, including housing.[31] From the individual perspective, being involved in a spin-off thus offers the best of both worlds—the benefits and higher income afforded by the success of a commercial venture, with the security of state employment.

The Role of Parent Units

Incubation Parent units play a critical role in the development of commercial spin-offs through the provision of facilities, staff, and resources and by providing spin-offs with an incubation period during which they can launch the venture. Individual scientists or researchers may have the commercial ideas and the familiarity with the technology, but they generally lack the capital, laboratories, infrastructure, and other resources to carry them to the product stage, thus presenting obstacles to strictly private commercial ventures. The incubation period provides fledgling commercial ventures a window of opportunity during which they can progress on product design and develop market strategies. During the incubation period employees generally continue to have access to their labs, offices, and other university facilities. Companies are often only officially established after an indefinite incubation period, and often just at the point when they are ready to directly market their product. Even after they are formally registered, commercial ventures often continue to operate out of the compound of their parent unit.[32] The Beijing University New Technology Company (later renamed the Beida Fangzheng Co.), for instance, grew out of a commercial venture first proposed by a group of professors from the physics and wireless electronics departments of Beida in the early 1980s. These professors proposed their idea for commercializing a software technology which they had been involved in developing to the University's Division of Scientific Development (*keji kaifa bu*), which oversees the scientific and technological research projects conducted by university staff.[33] The development division supported the project, allowing them to proceed with the work of product design. However, the company was only formally established after a couple of years when the product was closer to being ready to be sold on the market.[34]

Technology transfers Parent units play a key role in the formation of their spin-offs through the provision of start-up technology. The fact that spin-offs are usually established after or just at the point when the research and development on a particular product or technology is complete suggests they are a mechanism by which institutions commercialize an existing technology, and are not, at least initially, engaged in developing new technologies. The technology used is often complete from the perspective of basic research, requiring only the secondary stages of commercialization in order to develop marketable products.[35] The Capital Computer Co. (*Hanjing Diannao*) was established by a group of computer scientists from the Software Institute of CAS with the specific goal of commercializing a software and hardware system which had been developed over many years by institute researchers.[36] The task was to design, package, and market profit-making commodities from it. The Beijing University New Technology Company was set up for the specific purpose of commercializing a computer software technology developed by Beijing University researchers with applications for the editing, layout, and printing of Chinese characters. The company developed computer software packages for publishing houses, newspapers, and other printing companies designed for doing the layout of Chinese characters in different fonts and sizes.

The technologies that are commercialized through spin-offs are often developed by specifically funded government projects, many of which may last for years and involve large numbers of researchers. The Chinese character printing software adopted by Beijing University's New Technology Company was specifically funded by the government's "748" project (indicating the date on which it began: August, 1974) and received years of government funding.[37] The Chinese medicine software technology used by the Capital Computer Co. was developed by a large team of researchers in the Software Institute of CAS over an eight to ten year period, also with government funding. Making these technologies available for commercialization is a way for individual state entities to exploit existing technologies for their own benefit, many of which are the product of government-funded research. The products derived from these technologies are often a spin-off's primary money-maker for a number of years. The Beijing University New Technology Company was established in 1986, but the products developed from this technology continued to be the company's primary source of revenue as late as 1993.[38]

Capital transfers Parent units are vital to the formation of spin-offs through the provision of capital. Spin-offs have thus been a critical mechanism for fostering entrepreneurial activity and for transferring capital resources from the state sector to the commercial sector, a particularly important role given the underdeveloped state of China's capital markets. Parent units provide capital either from their own resources or from loans. Of the 4,000,000 rmb Beijing University provided the New Technology Company, 1,500,000 rmb was said to be from the university's own assets, while 2,500,000 rmb was borrowed from a local farmers' collective.[39] Even when a parent unit does not directly provide capital, it can facilitate the mobilization of capital from either domestic or foreign sources through its institutional prestige. Scientists affiliated with a prestigious institution such as the Chinese Academy of Sciences have advantages over individuals lacking such an institutional affiliation in attracting foreign partners for a joint venture and setting up a commercial firm. In the firms investigated, parent units were in all but one case the primary provider of start-up capital, and in the one exception the parent unit nevertheless played a critical role in helping the scientists get capital from foreign businessmen. According to a manager of this company, the fact that the founders of the company came from CAS was very helpful to them in finding a foreign investor. It was because of the CAS connection that they were able to gain the trust of the foreign partners.[40]

The capital provided by parent units is usually used for product design, market research, and manufacturing, since basic technology has usually been provided. In some cases even the product design and market research stages have been carried out prior to the formal establishment of the company. The initial capital transfer is used primarily for the costs of actual production and manufacturing, the construction of office buildings, hiring of outside employees, etc. Most parent units were reported to have provided a one-time capital transfer to newly formed companies. The Computer Institute of CAS was said to have provided only 200,000 rmb to set up the Lian Xiang Co. in 1984 and subsequently not to have provided any more direct capital transfer after that. In some cases, however, parent units do provide more than one round of capital infusion. Beijing University provided 400,000 rmb to its New Technology Company in 1986. Two years later it provided another 4,000,000 rmb to the company. The initial investment was used to pay for initial stages of product design and market research, while the second round of investment paid for the purchase and retooling of a factory site where the products would actually be manufactured.

Personnel Parent units facilitate the development of spin-offs through the provision of skilled personnel. Public sector employees, particularly professionals, are still hesitant to give up the benefits of state employment, and are thus hesitant to launch commercial ventures entirely on their own. Furthermore, public research institutions, particularly nationally renowned ones such as CAS, Beijing University, etc., employ some of the nation's top-notch scientists and researchers. Spin-offs facilitate the shift of skilled personnel into more commercial activities by allowing them to simultaneously retain their public-sector status. Parent units encourage employees to go into commercial ventures by allowing them to remain employees of the parent unit and to retain their status as civil servants, together with all the associated state employment benefits.

The number of employees involved in launching a commercial venture varies, but in the case of Haidian's computer and electronics firms it was usually between half a dozen to around 20. Twenty employees from the Computer Institute of CAS were involved in setting up the Lian Xiang Company. The Software Institute similarly initially transferred around 20 of its employees to set up and manage the Capital Computer Company, one of its spin-offs. Spin-offs are thus not typically a method for institutions to "dump" excess and less-qualified personnel, although this may sometimes be the case. Rather, they represent an important investment by these institutions of some of their top-notch personnel, giving these firms a competitive edge in the marketplace. An increasingly large proportion of university and research institute staff is working in commercial ventures.[41]

Spin-offs as Non-Private Market-Oriented Firms

Despite their nominal status as public firms, and the important role that parent units play in their formation, spin-offs share important features with firms operating under capitalist market conditions and differ significantly from traditional state-owned enterprises. They enjoy considerable managerial autonomy and are less administratively integrated into the state bureaucracy than traditional state-owned enterprises. Spin-off managers enjoy relative autonomy in personnel, production, finances, and investment decision-making. Spin-offs are relatively less burdened by social responsibilities than state enterprises and operate more strictly as profit-making enterprises rather than social welfare institutions. They operate under tighter budget constraints and are more profit-oriented than traditional state-owned enterprises, and they contribute to the competitive structure of the market.

Managerial autonomy Spin-off managers exercise relative autonomy in personnel management. Among the firms investigated, managers claimed to be able to make hiring decisions with little interference from their parent unit. Reforms in the personnel system operating in Haidian's High Technology Zone have helped to expand the role of the labor market and the autonomy of firms over hiring. A newly created Human Resources Exchange Center now functions as a clearinghouse for personnel where firms and employees can independently link up according to their needs.[42] The new system allows employees to "park" their dossiers (*dang'an*) with the Exchange Center and enables companies to browse through the list of job-seekers registered with the center and hire them directly, without intermediation by the state or their parent units.[43] This contrasts with traditional state-owned enterprises, which are assigned their pool of personnel by the state. While the individuals who form the initial management team of a spin-off are typically transferred from the parent unit, employees tend subsequently to be hired through the new market-oriented system. According to a manager in charge of personnel at the Lian Xiang Co., 80–90 percent of the company's *new* employees are hired through market channels, rather than the traditional state channels.[44] Managers claimed to have almost complete discretion in hiring through the new personnel system. They were only expected to draw up a list of the type of personnel they required and to have this approved by the Haidian Human Resources Exchange Center. The only constraint that spin-off managers said they experienced was the restriction on the percentage of employees they could hire who did not have Beijing residency.[45] Spin-off managers also exercised relative autonomy in promotions, layoffs, and the distribution of bonuses. This gives spin-off managers greater latitude to hire and fire, since hired employees (as opposed to transferred employees) are employed on the basis of a labor contract and do not enjoy the same job security as state employees.

Spin-off managers also exercise considerable autonomy in production, finance, and other business decision-making. And it is generally expected that the greater autonomy and control exercised by agents, the greater their incentive to productivity and efficiency.[46] Unlike traditional state-owned firms, spin-offs do not operate within a state plan and receive no quantitative or profit quotas from the state. This gives spin-off managers greater freedom to determine and change their firm's product mix without having to get approval from their parent unit or the state bureaucracy. Spin-off managers exercise considerable autonomy in setting up new subsidiary business ventures including second-tier spin-offs, semi-autonomous subsidiaries, joint ventures with domestic or foreign partners, and autonomous regional branches. In fact, spin-offs tend to diversify widely, particularly into speculative ventures including real estate, securities, and trade ventures. Beida Fangzheng, one of the largest companies in Haidian, had, by 1993, diversified into a complex of 20 different companies. Larger spin-offs thus often take the form of holding companies with numerous subsidiaries operating semi-autonomously. Some companies diversified so much that their core identity as computer or high-technology firms was doubtful.[47] Some appeared to simply maintain a high technology "front" store in Haidian, while the bulk of their activities was in more speculative ventures. One suspects this enabled companies to qualify for the tax breaks and other preferential policies given to high-technology firms by the local government while they actually engaged in a wide variety of other businesses and speculative ventures.[48]

Managers exercised relative autonomy over the use of after-tax profits, at least within the scope of municipal government regulations. Until 1993 Beijing municipal regulations stated that at least 50 percent of profits had to be reinvested, that no more that 30 percent could be used for individual employee benefits (*fuli*), and that no more than 20 percent could be used as individual bonuses.[49] However, within the scope of government regulations, managers claimed to exercise discretion over the use of profits—how they were reinvested, distributed as bonuses, etc.[50] Managers were free to structure individual bonuses as they wished, the only constraint being that employees had to pay individual income taxes over a certain fixed income. The de facto right that spin-off managers exercise over profits would be expected to further strengthen their incentive to fully exploit the resources under their control.[51]

Tightened budget constraints Spin-offs, despite their non-private status, share with market-based firms the fact that they operate under relatively tight budget constraints.[52] None of the companies investigated in this study reported receiving subsidies either from their parent unit or the state. They were, on the contrary, generally a source of revenue for their parent units. And the profits remitted by spin-offs to their parent units are in some cases larger than the latter's total budget.[53] Several features account for the relatively tight budget constraints among spin-off firms. First, spin-offs are not integrated into either central or local government budgets, so there are no state funds specially earmarked for subsidizing their employee wages, benefits, and other operating expenses. Only the jobs, wages, and benefits of the "transferred personnel" continue to be subsidized by the state during their employment with a spin-off because of their continued affiliation with the parent state work unit. *Hired* personnel, who are a majority of employees, do not receive such guarantees. The state thus does not have the same obligation to bail out failing spin-offs as it does with traditional state-owned enterprises, and spin-offs thus cannot count on the state to subsidize wages, benefits, and other operating expenses.

Parent units, for their part, are constrained in their ability to subsidize their spin-offs, at least relative to the state, due to their smaller budgets. This is particularly the case with universities and scientific and academic research institutes, many of which have had their budgets cut, making them even more dependent on revenue generated by their spin-offs. Parent units are particularly constrained in the case of large firms, some of which have extremely large operating budgets. Lian Xiang, which had been set up by the Computer Institute of CAS with only 200,000 rmb capital investment, had by 1993 grown into a company with 1,400 employees and annual gross profits of 1.5 million yuan (in 1992). With a budget many times smaller than that the Computer Institute was in no position to subsidize this company in the event of financial collapse. And the large number of spin-offs that parent units often set up makes it even more difficult for them to bail them all out in case of financial failure.

Competitive market structure Spin-offs contribute to the competitive structure of the market despite the fact that they are non-private enterprises. The decentralized and dispersed ownership structure of the spin-off political economy, in which "state" ownership is ownership by myriad dispersed state entities, contributes to competition between individual firms. Because "the state" is divided into a plurality of individual entities and ownership is exercised by these distinct entities, competition is not quashed. Computer companies set up by Beijing University compete with those set up by other institutions including ministries and the Chinese Academy of Science. There is even evidence of competition between firms set up by institutes within CAS, such as computer software companies set up by the Software Institute and the Computer Institute of CAS.[54]

Ownership and Contested Property Rights

Legally, Haidian's computer spin-offs register with the local government as state enterprises, or "enterprises owned by the whole people" (*quanmin qiye*). However, this ownership category fails to answer many questions relating to the exercise of property rights over these firms. Who actually owns these firms? Who can claim the income generated by these firms? The concept of "state" ownership is a problematic one even in the context of state socialist economies. Some have argued that state ownership (in the context of state socialism) is more accurately described as a communal form of property, in which the state lacks clear title and control over its nominal assets and managers and corporate groups have de facto access to these assets.[55] Others have suggested that the concept of ownership by the state has to be differentiated according to ownership by different levels of the state, such as regional governments.[56]

The concept of state ownership is even more problematic in the context of the rapid changes in China's property rights regime during the reform era. Conventional categories of ownership do not adequately capture the complexity of the property rights structures of many of the hybrid enterprise types which have emerged during the course of market reforms in China. The analysis of the exercise of property rights over spin-off firms is complicated by two facts. First, there were no legally defined owners of the capital assets of the firms in the study, through either shareholding among the firm's founders, investors of inputs, or outside investors.[57] China is only slowly developing a capital market and the government has tightly controlled which firms are allowed to sell stocks or to have internal shareholding. Up until 1993 the Haidian District High Technology Zone had prohibited firms registered in the district (both state firms, quanmin, and collective firms, minban) from instituting shareholding arrangements.[58] As a consequence, there were no legal owners of the capital assets of these firms.[59] Second, the property rights over the assets of these firms are not defined by legal contracts but tend to be worked out through informal mechanisms, norms, and bargaining.

Dividing the profit. One of the key components of the bundle of property rights associated with ownership is the right to the benefits of or the income derived from an asset. Given the absence of shareholding among Haidian's high-tech firms, the key issue regarding the right to returns is the distribution of enterprise profits. An analysis of who gets the profits is thus key to determining de facto property rights. Who in practice lays claim to the profits of Haidian's computer spin-off firms? Several features characterized the distribution of the profits of spin-off firms in this study. First, they were shared and fought over by two primary sets of actors—the governmental parent units and the firms themselves, although each of these sets was divided into sub-groups and in practice they were not always clearly divided. Second, profit-sharing arrangements between these sets of claimants were not contractually determined nor determined ex ante, but worked out through an informal process of bargaining and negotiation. As a consequence, this process was generally a contentious one and a primary source of struggle between spin-off managers and their parent units. Finally, no uniform formula was applied among spin-off firms as to how to divide the profit. Rather, this was worked out on the basis of specific conditions which applied in each case. Despite this, parent units and spin-offs worked out specific agreements regarding how profits were to be divided between them.

Spin-offs were generally given an initial incubation period by their parent unit during which they do not have to turn over any of their profits (often they were not yet making any).[60] However, once they started turning a solid profit, companies would have to turn over a portion of it to their parent units. Among the firms investigated this portion varied widely and also varied over time in the case of any particular firm. It ranged between 16.66 percent to 70 percent of net profits among the firms in the study, although the average was between 25 and 35 percent.

One company reported having had to turn over 70 percent of its net profits to its parent unit, the Industrial Control Institute of the Ministry of Aeronautics and Aviation, in its earliest days. According to a deputy company manager, "the institute didn't trust us, so they took all our profits."[61] The Lian Xiang Company turned over to the Computer Institute of CAS in 1992 a sum that a top manager said was equivalent to the amount the company paid to the state in taxes, which he said was roughly 30–35 percent of net profits.[62] Over time spin-offs appeared generally to be able to bargain to turn over a *reduced* portion of their profits. The Capital Computer Co., which had turned over upwards of 70 percent of its profits to the Industrial Control Institute of the Ministry of Aeronautics in its earlier days, was able over time to bring that proportion down to around 30 percent. In 1993 the company turned over 30 percent of its profits to the institute.[63] And company mangers expressed optimism that this proportion could be lowered even further.[64] Beida Fangzheng paid a relatively low proportion of its profits to its parent unit, Beijing University, after several years of operation, 10 million rmb out a total net profit of 60 million in 1992.

Parent Unit Leverage

Because of the lack of legal contracts between firms and their parent units, profit sharing was determined through informal bargaining between them based on their respective leverage. Parent units derived bargaining leverage from the legitimacy of their role in providing vital inputs in the formation of their spin-offs. Their degree of legitimacy appeared to be a function of the comprehensiveness of their support in providing these inputs, including capital, personnel, technology, institutional, and bureaucratic support. The more comprehensive their support was the more legitimate their proprietary claim appeared to be. Parent units which were the sole or primary investor of capital, personnel, technology, and infrastructure tended to have a stronger proprietary status vis-à-vis their spin-offs, assessed both by the proportion of profits they receive as well as by the attitudes of company employees toward their claim on profits. Spin-off employees (particularly managers) were more likely to feel beholden to their parent unit if it had contributed comprehensively to their company's formation. Managers of Beida Fangzheng, for instance, expressed relative acceptance of the ownership rights of Beijing University, which had been comprehensively involved in the company's development.[65] This contrasted with the attitudes of managers and employees of the Capital Computer Co., who denied that the Software Institute of CAS had any legitimate claims to company profits since it had not provided any capital in the formation of the company, despite the fact that the institute had provided the company's key start-up technology, that its initial personnel all came from the institute, and that the institute's institutional prestige and connections had been critical in the company's ability to the attract foreign investors. According to a company manager, "since the Institute did not invest any of its own money, it can't claim any of the profits."[66] This company did not turn over any profits to the Institute.[67]

Technology transfers did not appear to establish as strong a proprietary claim on the part of a parent unit as capital investment, as parents were not given much credit for "providing" start-up technology. This in part reflects the legacy of the old centralized economy which created an incentive for individuals and institutions to get government research grants to carry out scientific research, but in which the final technology had little value by itself due to the inability of individuals and organizations to exploit it commercially and the lack of an open market for scientific knowledge and technology.[68] It also reflects the lack of clear rights over intellectual property. Universities and research institutes were not generally recognized as having clear ownership of the technology and knowledge produced by scientists and researchers employed by them, who thus give them little credit for "contributing" it to their commercial ventures.[69] The lack of a well-enforced intellectual property rights regime also makes it more difficult for universities and research institutes to sell their technology on the market, giving them greater incentives to make it available to their employees for commercial exploitation, particularly since the latter might otherwise take it and launch a venture on their own.[70]

Parent units derive leverage over their spin-offs through their control over management appointments. Nominally, parent units have the authority to appoint, dismiss, and promote top-level managers. Some institutions, including Beijing University and CAS, set up special committees to handle managerial appointments. In some cases parent units appoint one of their own officials to a managerial position in the spin-off.[71] One former general manager of Beida Fangzheng was removed from his position by the University and was replaced by a former division head within the company.[72] This, according to behavioral theory, should insure a degree of managerial responsiveness to the interests of the parent unit. Managers who perform badly may be demoted or transferred. The desirability of the position of spin-off managers—due to its high salaries and generous perks and benefits—could be expected to establish a dependency on the part of spin-off managers on their parent units and be a mechanism of discipline over them. As unitary organizational actors, parent units, it could be argued, are not as dispersed as shareholders of a large modern corporation and could thus be expected to be more coherent and effective as owners (or principals).[73]

The lack of a well-developed managerial market further strengthens the hand of parent institutions, as it reduces the options that managers exercise. Managers who leave cannot easily find equivalent positions elsewhere. Spin-off managers in fact appear to be mostly employees of the parent unit, rather than outsiders. Managers are further dependent on their parent unit because if they quit their job they lose their state cadre status and all the security and perks associated with it.[74] Despite the development of a private sector in China, professionals still value public sector employment for its security, benefits, and prestige. The managers of the firms in this study appeared generally unwilling to cut their ties with their parent unit.[75]

Parent units gain leverage through their political and institutional connections, which can be critical to spin-offs in securing contracts, market share, and foreign investment. According to managers at Beida Fangzheng, the University's bureaucratic connections and institutional prestige were helpful to the company in getting contracts for their software product from large, national-level publishing houses and newspapers around the country.[76] CAS similarly played a role in helping its spin-offs obtain contracts for their products, particularly from government agencies.

Parent units, finally, gain leverage from the informal norms of behavior imposed by the local political economy. Spin-off managers said they felt unable to "buck the system"—to refuse to turn over a portion of their profits to their parent units—due to concern over reprisals from local bureaucrats. In their view local bureaucrats would punish a "renegade" company—one that broke its informal agreement with its parent unit—by disfavoring them on a range of vital issues, including the disbursement of loans, licenses, services, and facilities. Discretion on the part of local political authorities thus functioned to informally enforce property rights exercised by parent units.

Parent units tend to set up numerous small, separate companies, which are easier to control, instead of allowing existing companies to expand. Smaller and new firms also have the advantage that they generally perform better in terms of productivity and market responsiveness. They provide a more effective incentive structure for managers, who have a greater stake in their success. And new companies have the benefit of the preferential polices offered by the local authorities, notably the three-year tax break. This helps explain the large number of spin-offs that parent units tend to set up, with single research institutes even setting up numerous companies working in similar areas.[77] Universities and research institutes in fact tend to set up a separate company for each technology item or product knowledge they want to commercialize, instead of making these available to existing companies.

Spin-off Leverage

What leverage do spin-offs exercise with respect to their parent units? The proprietary claim of parent units is weakened by their lack of clear ownership over the firm's capital assets, which allows company managers and employees to develop an informal proprietary sense over these firms. This informal sense of proprietorship is often intensified by the fact that individual scientists and researchers have the initial idea for launching the commercial venture. These individuals, who often become the firm's managers and build the company from scratch, tend to exhibit a strong sense of proprietorship, albeit an informal sense, over these companies.[78]

Spin-off managers derive a degree of leverage vis-à-vis their parent units from their managerial autonomy and their firm's market orientation. Spin-offs do not rely on the state for the same range of production inputs as traditional state-owned enterprises, nor do they rely on the state to market their products.[79] And while spin-offs often rely on their parent unit in their formative stages of development, they generally become increasingly independent as time goes on. The more profitable, financially successful, and reputable a spin-off is, the more independence it gains from its parent unit. Furthermore, over time the memories of and the relative importance of the initial contributions made by parent units tend to fade and the proprietary sense on the part of firm managers intensifies. The initial capital that parent units contribute tends increasingly to be viewed as insignificant, particularly in relation to the company's growing profits.

Managers of the largest and most reputable computer firms in Haidian exercise considerable leverage in bargaining over the terms of profit sharing with their parent units. This is evident in the case of You Quanzi, the manager of the Lian Xiang Co.[80] Lian Xiang was, as we saw above, established in 1984 by a group of around 20 members of the Computer Research Institute of CAS and the company received an initial capital investment of 200,000 rmb from the institute. Under the leadership of You Quanzi, Lian Xiang's manager, the company grew rapidly and became very profitable.[81] In 1987 tensions mounted between You Quanzi and the Computer Research Institute over the division of the profit. You felt the institute was demanding too large a portion of the profits, and he wanted a larger share to be retained by the company. The Computer Institute rejected his demand and threatened to fire him if he did not cooperate. You, in turn, launched an all-out battle against the institute. He used the press to build up his personal case and to establish his role in the success of the company. He also threatened to use the press to tarnish the firm's reputation by reporting on its internal problems in the event he was fired. Unlike traditional state-owned enterprises whose public relations were relatively unimportant, spin-offs are more dependent on their public image and reliant on the public's confidence.

You Quanzi also used more traditional bureaucratic strategies in his battle with the Institute. He appealed for support to the leadership of CAS, the leadership level above the institute, going over the head of his immediate parent unit. You made his case that the Computer Institute was demanding an excessive portion of the company's profits, and that the resources could be better spent by the company. You secured the backing of top CAS officials, and with this gained an edge over the Institute. In the end You managed not to be fired and he was said to have won a more favorable profit-sharing arrangement for the company. You's case was among the more contentious and publicly contested. It shows the considerable leverage that the managers of large profitable spin-offs exercise vis-à-vis their parent units, and the informal way that profit-sharing agreements are worked out.

It further shows that even in one of the most contentious cases, parent units and firms worked out agreements through informal bargaining.

Bargaining over shareholding. By 1993, in response to changes in local Haidian regulations which allowed shareholding among firms registered there, some companies began negotiating shareholding arrangements with their parent units which would legally determine the ownership of company assets. Demands for such arrangements were generally initiated by spin-offs, not parent units. In the bargaining over shares, three types of shares were defined: shares belonging to "the state" (in effect the parent unit), shares belonging to the company, and shares for individual employees. Regulations in place in 1993 limited the portion of individual employee shares to 15 percent, leaving 85 percent of the shares to be divided between the parent unit and the company as a collective. And in some companies no individual shares were to be assigned, with all shares being assigned either to the parent or the company collectively. Company shares were in principle to be used for collective goods which would benefit all employees, including a variety of services, cafeteria and food services, recreational facilities, building improvements, etc.[82] Shares belonging to the parent unit were also in principle to be used for similar types of collective benefits for the employees in the parent institution.

The fact that negotiations were being conducted at all confirms that spin-offs and their employees were recognized by their parent units as having a legitimate proprietary claim over the assets of these companies. The specifics of the negotiations did reflect significant variation in perceptions of how shares should be divided. The managers of Lian Xiang, for instance, initially suggested that 60 percent of the company's assets be assigned to "the state," meaning CAS, while 40 percent be assigned to the company. CAS officials rejected this proposal, offering only 5 percent of shares to the company. They argued that CAS's initial investment of 200,000 rmb in the company gave it primary ownership rights over company assets. Company managers, on the other hand, argued that the ideas, labor, ingenuity, and effort invested by the individual employees of the company should count for more than that.[83] In the case of Lian Xiang all shares were to be divided between CAS and the company, with no shares being assigned to individual employees. However, negotiators for Beida Fangzheng had, by the end of 1993, reached agreement that 15 percent of shares should be assigned to individual employees. However, they were unable to agree on how to divide the remaining 85 percent of the shares, with both sides demanding a majority of 50 percent, leaving 35 percent to the other side. No compromise had yet been worked out, although the two sides appeared to be closer to an agreement than Lian Xiang.

Property Rights over Haidian's Computer Spin-Offs

It is obvious that ownership of the capital assets of Haidian's computer spin-offs was not legally clearly defined. The process of negotiating shareholding arrangements was precisely an effort to bring legal clarity to this issue. And the determination of the ownership of fixed shares would, to the extent this was accomplished, help to clarify the question of ownership of spin-off firms. However, no firms in the case study had yet succeeded in doing so, and the majority of spin-offs remain without legal clarification of the ownership of their assets. In light of this, how can the "ownership" over this type of firm be understood? To address this we can return to the framework provided by the property rights literature and asses who actually exercises the range of rights associated with ownership, including the rights to utilization, the rights to returns from assets, and the right to sell or exchange the resources of spin-off firms.

The first right associated with ownership is the right to utilization. In the case of the firm this right is associated with the right to conclude contracts for the company, including employment and other types of business contracts, and the right to "unilaterally fill in the details of incomplete contracts with some agents," e.g., employees.[84] The case study shows that managers in practice exercise considerable autonomy in determining contracts and business deals, including those with employees, foreign and domestic business partners, and others. Managers exercised considerable autonomy in determining employment contracts, in setting up joint ventures with foreign partners, in setting up domestic subsidiaries, in contracting with suppliers, etc. However, managers do not entirely monopolize this right. Parent units retain authority over certain vital decisions, including managers' own employment contracts. So with regard to the right of *utilization*, spin-offs are thus characterized by a significant degree of managerialism—de facto control exercised by managers over a wide range of key decision-making, coupled with selective rights retained by parent organizations.

The rights to the returns from the assets of the firm, the second key component of the ownership bundle, are similarly shared between companies and parent units. While legal ownership of capital assets was not clear, informal bargaining between spin-offs and their parent units produced specific arrangements regarding how the profits would be divided. While these arrangements were often contested and contentiously bargained over, they nevertheless produced mutual agreements regarding which side would get what. This points to the fact that in the absence of legal regulations, informal mechanisms served to determine the de facto exercise of rights.

Regarding the third major type of property right, the right to sell, exchange, or transfer the assets of property, what is striking about spin-offs is the fact that this right is restricted for all interested agents. No set of agents has a clear right to sell these firms, or even to sell portions of their capital assets. Without clear ownership of capital assets, there is no way for these firms or for portions of their assets to be transferred or sold. There were also restrictions on the right to sell or exchange assets even in the event of the successful resolution of shareholding arrangements because of the collective nature of a large portion of the shares. While employees could enjoy the collective shares of the company as long as they were members of this community, no one had the right to sell or exchange the collective shares. According to Haidian District Zone regulations, the shares belonging to either parent units or companies were not subject to alienation from the firm.[85] At the time it was also reported that individual shares would not be allowed to be sold or exchanged and could only be held by current company employees. This significantly restricts ownership's key right of sale or exchange of property and means that a large portion of the assets of spin-off firms cannot be transferred.

Conclusion

To return to the broad question set out at the beginning of this article, How does this case study of one set of computer and high-technology firms contribute to our understanding of China's unorthodox market economy? The evidence suggests that Haidian's computer spin-offs help explain three features of China's unorthodox market economy. First, they help explain the successes of China's market transition despite the proliferation of enterprise types with apparently "vaguely defined" ownership and the absence of a well-enforced property rights regime. In the absence of clearly defined ownership of capital assets, informal norms and mechanisms served in practice to define the de facto exercise of property rights. Firm managers worked out customary arrangements with their parent units regarding the exercise of property rights through informal norms and bargaining. These permitted the various claimants to know specifically the extent of their customary rights over the returns from the firm. This accords with the property rights framework, which posits that property rights may be determined through custom, culture, historical precedent, informal norms, or legal structures.[86] Property rights are not necessarily more "fuzzy" simply because they are informally determined. As in the case of Haidian's computer firms, ownership was vaguely defined in a legal sense, but very specifically determined through customary practice. While this process was often contentious, it nevertheless produced specific arrangements which determined the distribution of profits between the major claimants.

This further suggests that the central state need not be the primary mechanism for enforcing property rights arrangements, as is often assumed in Western economic thought, but that these may also be enforced through less formalized, less systematized regulations operating in local political economies. In Haidian the threat of sanctions by local officials served to deter spin-off managers from breaking their informal customary arrangements with their parent units. The case more functional in determining and enforcing property rights in the absence of a nationally integrated, legally clarified, and formally enforced property rights regime than idealized Western models of well-functioning market systems allow for.

Second, the case study helps explain how China's economy has had successes typically associated with a market system based predominantly on private enterprise, despite the small size of its private sector. Spin-offs, despite being non-private enterprises, capture some of the critical advantages typically associated with private enterprise. Spin-offs operate under relatively hard budget constraints, since they are not integrated into state budgets, and their wages, benefits, and other operating expenses are not guaranteed by state funds. These budget constraints are further tightened by the limitations of the budgets of parent units in contrast to governments, particularly in the case of universities and research institutes. Spin-offs, through the system of decentralized property rights exercised by myriad and diverse state entities, contribute to the competitive structure of the market, a feature also typically associated with private enterprise. The decentralized property rights, by granting considerable autonomy and controls to spin-off managers, also create an effective incentive structure which further captures some of the features associated with private enterprise. It might even be argued that because spin-off managers exercise both wide-ranging controls and considerable property rights, that spin-offs as an enterprise type are not as hindered by the problem of the separation of control and ownership as the large modern corporation (or the state socialist enterprise). Spin-offs, finally, are not, despite their non-private status, primarily social and welfare institutions one of whose key functions is sustaining the livelihood of their employees, but are market-oriented enterprises driven by the profit motive. For these and other reasons, spin-offs are able to capture important features associated with private enterprise, and thus help explain China's market transition in the absence of systematic privatization of the economy.

Finally, the case study helps explain the unorthodox blurring of the boundaries between the state and the market which has been one of the primary features of China's market system. Conventional views treat "the state" as a unitary and homogeneous entity and argue from this basis that if "the state" is closely involved in the market and is a key property holder that it will be unable to neutrally enforce the property rights regime more generally. However, the political economy of Chinese government spin-offs shows how a highly differentiated state allows for sub-units of the state to act as property holders while higher levels of the state can continue to perform more traditional state functions such as taxation and regulation. And the devolution of property rights to component sub-units of the state does not, as we saw above, bring with it the kind of negative features typically associated with state ownership such as monopoly and lack of competition, poor incentive structures, and lack of managerial autonomy. China's spin-offs thus contribute to an understanding of China's rather unorthodox approach to the transition to a market economy.

Notes

1. The Chinese have a variety of terms which refer to semi-autonomous commercial ventures established by government units, but there has not been a single term which has been used consistently. One term used particularly in relation to the urban economy has been "jingji shiti," or economic entity. Another term is "shiye danwei," which has been used to refer to small-scale ventures set up by individual state employees. For instance, see "Geji Renmin Yinhang yu Suoban Jingji Shiti," *Renmin Ribao*, August 30, 1993.

2. For one variety of spin-off, the township and village enterprises (TVEs), there is considerable empirical evidence that shows that this category of firm has been one of the primary engines of China's rapid economic growth. A number of studies have demonstrated the contribution of TVEs to economic growth and suggest that they have performed as efficiently as private firms. See Dilip Ratha, Inderit Singh, Geng Xiao, "Non-State Enterprises As an Engine of Growth: An Analysis of Provincial Industrial Growth in Post-Reform China," World Bank Research Paper Series, CH-RPS-20. Mark M. Pitt and Louis Putterman, "Employment and Wages in Township, Village, and Other Rural Enterprises," mimeo, Brown University, 1992. Jan Svejnar, "Productive Efficiency and Employment," in W. Byrd and Q. Lin, eds., *China's Rural Industry: Structure, Development and Reform* (New York: Oxford University Press, 1990). However, most of the literature includes TVEs in the category of "collective enterprise," while urban spin-offs are still treated as state-owned. Here I am treating both TVEs and urban varieties of spin-offs in the same category, despite the fact that they may have important differences.

3. This view fundamentally characterizes classical and neoclassical economic thought and it has pervaded the literature on post-communist market transitions. For discussion and critique of this neoclassical framework see Louis Putterman and Dietrich Rueschemeyer, eds., *State and Market in Development* (Boulder, CO: Lynne Rienner Publishers, 1992) and Jan-Erik Lane, ed., *State and Market: The Politics of the Public and the Private* (London: Sage Press, 1985).

4. For a critique of this view see Paul Streeten, "Against Minimalism," in *State and Market in Development*, as well as other chapters in that volume. See also Michael Moran and Maurice Wright, *The Market and the State* (New York, NY: St. Martin's Press, 1991).

5. For this argument see Douglas North, *Institutions, Institutional Change and Economic Performance* (Cambridge: Cambridge University Press, 1990). See also Arman A. Alchian and Harold Demsetz, "Production, Information Costs, and Economic Organization," *American Economic Review* 62 (December 1972): 777–795. This view is also characteristic of most of the literature on market transitions in Russia and Eastern Europe. For example, see Ellen Comisso, "Property Rights, Liberalism, and the Transition from 'Actually Existing' Socialism," *East European Politics and Society* 5, no. 1 (Winter 1991): 162–188.

6. For one view of the blurred boundaries between state and market in China see Dorothy Solinger, *China's Transition from Socialism* (Armonk, NY: M. E. Sharpe, 1993), especially chapter 11, "Urban Entrepreneurs and the State: The Merger of State and Society."

7. Walder sees China's economic reforms as having brought about a reassignment and devolution of property rights as well as a reassignment of rights to returns from assets. See Andrew Walder, "Corporate Organization and Local Government Property Rights in

China," in Vedat Milor, ed., *Changing Political Economies* (New York, NY: Lynne Rienner Publishers, 1992).

8. Douglas North and Barry Weingast argue that secure property rights are critical for economic development. See North and Weingast, "Constitutions and Commitment: The Evolution of Institutions Governing Public Choice in Seventeenth-Century England," *England Journal of Economic History*, no. 49, December, 1989. For further discussion of the property rights paradigm see Alchian and Demsetz, "Production, Information Costs, and Economic Organization"; Harold Demsetz, "Toward a Theory of Property Rights," *American Economic Review* 2, no. 57 (May 1967): 347–359; Eirik G. Furubotn and Svetozar Pejovich, "Introduction: The Property Rights Literature," in E. Furubotn and S. Pejovich, *The Economics of Property Rights* (Cambridge, MA: Ballinger Press, 1974), pp. 1–9.

9. The portion of national output value contributed by state-owned enterprises fell from 77.6 percent in 1978 to 48.3 percent at the end of 1992, while that of collective industrial enterprises rose from 22.4 percent to 38.2 percent. Foreign-invested, self-employed, and private industrial enterprises accounted for only 13.5 percent of output value in 1992. See "Progress in Property Rights System Reform," *Renmin Ribao*, 25 February 1994, p. 2. From FBIS-CHI-94-053, p. 53.

10. The literature on this is immense. Some examples include most of the contributors to Bruno Dallago, Gianmaria Ajani, and Bruno Grancelli, eds., *Privatization and Entrepreneurship in Post-Socialist Countries* (New York, NY: St. Martin's Press, 1992); and Anthony Jones and William Moskoff, eds., *The Great Market Debate in Soviet Economics* (Armonk, NY: M. E. Sharpe, 1991).

11. For a discussion of the arguments against public property see Carol Rose, "The Comedy of the Commons: Custom, Commerce, and Inherently Public Property," *The University of Chicago Law Review* 53, no. 3 (Summer 1986): 711–781.

12. Carol Rose, "The Comedy of the Commons," p. 711.

13. For a variety of views on state-owned enterprises in the context of advanced industrial societies see Raymond Vernon and Yair Aharoni, eds., *State-Owned Enterprise in the Western Economies* (New York, NY: St. Martin's Press, 1981).

14. The research is based on a case study of eleven computer and high-technology firms located in Haidian District, Beijing which was carried out over a two-and-a-half-month period from July to September, 1993. The size of the companies ranged from around 20 to 3,100 employees. One of the firms had 3,100 employees, one had 1,400 employees, four others had 6–400 employees, and three had fewer than 50 employees. Interviews were conducted with 25 employees or former employees of these companies, which included top and middle level managers, managers in charge of personnel, public relations, and sales, technicians, sales people, regional branch representatives, division heads, and rank and file staff members.

15. Blackstone, Commentaries. For reference see Carol Rose, "The Comedy of the Commons," footnote no. 2, p. 711.

16. It has been pointed out that from this perspective there can only be "private property," in the sense of there being a singular, identifiable owner (whether an individual or single collective of individuals). "Public property," in the sense of resources being left to or "owned" by the public, would in this perspective be an oxymoron. See Carol Rose, "The Comedy of the Commons: Custom, Commerce, and Inherently Public Property," p. 712.

17. For one of the earliest statements of the property rights paradigm see Alchian and

Demsetz, "The Property Rights Paradigm," and Harold Demsetz, "Toward a Theory of Property Rights."

18. Much of the property rights literature also makes the argument for the "nexus of contracts" model of the firm. See Jensen and Meckling, "Theory of the Firm: Managerial Behavior, Agency Costs, and Ownership Structure," *Journal of Finance and Economics* 305, no. 3 (1975); Oliver Williamson, "Corporate Governance," *Yale Law Journal* 1197 (1984).

19. Martin L. Weitzman and Chenggang Xu, "Chinese Township-Village Enterprises as Vaguely Defined Cooperatives," *Journal of Comparative Economics* 18 (1994): 121–145. Jean Oi, "Fiscal Reform and the Economic Foundations of Local State Corporatism in China," *World Politics*, October 1992.

20. Alchian and Demsetz, "The Property Rights Paradigm," p. 17.

21. For an overview of how this has operated in the context of regional governments see Jean Oi, "Local State Corporatism," *World Politics.*

22. For an overview of science and technology policy in the reform era and the state's encouragement of universities engaging in the high-tech industry, see Richard Suttmeier, "China's High Technology: Programs, Problems, and Prospects," in *China's Economic Dilemmas in the 1990s: The Problems of Reforms, Modernization, and Interdependence*, edited by the Joint Economic Committee, Congress of the United States (Armonk, NY: M. E. Sharpe, 1993), pp. 546–564, and Erik Baark, "Fragmented Innovation: China's Science and Technology Policy Reforms in Retrospect," ibid., pp. 531–545.

23. According to one report the government slashed by 70 percent the operating budgets of the 117 institutes of CAS (Chinese Academy of Sciences). See "Dozens of High-Tech Firms Started by University Researchers," *Science*, October 15, 1993.

24. "Li Peng on Trading Autonomy for Research Groups," *Beijing Xinhua News*, 12 March 1993, see FBIS-CHI-93-047, 12 March, 1993, p. 11. "Colleges Developing High-Tech Industry," Beijing, Xinhua News Agency, 28 March 93. FBIS-CHI-93-059, 30 March 1993, p. 66.

25. Author's interviews, 1993.

26. The other kind of firm operating in Haidian's high-tech zone is the "min-ban jiti" firm, a type of collective firm set up through a partnership among several or more founders. *Minban* firms operating in Haidian were not allowed, up until 1993, to establish a shareholding system that determined the portion of assets which each partner owned. In fact, up until 1993 the founders of minban companies were not allowed to extract profits based on their initial investment. For elaboration on the operation of *minban* firms in Haidian see Corinna-Barbara Francis, unpublished manuscript "The Reproduction of *Danwei* Institutional Features in the Context of China's Market Economy: The Case of Haidian District's Hi-tech Sector."

27. According to one report the Ministry of Machinery and Electronics Industry operates some 150 commercial spin-offs. "Dozens of High-Tech Firms Started by University Researchers."

28. "Dozens of High-Tech Firms Started by University Researchers."

29. In his overview of science and technology policy in post-Mao China, Suttmeier discusses the legacy of the failure of the old system to link basic research with the civilian economy, despite successes in the research in certain strategic technologies. See R. Suttmeier, "China's High Technology."

30. This differs from the situation in Western corporations, where employees tend not to want to work in regional branches or subsidiaries because it is difficult for them to subsequently get back into the company's mainstream career track which leads to top-level management positions. As a consequence, second-tier employees tend to be assigned to work in subsidiaries or regional branches. The situation in Chinese computer and electronic spin-offs is thus the opposite, where employment in the subsidiary offers more attractive career options and remuneration.

31. The Chinese refer to this as maintaining their "personnel links" (*renshi guanxi*) with their parent unit. In this case the dossiers of the transferred personnel generally stay with their original unit. Author's interviews, 1993.

32. Because of this, informants often had difficulty pinpointing exactly when a company was established.

33. This division was said to be supervising more than 1,000 scientific and technological research projects carried out within the university. Author's interviews, 1993.

34. One rationale for delaying the formal establishment of a company may be to make the best use of the three-year tax shelter that Haidian's High Technology Zone offered newly established companies. If a company is registered too early, before it is making much of a profit, it does not get the most benefit out of this tax shelter. The most logical strategy is to receive the tax shelter when the company is already making a sound profit. Delaying the formal establishment of a spin-off may also allow the parent institution to maximize the use of "public" funds to nurture its commercial venture. Many of the technologies used by spin-offs are developed through government-funded projects. By using state-allocated monies to develop technologies, and then commercializing them through spin-offs, parent institutions are able to minimize their direct costs and maximize the returns to their corporate interests.

35. This does not minimize the difficulty of nor the time and resources required for the develop process, but simply stresses the advantages of the availability of some primary technology.

36. This was a software system with potential applications for use by doctors in prescribing Chinese medicine, and could be used by hospitals to reduce the man-hours put in by doctors.

37. While the project had initially been approved and funded by the government in 1974, it is not clear how many years of research it subsequently entailed. According to informants, however, it was a "major" project that involved numerous researchers over many years. Author's interviews, 1993.

38. This could also also be seen as the failure on the part of many spin-offs to generate new technologies and derivative products.

39. This was the same farmers' collective that owned the factory that the university was purchasing for setting up the manufacturing process. Author's interviews, 1993.

40. Author's interviews, 1993.

41. For instance, 50 out of the 250 faculty members of Beijing University's chemistry department were reported to be involved in a commercial venture to produce chemicals and machinery. According to the same report CAS plans eventually to transfer two-thirds of its staff to spin-off companies, leaving only 15 percent of staff to do basic research. "Dozens of High-Tech Firms Started by University Researchers."

42. The method of hiring employees from the labor market is referred to as the "zhaopin" method (meaning to hire). This is in contrast to employees who are *transferred*

from the parent unit. Employees are clearly divided between *transferred* and *hired* personnel. Author's interviews, 1993.

43. Author's interviews, 1993.

44. Author's interviews, 1993. The remaining 10–20 percent are either transferred from CAS or from other state work units through the traditional personnel system. Interviewee No. 3.

45. The question of whether a new personnel system is being constructed which would tighten the controls over personnel is an interesting one. This is considered in Corinna-Barbara Francis, "The Reproduction of Danwei Institutional Features in Haidian's High Technology Sector." Interviewee No. 3.

46. This is the basis of Jean Oi's argument regarding the emergence of local state corporatism. Oi, "Fiscal Reforms in Post-Mao China." Andrew Walder makes a similar point with regard to TVSs. See "The Industrial Organization of Zouping County: Agency and Ownership in Local Public Enterprise," prepared for presentation for the panel "Property Rights, Institutions, and Social Structure in China," Annual Meeting of the Association for Asian Studies, Boston, March 24–27, 1994.

47. In some companies it was difficult to arrange to meet the managers because they were always gone, either in Shenzhen doing real estate deals, or in Hong Kong planning joint ventures with foreign investment firms, or some other sort of activity which had little to do with the supposed activities of the company.

48. Real estate appeared to be one of the most popular investments. One company had invested in a Shenzhen real estate company together with six other companies. Interviewee No. 10.

49. According to one manager, these regulations were relaxed after 1993. New regulations required only that 10 percent of after-tax profits be reinvested. In response, at least some companies raised the proportion of profits used for benefits, while decreasing the proportion that they reinvested. Lian Xiang increased the percentage of profits used for benefits and bonuses to 60 percent, and decreased the proportion going to reinvestment to 40 percent. Interviewee No. 3.

50. Author's interviews, 1993.

51. The hypothesis that enhanced managerial autonomy tends to lead to higher productivity is the basis of China's reform of state-owned enterprises. The empirical evidence suggests that the relationship holds. See Theodore Groves, Yongmiao Hong, John McMillan, and Barry Naughton, "Autonomy and Incentives in Chinese State Enterprises," *Quarterly Journal of Economics*, 1994, pp. 183–209.

52. For discussion of the soft budget syndrome in state socialist economies see Janos Kornai, The Economics of Shortage (Amsterdam: North-Holland, 1980); Janos Kornai, "The Soft Budget Constraint," *Kyklos* 39, no. 1 (1986): 3–30; and Janos Kornai, "The Hungarian Reform Process: Visions, Hopes, and Reality," *Journal of Economic Literature*, no. 24 (December 1986): 1687–1737.

53. The revenue generated by subsidiary companies may be several times that of a research institute's core funding. For example see "Shanghai Enlists Scientists to Foster Economic Growth," *Science*, Vol. 265, August 12, 1994.

54. At the national level, the Ministry of Electronics had had a dominant position in the developing electronics industry in the earlier stages of reform. Now, however, a range of ministerial level state agencies are involved in this sector, including the Ministry of

Aviation and Aerospace, the Ministry of Light Industry, the Ministry of Defense, and the Ministry of Machine Building and Electronics.

55. Ellen Comisso makes this argument in "Property Rights, Liberalism, and the Transition from 'Actually Existing' Socialism," *East European Politics and Societies* 5, no. 1 (Winter 1991): 162–188.

56. David Granick, *Chinese State Enterprises: A Regional Property Rights Analysis* (Chicago: The Chicago University Press, 1990).

57. Up until 1993 local regulations operating in Haidian's High Technology Zone prohibited shareholding and the payment of dividends. Author's interviews, 1993.

58. In late 1993 informants reported that regulations had changed and that firms were then allowed to implement shareholding arrangements among the interested parties, although enterprises still needed permission to publicly sell shares. Author's interviews, 1993.

59. So in the case of spin-offs there is not only the problem of the separation of ownership (of securities) and (managerial) control that is characteristic of the corporation, but the more basic problem of the lack of owners of securities.

60. How long this grace period lasts varies depending on how quickly the company begins making a profit and how solid the profit base is.

61. Author's interviews, 1993. Interviewee No. 11.

62. Because of the sensitive nature of this question, management was very hesitant to divulge this sum, and so it was not always possible to find out exactly how much firms turned over to their parent unit.

63. The company was founded in 1989. Interview No. 10.

64. Author's interviews, 1993.

65. Interviews Nos. 13 and 16. Beijing University as we saw above invested comprehensively in Beida Fangzheng's initial personnel, start-up capital, and start-up technology.

66. As we saw above the company had been set up as a joint venture with a group of overseas Chinese who had invested 60 percent of initial equity, with the remaining 40 percent from individuals. Interview No. 19, Beijing, September 8, 1993.

67. It did, however, pay a "management fee" of 6,000 rmb/year per head for each institute employee working in the company. Since around 20 or so institute employees worked for the company, it paid roughly 120,000 rmb/year to the institute. Author's interviews, 1993.

68. Much of the technology produced under the socialist system went unexploited, gathering dust on university shelves. While the market now allows and provides the incentive for the commercialization of technologies, attitudes are perhaps slower to change. See Suttmeier, "China's High Technology Programs."

69. For a survey of the literature on property rights in developing countries see Wolfgang E. Siebeck, editor with Robert E. Evenson, William Lesser, and Carlos A. Primo Braga, *Strengthening Protection of Intellectual Property in Developing Countries.* World Bank Discussion Paper, No. 112, 1990.

70. The weak intellectual property rights regime in China should at the same time be seen as facilitating the commercialization of technology since it allows the privatization of what could justifiably be said to be public property. Since universities and research institutes are public institutions, the product of their research should be considered public goods.

Certain technologies, furthermore, are funded by specific government grants, which makes a unversity's or research institute's property claim even less justified.

71. The director of the Software Institute, for instance, concurrently held the position of president of the Capital Computer Company, one of companies set up by the institute. Interviewee No. 10.

72. In a pattern somewhat reminiscent of the traditional state enterprises in which top managers are very often promoted from within the enterprise, the new (and current) manager of Beida Fangzheng, a former head of division, was promoted from within the company. Interview No. 13.

73. For a discussion of the agency problem see Eugene F. Fama, "Agency Problems and the Theory of the Firm," *Journal of Political Economy* 88, no. 21 (1980): 288–307, and Victor Brudney, "Corporate Governance, Agency Costs, and the Rhetoric of Contract," *Columbia Law Review* 85, no. 7 (November 1985): 1403–1444. According to arguments made by property rights theorists, owning stock in a corporation does not make security holders the "owners" of "the firm."

74. Employees who leave their jobs without the permission of their parent unit (and as a consequence do not have their dossier released by their work unit) irrevocably lose their civil servant status and all the benefits that go along with this. Author's interviews, 1993.

75. This is particularly so among employees of Beijing University or CAS, who are even less likely to cut their ties with their parent unit than employees from smaller, less prestigious work units. According to a manager in the personnel department overseeing personnel matters in the spin-offs set up by CAS, very few of the 3–4,000 employees who transferred to these spin-offs from CAS have left these companies. To do so would mean severing their ties with CAS and losing all the benefits of their cadre status. Furthermore, there is a strong institutional culture and sense of familiarity which keeps people within the system. According to this manager, these people have personal ties to CAS—"dui kexueyuan you ganqing." Interviewee No. 2.

76. Beida Fangzheng had been able to capture much of the publishing market. Out of a total of 1,850 major newspapers in China, 1,200 had by 1993 adopted Beida Fangzheng's software technology. Of the 5,000 large and medium sized publishing houses, 3,000 had adopted this technology. Interview No. 16. Author's interviews, 1993.

77. Another reason that numerous small companies are set up is that individual scientists and researchers would rather set up their own commercial venture rather than attaching themselves to an existing one. Informants frequently stated that most people would rather be their own boss. Author's interviews, 1993.

78. Author's interviews, 1993.

79. As we saw above, however, parent units can be helpful in securing government contracts and other institutional customers.

80. Author's interviews, 1993.

81. In 1992 it reported 1.5 billion yuan in gross profits. Author's interviews, 1993.

82. Since no shareholding arrangements had yet been finalized, there was uncertainty among informants regarding exactly how these shares would in the future be used. Author's interviews, 1993.

83. Author's interviews, 1993.

84. Louis Putterman, "Ownership and the Nature of the Firm," *Journal of Comparative Economics* 17 (1993): 243–263, see p. 246.

85. While it was not clear at the time what the laws regarding shares would be, it appeared that there would be restrictions on the sale of individual shares. Author's interviews, 1993.

86. Carol Rose, "The Comedy of the Commons: Custom, Commerce, and Inherently Public Property." David Granick refers to regional government property rights, even though he recognizes that these do not "factually," or legally, exist. See Granick, *Chinese State Enterprises*.

16

Financing the Commercialization of Russian Defense Technologies: Venture Capital and Related Transactional Structures

Larry W. Schwartz

Introduction

That the success of Russian defense conversions is largely dependent upon the development of a fully functional free market economy is a truism. Correspondingly but perhaps less conspicuously, the entrepreneurial entities arising out of these conversions help to set the stage for venture capital to play a more prominent role in the commercialization of Russian defense technologies. While venture capital is certainly not a panacea for all cash-strapped companies, it is nonetheless an increasingly viable alternative for financing the growth of spin-offs and other emerging growth companies in the Russian defense industry.

In the wake of Russia's free market reforms, an increasing number of Russian companies are experimenting with a variety of financing alternatives. Many of these companies have already been frustrated by their inability to obtain favorable financing terms from Russian banks. In Russia as in the United States, this financing gap is particularly prevalent among emerging-growth companies in technology-driven industries.

Nowhere is this financing void more noticeable than in the recent spin-offs of entrepreneurial groups from larger defense companies. Indeed, venture capital may be one of the few means of financing growth for spin-offs and other newly formed, technology-driven companies in Russia. While these spin-offs constitute an integral part of the first step of the defense conversion process, a logical second step should include a primary role for venture capital to nurture the growth of entrepreneurial seeds planted by this conversion process.

This paper begins with an overview of the general role of venture capital and an analysis of the factors that will help increase venture capital activities involving Russian technology companies. Certain transactional structures that involve venture capital or venture-backed companies are then proposed. Next, venture-related issues that Russian technology companies should anticipate addressing in their attempts to solicit such investors are examined. While this paper focuses on efforts to attract Western venture capital, much of the discussion is also relevant to indigenous, Russian venture capital. As will be evident by the following discussion, there are certain concerns that arise in any venture financing, regardless of where such funding originates.

The Role of Venture Capital in the Marketplace

The European Venture Capital Association defines a venture capitalist as an individual or group that provides equity or quasi-equity financing for start-up companies and for the development of small and medium-sized privately held enterprises that have significant growth potential in terms of products, technology, business concepts, and/or services. The investment is combined with active management support to investees. This management support is a value-added role of the venture capitalist that distinguishes a venture investment from other forms of direct investment. A venture capitalist may, in addition to his equity investment, become a member of the investee's board and may also provide the investee with access to outside experts, additional investors, and other contacts for strategic development purposes. The main objective of the venture capitalist is to recognize a long-term capital gain as remuneration for his risk.

Notwithstanding that the term "venture capitalist" may connote that such investor is an individual, that is usually not the case. The typical venture capital investor is a firm that sets up a fund (generally formed as a limited partnership) and solicits institutional investors to contribute capital to the fund in exchange for interests therein (i.e., limited partnership interests). The venture capital firm will serve as managing partner of the fund and identify prospective opportunities for investing the fund's money.

The role of venture capitalists in the marketplace is to provide entrepreneurial firms with an access to cash that would not be available elsewhere. William D. Bygrave and Jeffrey A. Timmons, in *Venture Capital at the Crossroads*, describe venture capital as "patient and brave" money. Their reason for this description is that young companies have voracious appetites for cash but lack solid customer bases and have little or no revenues. An additional factor that is especially relevant to technology-driven companies is that the limited tangible property and often speculative valuations of intangible property held by such companies discourage lenders seeking to have their loans collateralized. Since traditional lenders are usually reluctant to view these companies as desirable credit risks, and founders usually do not have sufficient cash of their own, venture capitalists fill this financing gap.

There is a direct relationship between venture capitalists and entrepreneurial enterprises. Venture capitalists serve as catalysts to the entrepreneurial process by seeking out, investing in, and assisting privately held enterprises that have significant growth potential. The two groups feed off of each other and are mutually dependent. In fact, some of today's most well-known companies (including Federal Express, Microsoft, Blockbuster Entertainment, Apple Computer, and Sun Microsystems) were venture-backed companies.

To attract and keep venture capital activity, a market must provide an environment that nurtures and is supportive of entrepreneurial businesses. Without such a climate, a venture capital firm will not be able to develop an adequate deal flow. "Deal flow" refers to the process by which a venture capitalist identifies entities that are prospective recipients of venture funds. For example, New Enterprise Associates (one of the largest U.S. venture capital firms, with approximately $680 million under management in several funds) reportedly reviews about 250 businesses per year and typically invests in only fifteen to twenty of these. The inability to provide a healthy entrepreneurial climate leads to a shortage of entrepreneurial businesses. The resultant lack of deal flow discourages venture capital activities and further harms the entrepreneurial process and the economy as a whole.

Factors Influencing the Growth of Venture Capital in Russia

There are a number of economic and financial trends that will help to increase venture capital activities in Russia: first, the recognition that entrepreneurial firms and innovation are keys to economic growth; second, the overall escalation in strategic and financial investments in Russia; third, recent developments in applicable intellectual property laws; fourth, the increased use of spin-offs as an integral part of the defense conversion process in both Russia and the United States; and fifth, the globalization of venture capital activities.

Entrepreneurs and Innovation

In both Russia and the United States, there is an increased acceptance of the proposition that a modern economy can not develop to its full potential unless it nurtures and encourages the innovation of its entrepreneurs. For a start-up firm or a conglomerate, and for a developed country or a developing one, innovation and technology play critical roles in economic development. The driving force behind innovation and technology is the entrepreneurial process. John Naisbitt, author of *Megatrends*, has recently emphasized that as the world economy becomes bigger and more open, small and mid-sized entrepreneurial companies will dominate. As if on cue, a number of large corporations, including AT&T, General Electric, DuPont, Pepsico, MCI, IBM, Chevron, and Xerox, have publicly announced that they are attempting to emulate smaller, more entrepreneurial businesses. By that, they mean providing quicker times to market, eliminating bureaucracy, and creating greater efficiency.

In short, economies of scale have been replaced in many cases by a concept of diseconomies of scale as a result of fragmentation of markets and the accelerated pace of technological change. While there still remain certain capital-intensive industries where the power of size is critical, many other industries no longer reflect the conventional wisdom. Instead, the emphasis is on being innovative, dynamic, quick to market, and close to the customer.

Strategic and Financial Investments

With the development of a free market economy, strategic investments and financial investments in Russia have both increased at astonishing rates. "Strategic investments" refer to acquisitions, joint ventures, alliances, licensing arrangements and the like that have been entered into by corporations as part of their global strategy. "Financial investments" refer to investments made by institutional investors in the listed securities of companies based in Russia or in funds that, in turn, are making such investments.

In the past three years, more than 18,000 new private companies have been formed and more than 70,000 state enterprises have been at least partially privatized in Russia. As a result, corporations from developed countries have identified Russia and certain of the other former Soviet republics as markets that offer fertile opportunities. For example, in the eighteen months ended March 1993, foreign companies struck 1,700 deals in Eastern and Central Europe, valued at $42 billion. The number of ventures, start-ups, and purchases entered into in that region by Western companies nearly tripled in the six months ending in March 1993, to 979 from 374 in the prior six months.

The increase in strategic investments provides greater opportunities for venture capital exit strategies. In a developed market, a venture capitalist's primary exit strategy is generally via an initial public offering (IPO) or via an acquisition. Unfortunately, the lack of liquidity currently offered by existing Russian stock exchanges (discussed below) precludes the venture capitalist from considering an IPO as a presently available means of exit. As a result, venture capitalists in Russia place greater emphasis on exiting by way of an acquisition. Hence, the increase in strategic investments by companies in Russia causes a corresponding increase in the feasibility of using an acquisition as an exit strategy in this market.

The amount of financial investments in Russia by institutional investors has also escalated. Beginning in 1990, high levels for U.S. stock values, low interest rates, and slow U.S. growth, contrasted with free market reforms and rapid growth in Russia and other developing markets, pushed many U.S. institutions (including pension funds) to look further afield to try to maintain their accustomed gains. For example, U.S. pension funds increased their foreign holdings by $53 billion in 1993 and approximately 25 percent of these investments were in emerging markets. By 1996, U.S. pension funds are expected to increase their overseas investments by an average of 50 percent.

The increase in financial investments is important insofar as many of the institutional investors that are making these investments are the same ones that invest in venture funds. This will facilitate the ability of a venture capitalist to raise capital for such funds following their formation. For example, the U.S. pension and mutual-fund system is the largest source of venture money in the world, representing approximately 40 percent of new committed capital for U.S. venture funds. Although most investments to date in Russia and other developing countries by pension funds have been in exchange-listed securities, a number of these investors have expressed their intention to increase their private investments in developing countries. This would be consistent with previous investment trends. Pension funds and other institutional investors often take a first step in developing countries by making investments in the securities of listed companies and then move on to make investments in venture funds formed to invest in those regions. For example, institutional investors first entered Mexico by investing in companies listed on the Bolsa and in the past few years funds have been formed with institutional investors' backing to make venture capital investments in that market.

These investment trends help to increase the future viability of venture capital activities in Russia. As a general matter, the strategic and financial investments being made now are made by many corporations and institutional investors who, until very recently, would have refused to even consider making such investments out of lack of understanding of such markets and fear of political risks. Now these investments are not only being considered, but are becoming an integral part of global investment strategies.

Intellectual Property Law Developments

Recent measures undertaken to clarify intellectual property rights will also encourage an increase in venture capital activity in Russia. The Patent Law of the Russian Federation, the Law of the Russian Federation on Trademarks, Service Marks and Identification of Place of Origin of Goods, and the Law of the Russian Federation on Legal Protection of Computer Programs and Data Bases were enacted in 1992, and a comprehensive copyright law was enacted in 1993. In addition, Russia and the other members of the Commonwealth of Independent States (other than Turkmenistan) have entered into an agreement with the World Intellectual Property Organization setting up the Eurasian Patent Convention. The Convention provides the means by which a patent applicant desiring to register a patent that covers all participant countries can do so with a single filing in the Eurasian Patent Office in Moscow.

As evidenced by the multitude of Russian-based technology brokers from the United States and elsewhere, these measures have helped to trigger a search for

Russian intellectual property for commercialization. While the significance of these developments to technology brokers may be self-evident, they are also central to piquing the interest of Western venture capitalists. As a general rule, venture capitalists focus on investments that can provide the potential for explosive growth in a four- to seven-year time frame. In the United States and other mature markets, such growth can often be found in high-tech industries. As a result, investments in high-tech, bio-tech and other emerging growth industries typically account for approximately 80 percent of all U.S. venture capital investments. Until recently, one of the biggest hurdles to attracting Western investment in Russian technologies was the uncertainty surrounding intellectual property rights. These new laws and conventions are a critical step in the effort to settle ownership and assignability issues and help pave the way for investors who recognize the technologies' commercial potential.

Spin-Offs

The increased emphasis on spin-offs from defense companies in both Russia and the United States will bring venture capitalists to the forefront of the defense conversion process. Insofar as spin-offs are recognized as technologically driven, emerging growth companies, they are prototypical candidates for venture backing. The reason for this is threefold. First, the value of these companies often lies in their technology, rather than in their tangible property, and many such companies have minimal revenues. As previously noted, traditional lenders are thus reluctant to lend funds, forcing these companies to consider private equity alternatives to satisfy their tremendous needs for funds. Second, to the extent that new technologies offer commercial application, they may be perceived to be in emerging growth industries. Third, entrepreneurial aspirations of these companies are often hamstrung by a lack of expertise in areas that venture capitalists can add value, such as in connection with marketing and distributing strategies.

Certainly the foregoing has not been lost on Lockheed Marietta Corp. When the Department of Energy selected the company in 1993 to manage Sandia National Laboratories, Lockheed was quick to emphasize an intention to form Technology Ventures Corporation (TVC) as an integral part of its emphasis on spin-offs. TVC's mission is to help attract venture capital to technology-driven start-ups. Lockheed's commitment to TVC (to date, Lockheed has invested close to $10 million in TVC) is consistent with venture capitalists' increased focus on finding commercial applications for defense technologies. For example, BDM Federal Inc. (which provides information technology, systems and services to federal agencies) and ARCH Venture Partners L.P. (a venture capital firm) recently formed AB Ventures to focus on commercializing technologies from Los Alamos and Sandia national laboratories. Start-up capital for promising ventures identified by AB Ventures will be funded by ARCH Venture Partners.

Global Venture Capital

Perhaps the most critical reason for the flow of venture capital to developing countries to increase in coming years is the fact that venture capital firms have been increasingly adopting global perspectives when contemplating fund formation and investment decisions. In 1980, venture capital markets were practically nonexistent outside the United States. By 1990, more than half of the $80 billion of new capital commitments worldwide was outside the United States. European venture capital commitments increased from $1.85 billion in 1985 to $6.28 billion by the end of the decade, and Japanese venture capital commitments jumped from $600 million to $1.3 billion in the same period. Venture capital disbursements increased in Europe from $1.9 billion in the same period. Venture capital disbursements increased in Europe from $1.9 billion to $4.6 billion and in Japan from $500 million to $1.8 billion during this period.

Nor is this growth limited only to developed countries. The International Finance Corporation (IFC), established by the World Bank in 1956 to help strengthen the private sector in developing countries, led the way into developing markets by forming Southeast Asian Venture Investment in 1983 (the first Asian venture fund). Since then, venture funds have been formed for a number of Eastern and Central European countries (including Russia, Ukraine, the Czech Republic, Poland, Kazakstan, and Hungary). Elsewhere, venture funds have been set up throughout Asia (including Singapore, Indonesia, Malaysia, the Philippines, Papua New Guinea, Sri Lanka, India, China, Thailand, and Vietnam), Central and South America (including Mexico, Chile, Brazil, and Argentina), and Africa (including the Africa Growth Fund, the Africa Fund, and the Mauritius Fund).

Alternative Transaction Structures Involving Venture Capital and Venture-Backed Companies

Set forth below is an outline of certain transactional structures that involve venture funds or venture-backed companies. It should be noted that this is not an exhaustive list of venture-related deals, and variations of structures described below are virtually infinite. Nonetheless, these structures exemplify the types of deals worthy of consideration by Russian technology companies seeking funds for the commercialization of their technologies and products.

Direct Investment by Venture Capital Firms

The traditional means by which a company obtains access to venture capital is via direct investment. Typically, the venture capitalist will receive an equity stake in the company, along with board representation, in exchange for cash consideration. In the course of negotiating this transaction, a variety of ancillary issues will also need to be addressed, including, e.g., minority investor protection provisions, rights of first refusal or first offer, registration rights (relating to public offerings of the company's stock) and other matters relevant to either the company's charter documents or to agreements among shareholders.

While a company may have rights to technology with commercial potential, this is certainly not the sole criterion by which the venture capitalist will assess a prospective investment. Of equal import is the viability of means by which the venture capitalist may exit his investment within a predetermined time frame. Although a venture capitalist is a long-term investor, his role is nonetheless a transitory one and usually lasts no more than seven years. At the outset, a venture capitalist will evaluate, among other things, the anticipated development of a public market for the company's securities and the expected level of acquisitions in the company's industry at the time of the desired exit.

Although a venture capitalist usually pays for his stock with cash, that need not always be the case. One possible scenario, for example, for a debt-ladened company would include a debt-for-equity swap. Under this approach, the venture capitalist first purchases the company's outstanding debt from its lender. More often than not, the venture capitalist will be in a better position than the existing lender to accurately assess the market value of the company's technology and its commercial potential. Thus, the venture capitalist should be able to purchase the debt for a negotiated price representing a portion of the amount outstanding. The company then issues shares of its stock to the venture capitalist in exchange for the forgiveness of this debt. If this swap would not provide the company with a needed cash infusion, then of course any combination of debt forgiveness and cash may also be used as consideration for the stock.

Alternatively, the venture capitalist may provide the company with an ability to restructure or refinance its existing debt to obtain more favorable terms for the company. For example, the venture capitalist may be able to arrange for the issuance of a guaranty or letter of credit that reduces a lender's credit risk. The services provided by the venture capitalist in this regard may be viewed as part of the purchase price for the shares.

Captive Venture Programs

An alternative source of venture funding for a Russian technology company may be found by approaching a corporation that has a captive venture program. A captive venture program (sometimes referred to as a corporate venture program) is one that has been formed by a corporation primarily for strategic investment purposes, rather than solely for financial gain. The captive venture investor seeks out investments that offer a potential synergy between the investor's and the investee's respective businesses. Most captive venture programs are managed by employees of the investing corporation and are either set up as internally managed subsidiaries or as functions within existing business units. The alternative form of a captive venture program involves hiring an unaffiliated venture capital firm to serve as general partner and to make investment decisions. The role of the investing corporation in this latter approach, which is referred to as a "dedicated" fund, is to serve as the sole limited partner.

The number of captive venture capital programs in the United States reached an all-time high of 100 in 1989. That number dropped to seventy-three by the end of 1993, largely due to the trend in a number of industries of de-emphasizing diversification. Nonetheless, the concept is one that continues to attract a number of companies. For example, Micron Investments Inc. was formed in 1993 by (and received $10 million in funds from) Micron Technologies Inc. to make equity investments in companies developing electronic technologies and products. And Adobe Systems recently contributed $25 million to a dedicated fund that has been formed with the help of Hambrecht & Quist for the purpose of making investments in the software industry.

From the perspective of a Russian technology company, there are a variety of reasons why a captive venture investor may be preferable to a more traditional venture capitalist. First, because the larger corporation is seeking access to new technology and products, it is often more receptive than a traditional venture capitalist to consider an investment in a start-up company. This helps to fill a gap created in recent years by many venture capitalists who have steered away from early stage financings in favor of management buyouts and other large investments (discussed below). Second, a captive venture investment can lead to a variety of strategic alliances between the Russian technology company and the investing corporation. Third, a captive venture investor generally asks for less equity in exchange for the same amount of funds than that insisted upon by a traditional venture capitalist in the investments that it makes. Finally, the lack of a public market for the Russian company's securities, which discourages traditional venture capitalists by limiting their exit strategies, is a moot point for a captive venture investor seeking a long-term strategic fit.

Corporate Partnering Deals with U.S. Venture-Backed Companies

Although there has been an increase in Western venture capital investments in Russia in recent years, the majority of institutional monies earmarked for venture capital funds and the majority of investments by these funds in technology companies continue to be based in the United States. As a result, Russian-based technology companies may be frustrated by their inability to attract the attention of these venture capitalists. Paradoxically, many of the U.S.-based companies into which venture funds are flowing are becoming increasingly active in Eastern and Central Europe. A corporate partnering transaction with a U.S. venture-backed company may help solve this conundrum for Russian technology companies.

The term "corporate partnering transaction" is used to refer to any ongoing interdependent relationship between two or more companies seeking to integrate at least two of the following elements: licensing, research and development, manufacturing, distribution, and equity or debt investment. The decision as to whether a corporate partnering deal should be done in the form of a contractual relationship, as opposed to via the creation of a new, separate legal entity (such as a corporation or a partnership), will be driven by a variety of accounting, tax, limited liability, cultural, managerial, and operational objectives. As a general rule, however, most of these alliances are set up without the creation of a separate legal entity. Most practitioners would agree that "except where the parties intend to create a new, truly independent business, separate entities frequently are more trouble than they are worth."

There are a variety of reasons why a company may desire to enter into a corporate partnering transaction. Chief among these may be a perceived need for funding, risk sharing, access to technology or expertise, access to distribution channels and customer bases, increased manufacturing capacity, or economies of scale. For a Russian technology company, additional advantages to such an alliance with a U.S. venture-backed company are that it helps to attract the interest of U.S. venture capitalists, provides a foothold in the U.S. market, and provides a reliable source of Western currency. The extent to which a company benefits from any of the foregoing factors will, of course, be dictated by the terms of the negotiated alliance.

Perhaps not surprisingly, emerging growth companies in Russia and in the United States share many of the same attributes. Both require tremendous cash infusions to support entrepreneurial growth, neither of them tends to own any significant real property, their value is largely derived from their technology, they have difficulty obtaining any significant debt financing, and they often recognize that their domestic market alone will not support their growth. An alliance allows each participating company to leverage its existing strengths in order to utilize the opportunities provided by the other. For example, the U.S. company is better positioned for distribution and sales in the U.S. and has greater access to venture capital. The Russian company offers an entree to Central and Eastern European markets and may provide needed services at a fraction of their cost in the West.

Venture Capital-Specific Issues Relevant to Investments in Russian Technology Companies

Notwithstanding that political and economic trends help to increase the viability of venture capital activities in Russia, a number of obstacles continue to impede the growth of this financing alternative. Generally speaking, Western individuals and institutions invariably face a host of concerns when contemplating a Russian-based transaction, regardless of the industry, the nature of the transaction, or the identity of the parties. Political risk, organized crime, burdensome bureaucracy, cultural and language barriers, inadequacy of adequate financial information, nonconformity to generally accepted accounting principles and countless other generic risk factors are well documented.

While Western investors and Russian technology companies certainly need to be comfortable with the manner in which these concerns will be addressed, they are not unique to venture capital-related transactions. Rather than rehashing a litany of those risks relevant to all emerging market investments, this section of the paper explores certain issues that are of particular concern to venture capitalists and to others involved in venture capital-related transactions. The harbinger of an increase in these deals for Russian technology companies will be their sensitivity to, and proposals to minimize, such venture capital-related concerns.

Limited Exit Strategies

The first rule of venture capital is to be mindful of exit strategies when making the investment. The typical means by which a venture capitalist will exit from an investment is via an IPO of the company's shares or an acquisition of the company by a third party. When both alternatives are feasible, the preferred route in the United States is usually by means of an offering. Experience in the U.S. market has shown that shares trade at more than ten times their original value when sold to the public in an IPO; in an acquisition, a venture capitalist looks to receive approximately five times the original value. While statistical evidence of the foregoing for Russia is hard to come by, the United States experiences of venture capitalists and venture-backed companies are indicative of the consideration given to the public offering process.

Although the number of stock exchanges registered in Russia was, at last count, seventy, most stock sales continue to be via over-the-counter transactions. For Russian technology companies, this creates a problem insofar as the exchanges do not currently offer a readily available source of liquidity and are not viewed as offering an exit strategy for venture capitalists. The absence of U.S.-style clearing and custodial systems, combined with deficient government regulation, have deterred most U.S. mutual and pension funds and other institutional investors from making significant investments in publicly traded stock of Russian companies.

This lack of a secondary market for shares has not gone unnoticed by Russian officials. A decree issued in November of 1994 by President Yeltsin ordered the formation of a federal securities and exchange commission; KPMG Peat Marwick has been developing a NASDAQ-like computer-based securities trading system in Russia; and KPMG and Deloitte and Touche are creating clearing and settlement organizations for Russian securities and currency trades and are creating broker-dealer associations in Moscow, St. Petersburg, and other cities. While these measures certainly bode well for future investors, they do not yet provide the exit strategy sought by venture capitalists.

What becomes evident in light of the shortfalls of domestic stock exchanges is that venture capitalists in Russia place a greater emphasis on exiting via an acquisition. However, that is not the only exit alternative. For example, a Russian technology company and a prospective venture investor would be wise to explore the possibility of an offering of American Depositary Receipts (ADRs). ADRs are negotiated certificates issued by U.S. banks and represent shares of stock in a non-U.S. company. They are traded on U.S. stock exchanges (e.g., NASDAQ, NYSE, and AMEX) like shares of stock and, as such, offer Russian companies an opportunity to access U.S. capital markets. Perhaps not surprisingly, the number of companies from emerging markets that have launched ADR or global depositary receipt programs has been escalating. For example, companies from emerging markets (including Russia) represented, in the aggregate, 56 percent of the depositary receipt programs undertaken in 1994, compared with only 11 percent in 1990. As a consequence of such programs, they raised $13.1 billion in 1994, accounting for 66 percent of all capital raised through depositary receipts. Of course, not all venture-backed companies will be able to undertake such an offering. Nonetheless, the depositary receipt alternative is indicative of the types of factors that should be addressed with venture capitalists.

Focus on People over Technology

"Build a better mousetrap and the world will beat a path to your door" is an adage of little relevance to technology companies seeking venture financing. Venture capitalists place a premium on the people behind the product, rather than on the product itself. That is not to say that venture investors de-emphasize the importance of a company's technology. Rather, they are conscious of the fact that commercial judgment, leadership, and like qualities will often be the deciding factors in a company's success.

This focus on people over technology is borne out by experience in a variety of industries. Many investors, for example, are mindful of Sony's failure in the videotape-recorder market in the last decade. Although Sony's Betamax was hailed by the company as the higher quality video format, the rest of the industry turned to VHS, which was a cheaper alternative and was marketed more successfully. In short, Russian technology companies should be prepared to show not only that they have valuable technology, but that the existing management has the experience and the judgment to make the company a success.

Ability to "Commoditize" Technology

Venture capitalists also value "commoditization" over technical complexity. The prospective venture investor will look to see whether the technology is highly dependent on the services or expertise of the company's employees. If a product that incorporates the technology can only be made on a customized basis or otherwise includes a significant service component, the attractiveness of the company is greatly diminished. Thus, Russian technology companies seeking private equity investors should seek ways to incorporate the technology in generic products that do not need to be tailor-made for each order.

As a corollary to the foregoing proposition, scientists and engineers that approach venture capitalists with proposals based on a single product tend to have the most trouble. As an equity investor, the goal of a venture capitalist is to recognize long-term capital gain. As a result, this type of investor will be attracted to technology and a management team that offer growth potential for a company as a whole, and he will shy away from situations that merely offer the potential for a one-time sale of the technology to a third party.

These factors highlight a predicament faced by Russian technology companies. Part of the reason for seeking venture capital is to be able to tap the commercial judgment and marketing and networking prowess of these investors. At the same time, to obtain venture capital a company must impress upon these investors that the management team already has the commercial and marketing savvy to undertake the entrepreneurial growth.

Increased Focus on Dividends

There is a greater focus on dividends in Eastern and Central European venture transactions than in similar U.S.-based transactions. For example, Claflin Capital Management, a Boston-based venture firm, recently formed a Ukraine Fund with $11.8 million of committed capital for the purpose of investing in consumer-products firms and is in the process of closing a second venture fund for the region with $40 million of committed capital. To date, Claflin has made equity investments in seventeen companies totaling approximately $4 million and has taken, on the average, 30–40 percent ownership stakes in the companies in which it has invested. Claflin has said that it expects to be receiving dividends from these companies within three years.

The desire of Claflin and other venture investors to receive dividends contrasts with their typical practices in the United States. In the United States, a venture capitalist looks to achieve his return on investment as a result of an increase in the capital value in the company, not from dividends. For those who invest in emerging growth companies in emerging markets, however, the insistence on dividends is a way to hedge the investment. While this is certainly understandable, it presents a frustrating picture for entrepreneurial-minded managers of Russian technology companies who may be eager to plow those funds back into research and development.

Reluctance to Invest in Start-Ups

A venture capital investment trend in the United States that may initially concern Russian technology companies is that venture capitalists have increasingly been moving away from making investments in start-ups and other early-stage financings. In 1983, financings of start-ups accounted for 43 percent of all venture deals done in the United States. By the end of that decade, they accounted for less than 10 percent. As the number of these start-up deals shrunk, the number of acquisitions, management buyouts, and later-stage financings involving venture capital rose. Although there are a number of reasons for this trend, the primary reason relates to the effect that pension funds and other institutional investors in venture funds have had on the market. As the role of these investors has increased, venture capitalists have become more hesitant to fund start-ups, which are riskier than later stage financings.

For Russian technology companies, many of which are spin-offs or otherwise at an early stage, the question is whether they might be similarly ignored as a result of being classified as start-ups. Although statistical evidence of a trend on this issue is not readily available for Russian venture capital transactions, anecdotal evidence indicates that this particular factor should not be a concern. To date, venture funds for Eastern and Central European companies have largely focused on early-stage investments. Arguably, a venture investor who knows that he will be exiting via an acquisition rather than via an IPO is more comfortable with early-stage investments. The reason for this is that the investor seeking a public offering of the company's stock recognizes that a precondition to such an offering is that the company achieve a sufficient level of maturity, whereas in the acquisition context an exit can take place at an earlier stage of the company's development.

As an aside, the percentage of U.S. venture deals that involve start-ups also seems to be on the rise. For the first six months of 1994, start-up and other early-stage financings accounted for 35 percent of all venture deals. While it would be premature to say that the downward trend of such investments in the last decade has been reversed, it does show a willingness of many investors to continue to provide "classic" venture capital.

Conclusion

The search for alternative means of financing the commercialization of Russian defense technologies coincides with the renewed, bullish sentiment of many Western investors toward emerging market investments. The explosion of emerging market investments (in the bond and stock markets as well as in direct investments) of the early 1990s was cut short in 1994 due to the U.S. Federal Reserve's decision to begin to raise interest rates in February of that year combined with Mexico's peso devaluation in the following December. Once again, however, investors are returning to these markets. For a variety of Western institutional investors, the desire to invest a portion of their funds in Russia and other Eastern and Central European markets is now viewed as a critical part of their hedging strategies. And for many Western corporations, entry into these markets is a sine qua non of a global corporate strategy.

These trends will help to increase the amount of capital committed to Russian venture capital funds and will increase the availability of these funds for emerging growth companies. For a Russian technology-driven company, the ability to attract these venture funds will be dependent upon evidence that its combination of management and technology provides the potential for significant capital gains.

Even in light of the foregoing trends, however, the actual amount of funds committed to venture investments in Russia will remain small relative to the many other forms of debt and equity that will be flowing into the market. As a consequence, venture capital will admittedly play a niche role in the Russian defense conversion process. Yet it would be precipitate to dismiss the effect that venture capital can have on technology companies and on the economy as a whole. The importance of venture capital lies not in its size but in its ability to seek out entities with explosive growth potential and help them to flourish. In this way, it catalyzes emerging growth companies and helps underpin Russia's free market reforms.

Bibliography

Anslow, Maurice. "Venturing in Central and Eastern Europe." *European Venture Capital Journal* (November 1994).

Arora, Ashish, and Maryellen Kelley. "Manufacturing Technology Centers and the Transfer of Technology to Small Firms." Paper presented at the Conference on University Goals, Institutional Mechanisms, and the Industrial Transferability of Research, Center for Economic Policy Research, Stanford University, March 18–20, 1994.

Barry, Christopher B. "New Directions in Research on Venture Capital Finance." *Financial Management* 23, no. 3 (1994).

Bartlett, Joseph W. *Venture Capital: Law, Business Strategies, and Investment Planning*. New York: John Wiley & Sons, 1988.

Barton, John, and Simone Shaheen. "Sharing the Wealth: The Role for Venture Capitalists in Russia's Economic Development." Unpublished paper presented at the Center for International Security and Arms Control at Stanford University, 1995.

Bernstein, David, ed. *Defense Industry Restructuring in Russia: Case Studies and Analysis*. Stanford, CA: Stanford University Center for International Security and Arms Control, 1994.

Bernstein, David, and William J. Perry. "Defense Conversion in Russia: A Strategic Imperative." *Stanford Journal of International Affairs* 2, no. 2 (1993).

Bygrave, William D., and Jeffrey A. Timmons. *Venture Capital at the Crossroads*. Cambridge: Harvard Business School Press, 1992.

Cohen, Ariel. "Restructuring Aid to Russia: Revising H.R. 4426." Heritage Foundation Reports, Issue Bulletin, no. 198 (1994).

"The Commercial Campus." *Intellectual Property*, March 1995.

Dobkin, James A. *International Technology Joint Ventures in the Countries of the Pacific Rim*. Stoneham, MA: Butterworth Legal Publishers, 1988.

Erickson, John W. "Investment in Russia: The Legal Environment." *Michigan International Lawyer* VI, no. 1 (1994).

Goldscheider, Robert. *Technology Management*. New York: Clark Boardman Callaghan, 1990.

Goldsweig, David N., and Roger H. Cummings. *International Joint Ventures: A Practical Approach to Working With Foreign Investors in the U.S. and Abroad*. Chicago: American Bar Association, 1990.

Gutterman, Alan S. *Counseling Emerging Companies in Going International*. American Bar Association, 1994.

"Interim Report: Redeploying Assets of the Russian Defense Sector to the Civilian Economy." Paper presented by the Committee on Enterprise Management in a Market Economy Under Defense Conversion, National Academy of Sciences, March 1993.

International Commercial Agreements. Practising Law Institute, 1991.

Joint Ventures and Privatization in Eastern Europe. Practising Law Institute, 1991.

Mansfield, Edwin. "Links Between University Research and Industrial Innovation." Paper presented at the Conference on University Goals, Institutional Mechanisms, and the Industrial Transferability of Research, Center for Economic Policy Research, Stanford University, March 18–20, 1994.

Moodie, Michael. "Beyond Proliferation: The Challenge of Technology Diffusion." *The Washington Quarterly* 18, no. 2 (1995).

Rahm, Dianne. "University-Firm Linkages for Industrial Innovation." Paper presented at the Conference on University Goals, Institutional Mechanisms, and the Industrial Transferability of Research, Center for Economic Policy Research, Stanford University, March 18–20, 1994.

"Russia's Emerging Market." *The Economist*, April 8–14, 1995.

Sagari, Sylvia B., and Gabriela Guidotti. "Venture Capital: Lessons from the Developed World for the Developing Markets." Discussion Paper No. 13, The World Bank, 1992.

Schwartz, Larry W. "Venture Abroad: Developing Countries Need Venture Capital Strategies." *Foreign Affairs* 73, no. 6 (Nov./Dec. 1994).

"Special Report: Focus on Start-up and Early Stage Investing." *UK Venture Capital Journal* (December 1991).

Sullivan, Ronald F. *Financing Transnational Projects*. New York: Matthew Bender, 1993.

Villeneuve, Thomas F., Robert V. Gunderson, and Daniel M. Kaufman. *Corporate Partnering: Structuring and Negotiating Domestic and International Strategic Alliances*. Englewood Cliffs, NJ: Prentice Hall Law & Business, 1994.

17

New Company Formation in Russia: Legal Regulation

Emily Silliman and Edward Kayukov

The perestroika period (1987–1991) unleashed a wave of entrepreneurial activity in Russia. The reforms of that period failed, however, to clarify property rights. The privatization period, begun in 1992, did address enterprise reform and sought to clarify property rights, but had the effect of dampening entrepreneurship, at least temporarily. There is a significant gap between the theory of the privatization process and the practice of privatization, which has brought about the preservation and in some cases the strengthening of centralized enterprise management.

The legal basis of new company formation is examined in this paper from the point of view of fostering entrepreneurship in general, and scientific-technical development in particular. An underlying question is whether the resources in the defense industrial and scientific research sectors will be available (for lease or purchase) to entrepreneurs who might further development and commercialization of valuable technologies.

The Legal History of New Company Formation Since 1987

Gorbachev's much-touted Law on the State Enterprise of 1987 states that the founding and cessation of operations of enterprises shall be regulated by extremely restrictive established rules on enterprise formation.[1] The basic rule was that such decisions were made at the ministerial or state committee level (Article 1). However, any enterprise formation linked to science and technology required Council of Ministers (i.e., the entire government) approval, in addition to backing from the relevant ministry (Article 4).

Under the enterprise formation decree, formation of any scientific-technical information "organs" (which could even include a shop within an existing enterprise) required approval of the USSR Ministry of Finance and the State Committee of the USSR on Science and Technology (GKNT) (7c), a process that entailed multiple, high-level approvals. These rules help explain the plethora of special decrees issued by the Council of Ministers to enable enterprise formation before June 1990 (when new registration rules came into effect).

The old rules on company formation were very slow to modernize in the perestroika period, although at the same time a quiet revolution was taking place. There had always been some limited exceptions to the above enterprise-formation rules. One category of exceptions were the housing cooperatives formed by local governments or by enterprises to provide housing for employees.[2] And a more significant movement began in 1987 when production cooperatives, or small businesses, were authorized for the first time, in public catering and consumer goods manufacturing.[3] In early 1988 a special decree was issued authorizing computer programming cooperatives, and cooperatives that provided computing and informatics services.[4] Finally, in May 1988, a general law on cooperatives was passed, authorizing any type of small business activity, unless expressly prohibited by legislation.[5]

Interestingly, cooperatives were not given the status of legal persons in the law,[6] but rather had the rights of "social organizations"(Article 5). Since a "social organization" does not necessarily have any rights to use, let alone own, property, the Law on Cooperatives spells out in detail that a cooperative may use and dispose of any property that is needed to fulfill its charter tasks (Articles 7–9). The relatively liberal 1988 law was followed at the end of the year by a list of restrictions on cooperative formation.[7] Contrary to the popular image, not all cooperatives were freebooting entrepreneurial companies. In fact, most were formed under the sponsorship of industrial state enterprises, a use of the cooperative form which was contemplated in the original Law on Cooperatives.[8] Subsequently, model charters for cooperatives, approved by the Council of Ministers, stated that cooperatives may be "attached" to local government organizations, or to enterprises or ministries.[9]

The cooperative movement was remarkable in that cooperatives were freed both from the planning system and from the system of state price controls.[10] Special mention is made in the Law on Cooperatives of the need to encourage small business to commercialize scientific-technical discoveries. Cooperatives were specifically permitted to make use of scientific-technical discoveries, and promote any related products through advertising and trade shows (Article 26).

Cooperatives became controversial in the eyes of the public, or at least in the eyes of the official press, because they charged higher prices than state enterprises, and they could make fast money by correcting imbalances in the economy.[11] A decree was issued on "small enterprises," which purported to promote within the traditional framework a new, more efficient type of small business.[12] Most cooperatives renamed themselves as "small enterprises," or reorganized themselves as a joint enterprise or some other type of company.

Coincident with the emergence of cooperatives and small enterprises, the government issued a string of decrees in 1987 permitting state enterprises to form joint enterprises with foreign companies and foreign individuals.[13] From the beginning, joint enterprises were a stronger organizational form than cooperatives, because a joint enterprise had the status of a legal person (Article 6). Joint enterprises required registration at a very high level—at the USSR Ministry of Finance—although subsequent liberalizations allowed for ministerial and even local registration.[14]

State enterprises were permitted to contribute assets to form the charter capital of a new joint enterprise.[15] Since the property of a joint enterprise was described as *sobstvennost*, or fully owned, formation of a joint enterprise became an effective means for spontaneous privatization. Some remarkable assets were transferred to joint enterprises under this system. Several joint enterprises were formed in the oil and gas sector which were 50 percent owned by American and other foreign companies. In the case of the Siberian-American Oil Company (the foreign partner was Pennzoil), the joint enterprise received a license to exploit the West Mogutlorsk oil field for a period of 25 years.[16] Although the oil field in question is a small one, it is exceptional in the Russian context to see that an oil field was effectively privatized.

Another source of spontaneous privatization was the advent in 1989 of the leasehold enterprise.[17] Managers or workers at selected enterprises were allowed to lease the property of the state enterprise where they worked. When the lease agreement permitted, any profits earned while the enterprise was under lease could be used to purchase the assets from the state, so that gradually the enterprise was privatized.[18] In order for this system to benefit managers or the employees, the enterprise had to be profitable in the first place. Accounting practices permitted purchase of the assets at 1989 or 1990 book value, or in other words for very small amounts of money. In 1994 the Supreme Arbitrazh Court[19] of Russia upheld the rights of employees or managers who obtained property under the leasehold rules.[20]

Interestingly, the area where state enterprises were given the most independence was in foreign trade. The decrees cited above that liberalized the joint enterprise regime also threw open the doors for state enterprises to carry out their own foreign trade transactions.[21] Initially, only the enterprises on a special list were granted foreign trade privileges, but later any enterprise could file with the foreign trade ministry and conclude contracts in its own name. This liberalization was remarkable, given that state enterprises still operated under a modified form of the planning system, and were domestically still subject to price controls.

Through 1989, legislation on economic liberalization skirted the core issues. The core issues during the perestroika period were ending the system of state prices, ending the system of centralized planning, and providing a unified, relatively simple system of enterprise formation. A major breakthrough in the enterprise formation area came in 1990 when both the USSR and the Russian Federation, which were competing entities at the time, passed comprehensive enterprise laws. Both laws provided for a universal registration system of all types of enterprises.[22] Until passage of the 1990 USSR Enterprise Law, state enterprises could only be formed and dissolved by state decree, whereas new company formation was encouraged under cooperative, small enterprise, and joint enterprise legislation. The Russian Federation law also provided for a unified registration system and provided enterprises with greater independence from ministerial and planning interference.[23]

In the period leading up to the failed coup of 1991, entrepreneurial activity increased, which seems to have had an effect in several areas including the computer software industry,[24] consumer goods trade, restaurants, and the encouragement of foreign trade activities, which made it easier to purchase goods and technology from abroad than to commercialize technology within Russia. After the coup, Boris Yeltsin permitted the open trade of goods, and issued a decree, effective in early 1992, freeing prices on most goods and services in Russia.[25] With these decrees, the system of state prices no longer applied for the most part to state enterprises.[26] The state pricing system had not applied to cooperatives, small enterprises, or joint enterprises, but until 1992 state enterprises were essentially prohibited from independent price setting. This protracted and rather late "emancipation" of state enterprises helps explain why reform in Russia has been troubled and slow. The three-year period between the liberalization of foreign trade and the elimination of the state pricing system encouraged corruption. During the three-year gap, powerful plant directors had access to goods at controlled state prices, and were able to sell them at world market prices. The windfall profits sometimes enriched the directors involved, and sometimes foreign or Russian trading partners. Meanwhile, state enterprises devoted themselves to getting around the restrictions placed on them.

The 1990 Russian Law on Enterprises was a major step in creating a company law that applied to both private and state-owned enterprises, and this law predominated in the aftermath of the failed coup of 1991. Russian enterprise law was not rewritten until 1994 with the passage of the new Civil Code.[27] The Civil Code became effective January 1, 1995, and superseded the Law on Enterprises.[28]

The 1990 Law on Enterprises does not clarify the relationship between the type of enterprise formed and the form of property. Since the shareholders (shareholders in some cases might be state enterprises) of some types of enterprises actually owned the property of the enterprise, enterprises were created that had a muddled, "mixed" form of ownership.[29] In contrast, the Civil Code states clearly that all enterprises (with the exception of state or municipal enterprises, and institutions formed by them) become the owners of the property contributed to them by founders or participants, as well as of the property they acquire by other means (Civil Code, Article 213, para. 3, and Article 66, para. 1). The 1990 Law on Property contained a similar provision, but the wording was weaker, and stood in direct conflict with the Law on Enterprises.[30] The Civil Code clarifies this issue.

In the period before adoption of the Civil Code, new enterprises formed with participation of state enterprises could theoretically be renationalized by decree. An edict of President Yeltsin of December, 1991 states that "concerns and corporations which were formed by enterprises of the defense and other industrial branches under the former Soviet ministries are considered state owned."[31] The decree does not explain what types of "concerns and corporations" were included in its provisions. By contrast, an enterprise that owns its property as defined in the new Civil Code is not subject to reorganization by decree.

The types of enterprises that may be formed, and their legal status, are outlined in the Civil Code. Enterprises are formed under implementing legislation for each type of enterprise. The following is a list of the types of enterprises contemplated under the Civil Code system, and the appendix provides greater detail concerning each organizational form: partnerships (general and limited), societies (limited liability, additional liability, joint-stock, daughter, and dependent), production and consumer cooperatives, state and municipal unitary enterprises, social or religious organizations, charitable funds, institutions, associations, and unions. All of the above enterprise types are legal persons. It is also possible for an existing enterprise to form a representation office or a branch (*filial*), but these types of organizations are not separate legal persons (Civil Code, Article 55).

The Privatization Program and the Effort to Define Property Rights

The Russian Privatization Program is one of the few relative successes of the Russian economic reform. The bold reformers who embarked on privatizing the medium to large sized state enterprises faced a task of enormous complexity. For that reason, the methods and results of privatization have to be seen in the context of the tremendous institutional resistance of the ministries that once ran the enterprises and of the entrenched managers of those enterprises who are loath to share control with outside investors. Thus, the privatization program in Russia has to be analyzed not only in light of its economic context, but more in terms of the enormous political and institutional changes it has wrought in the Russian industrial structure.

The initial legislation outlining the foundations for the Russian Privatization Program was the June 3, 1991 RSFSR Law on Privatization of State and Municipal Enterprises.[32] This law created the State Committee on Property (*Goskomimushchestvo* or GKI) which is primarily responsible for carrying out privatization and administering the proceeds. Perhaps in recognition of the important role that GKI has played in Russia's privatization effort, a recent presidential decree elevated the GKI from the status of a committee to the status of a federal ministry.[33] Another important national body created to facilitate privatization is the Russian Fund for Federal Property, which was designed to represent the Russian Federation as a stakeholder at shareholder meetings. Another function of this institution is to distribute the proceeds of privatization auctions and tenders (Law on Privatization, Article 5).

The architects of the privatization program had to balance the interests of the local authorities against the need for an internally consistent national privatization program. A measure of accommodation was achieved in the Law on Privatization, which provided for the creation of local property management committees (on the level of an autonomous republic, oblast, or city with a population of more than one million people) (Article 5). The GKI branches are responsible for privatizing federal property, while the local property management committees make decisions on municipal property within their borders. The local property committees were to develop privatization plans and to submit them to local authorities for approval. Unlike the federally supported GKI, the local property committees are allowed to retain preferred shares with a fixed dividend (usually five percent) on the assumption that the committees would be able to sell such shares at a profit in the future. Since shares are automatically convertible into common stock, the holdings of a local committee in a typical privatized enterprise (under Option Two) will amount to significant voting power (10 percent of the common shares).[34] The Law on Privatization also sets out the basic procedure for privatization (Articles 11 to 26). Under the regular procedure, once the management or the workers of the enterprise, or the GKI itself, had made the decision to privatize an enterprise, a privatization committee had to be appointed, which would be charged with drafting a privatization plan for the enterprise. Such a plan would recommend one of the privatization methods listed under Article 15 (liquidation, auction, public bid invitation, or transformation into a joint-stock corporation). The plan would also ascribe value to the enterprise, by determining the starting price of the enterprise in an auction context, or the required statutory capital if the enterprise were to be transformed into a joint-stock society.

The procedures set out in the Law on Privatization were undermined by the practical difficulties of ascribing an initial value to an enterprise. The procedures for privatization were modified by the June 11, 1992 State Privatization Program.[35] The program provided for a system of privatization vouchers to be distributed free of charge to all Russian citizens, and introduced the concept of an investment fund (Article 7 of the program). As a practical matter, investment funds were delegated the fiendishly difficult task of valuating enterprises to be privatized. But for those funds, an individual investor would be hard-pressed to properly analyze the value of a given enterprise. Article 5 of the program introduced three new options specifying methods of privatization to be applied to individual state enterprises based upon the size of the enterprise's workforce and its balance sheet value.

The State Privatization Program of 1992 set out three options for privatizing an enterprise. Under the first option, the employees of an enterprise to be privatized received 25 percent of preferred stock free of charge and the right to purchase 10 percent of the common stock, while the enterprise management in effect received a call option on common shares comprising 5 percent of the charter capital of the enterprise. Under the second option, the employees could buy common stock up to 51 percent of the charter capital of the enterprise. The third option of privatization gave the employees a call option on common stock limited to 20 percent of the charter capital and hence has been the least popular option.

The official end of the voucher privatization program on June 30, 1994 has sparked heated debates about the results of privatization. There is a great divergence of opinion on whether the voucher stage of privatization achieved its stated goals of (1) creating a wide stratum of private investors, (2) increasing the efficiency of enterprises, (3) minimizing subsidies, and (4) attracting foreign investment.[36] The conclusions that policymakers draw about the relative success or failure of voucher privatization depends directly on the assumptions they make about the relative importance of the privatization goals listed above. Thus on December 9, 1994 the Russian Duma adopted a resolution that characterized the results of the voucher stage of privatization as unsatisfactory.[37] The legislators mainly criticized the privatization program for failure to achieve a greater efficiency at the privatized enterprises, and for the failure to attract high levels of foreign investment.[38] Moreover, Duma members pointed out that voucher privatization was conducted in a top-down, "unitary" fashion in which the federal GKI imposed its policies on the local property committees. Some critics assert that the voucher privatization amounted to a huge giveaway of roughly 300 trillion rubles in 1994 prices to the new owners, due to extremely low valuation of the enterprises.

All these criticisms assume that the architects of the privatization program expected to obtain full replacement value of the enterprises to be privatized, and that enterprises would achieve higher efficiency very quickly. Yet Anatoly Chubais and the other leaders of the privatization drive understood from the beginning that turning around an industrial structure that had existed for 70 years was bound to take time. Their focus was on the rapid-fire destatization of the state-owned mid-sized and large enterprises to create the new stratum of investors, a result that has been achieved. Even the critics accept that the number of individual shareholders in Russia today is between 40 and 50 million people.[39] A system of relatively transparent voucher auctions gave ordinary citizens a much greater sense of inclusion and participation than under the alternative to voucher privatization: spontaneous privatization, which amounts to the transfer of state property to powerful managers.

The privatization program has put approximately 15 thousand mid-sized and large enterprises into private hands (most of those private shareholders are insiders—management and workers). Out of these 15 thousand, only a small percentage (around 2,500 enterprises) have been sold at auction.[40] Some 80 percent of Russian workers now work in privatized enterprises.[41] While this has not lessened the lobbying zeal of many managers, who continue to press the state for subsidies for their newly privatized enterprises, this at least gives the Russian government a stronger argument for resisting the pressure from the lobbying efforts of entrenched directors. Government agencies are no longer formally responsible for these enterprises; the new enterprise charters, instead of naming a ministry or other responsible government agency, mention only board members, shareholders, and managers as responsible parties. This gives government agencies the opportunity to ignore an enterprise's plea for help.

The privatization program has created the beginnings of capital markets in Russia by distributing some 146 million vouchers among Russian citizens. In addition to creating capital markets, the privatization program created some 600 voucher and investment funds. Those funds, as well as private banks with significant equity holdings, have become powerful outside investors, capable of enforcing shareholders' rights and monitoring corporate governance of the newly privatized companies.[42]

The critics of privatization point out that insider management, which has tied up workers' shares through voting trusts, five-year-long irrevocable proxies, and blatant violations of outsider shareholders' rights, still wields the real power.[43] Indeed, approximately 78 percent of privatized companies have chosen the second option, which gave the insiders (managers and workers) 60 percent of voting shares, while only 29 percent of shares went to outside investors.[44] Some 21 percent of privatized companies, mostly capital-hungry companies with high nominal share values, chose the first option, thus transferring control to the outside investors. Very few enterprises chose the third option.

If the government upholds its promise not to grant soft credits to newly privatized enterprises, the entrenched management of those enterprises will be forced to tap capital markets. To do that, the insider management will have to become more accommodating to the outside investors. Eventually the ownership and control over most enterprises will pass to strategic investors who will be able to transform those enterprises from value-subtracting to value-adding enterprises. The position of outside investors should also become stronger when the local property funds sell their preferred shares, which are automatically convertible for up to 10 percent of the authorized common shares. Thus, it appears that the position of the insider management will become more vulnerable over time. Their position will be further eroded by the advent of the post-voucher privatization.

The deep divisions in the Russian body politic over the results of the first stage of privatization were closely connected to the ongoing struggle over the direction that the post-voucher privatization should take. The Russian Duma engaged in a campaign to prevent introduction by presidential decree of the post-voucher privatization.[45] As a result of this pressure, the GKI was forced to negotiate with five Duma committees on the contours of the second stage of privatization. However, the draft legislation did not gain enough votes in the first hearing. While the details of the post-voucher privatization program remain undefined, it is clear that the government will concentrate on auctioning off its stake in various state enterprises. A new 1994 State Privatization Program provides for widespread auctions of the state holdings in privatized enterprises.[46] The potential for further privatization is still enormous—the government intends to market 10 to 15 percent of all the oil producers with the starting price of $3.5 billion. The state intends to sell a controlling stake in Russia's largest and arguably most profitable companies: oil-producing companies; Gazprom, the natural gas monopoly; and the Unified energy grid.[47] The size of the revenues from this second stage should dwarf the government's receipts from the voucher stage of the privatization and will hopefully enable the government to comply with its tight 1995 budget.

Specifics of Privatization in the Defense Sector

From the outset, the architects of the privatization program attempted to insulate sensitive defense enterprises from privatization. However, this goal has conflicted with the need to spin off or convert unnecessary or unprofitable enterprises. The ensuing legislation reflects that conflict. From the State Privatization Program of 1992 onward, the creators of the Russian Privatization Program recognized that strategic state property should not be privatized. Such property included the facilities that belonged to the Ministry of Defense and national security agencies.[48] On the other hand, military production facilities can be privatized, provided that the Russian government makes the decision. With respect to scientific and research institutions, the decision to privatize or separate off subdivisions has to be made by the relevant ministry along with the GKI. The 1994 State Privatization Program requires that all state enterprises open to privatization with a January 1, 1994 book value exceeding 20 million rubles must turn themselves into openly held joint-stock societies and float their shares on the capital markets. This provision is mandatory for defense plants and factories as well, provided that the share of state procurement orders accounts for less than 30 percent of their total turnover.[49]

A 1994 decree governs privatization of a broad class of research facilities typically held as federal property.[50] Under the decree, the relevant ministries are required to draw up (1) lists of facilities whose privatization is forbidden; (2) facilities to be transferred to other entities to be financed out of the federal budget; and lastly, but most importantly, (3) facilities to be turned into openly held joint-stock societies, although the state may initially retain up to 100 percent of equity. The primary responsibility for determining the terms of securitization and the terms of the "golden share" arrangements is left with the GKI.

The same decree also stipulates that State Scientific Centers (research centers financed by the federal budget) are to be privatized by the decision of the Russian government. If the government decides to privatize a State Scientific Center, the GKI may employ all privatization methods, except for direct auctions. However, if an investment tender is held for a State Scientific Center, the controlling shareholders will have to agree to maintain the same level of employment and the same research profile during the next three years. Such requirements, along with the difficulty of obtaining a governmental or ministerial approval, make the privatization of research institutions difficult at best. In addition, a subsequent 1994 decree on defense complex enterprises exhibits even more hostility toward privatization of defense enterprises than the previous decree.[51] It creates new State Scientific Centers from some defense research institutions and allows others to be transferred to the state fiscus or be auctioned off or privatized. The prime thrust of this recent decree is directed not at privatization, but at rationalizing the defense sector and keeping the core production in state hands: the decree charges various state agencies with developing a program for solidifying the national technological base, which would provide for identifying and promoting exportable defense products and technologies.

Practical Aspects of New Company Formation

Any new company organizing itself in Russia encounters serious barriers to entry. The first, most obvious barrier to entry is the process of registering a company. The number of documents required has been reduced under a new statute on registration, and registration agencies are prohibited from demanding guarantee letters or other documents other than the ones listed in the statute.[52] However, notarization of documents is only prohibited in the case of individuals forming enterprises (Statute, Article 3, h). This means that the extremely cumbersome process of notarizing documents in Russia is required for the formation of most new companies.

In addition, new rules require that documents be presented at the time of registration which prove that at least 50 percent of the charter capital has already been paid (Article 1). Logistically, this requirement can be difficult to meet. The simplest way to prove that the charter capital has been paid in is to show bank documents indicating deposits of a certain amount of money. In order to show such bank documentation, one has to open a bank account. Russian banks, like banks in many other countries, do not open a bank account just because you say you want to form a company. Either an existing organization has to use its influence at the bank to help open an account, or organizers of a new company have to obtain a provisional registration certificate, use the certificate to open the bank account, and then apply for registration.[53] In Russia, these procedures could take weeks or longer, and that is assuming that those wishing to form a company have experienced advisors, such as an attorney.

The documents required for registration are (Article 1):

- An application for registration prepared in the correct form and signed by the founders.

- An agreement of the founders, or a decision on the founding of the enterprise (except for some non-commercial enterprises).
- Documents proving payment of at least 50 percent of the charter capital.
- Proof of payment of the state registration fee.
- Charter of the enterprise if required for that particular type of enterprise.

The taxation system is a second source of problems both because it is administered poorly and because of the tax rates themselves. Any new company has to register with the local tax authority, and set up its system for filing required tax forms. This usually requires hiring an accountant who not only knows how to run an accounting system, but who is a good negotiator. The tax laws leave a number of gray areas within which, for example, certain types of benefits for employees may or may not be deductible. A good working relationship with the tax inspectorate is very important in order to work the gray areas in one's favor, or even just to file the correct forms.

The taxation rates create their own set of problems. A producer of electronic goods would pay effective taxes of roughly 45 percent of gross revenues.[54] High tax rates require that prices be raised, slow the start-up phase of a company, and encourage hiding revenue and other illegal practices. The need to fight with such a tax burden reduces the number of people who might otherwise be willing to start companies. From day one a new company needs leadership with strong negotiation skills and street smarts in order to survive. Companies that choose to operate illegally as their strategy to survive make themselves vulnerable in other ways.

Real estate is a third and serious handicap to new start-ups. In order to obtain space for offices or other uses, the new company manager could purchase space, but more often space will be leased. However, in Russia it can be difficult if not impossible to determine who might have the right to conclude a lease agreement. The supply of available space has been artificially low due to the uncertainties of property rights and therefore of real estate markets. Many of the defense enterprises and research institutes that control vast amounts of usable space have been prohibited from leasing the space without permission of the State Committee on Privatization.[55] Since lease payments are made to the State Committee on Privatization and not to the lessor, enterprises have little incentive to lease their premises.

Many local governments that have gained control over real estate have been slow to relinquish control and foster a genuine real estate market. The Moscow mayor's office, for example, recently required potential lessees to purchase at auction the "rights to lease" Moscow real estate. One such "right" to a lot in downtown Moscow was recently sold for $6 million.[56] Such requirements push commercial property rents further out of reach for small start-up companies. One obvious alternative is to avoid central Moscow, but local governments anywhere have interests to protect, and may choose to limit supply of usable space in order to extract higher revenues.[57]

For the entrepreneur, forming an attached company might help surmount some of the barriers to entry as the parent company might provide space on a favorable basis, help with bank accounts, or lend weight to negotiations with tax authorities. An attached company is any company formed under close collaboration with the management of an established enterprise or institute. For the parent company, the attached company may perform a specialized function for the parent, perform contract work, or facilitate avoidance of certain types of restrictive government regulations.

In the earlier perestroika period, state enterprises did not have the right to reorganize themselves, pay employees properly, fire employees if necessary, sell goods at market prices, and a whole host of other rights that new forms such as cooperatives, joint ventures, and others had.

Many state enterprises converted internal subdivisions into a new form in order to get around strict limitations on wages. In a perfectly legal maneuver, for example, the state enterprise Khokkei, which made sports equipment, formed a cooperative in 1989, and certain jobs were contracted out to the cooperative.[58] The cooperative proceeded to hire many of the state enterprise's employees at higher wages. Management liked this scheme, because it made it possible to reward good employees by paying them better, and the cooperative was very productive. In this particular example, the formation of the cooperative brought an interesting manager into the company, who had been a designer of military aircraft. The manager of the cooperative launched a high-tech research effort to use materials developed for the military to produce better hockey sticks and other sporting goods. His efforts at using high-tech materials were met with mixed success, but overall his company is profitable and is now becoming more independent of the parent company. This is an example of a scheme to avoid state regulations which led to some creative entrepreneurship, and launched the management career of a talented person.

Other parent firms engage in new company formation to circumvent Russian labor laws which are still very restrictive on firing employees.[59] It is easier to form a new company and shift all important business to the new company (leaving unwanted employees behind) than it is to simply fire an unsatisfactory employee.

Not all state enterprises were able to carry out foreign trade transactions in their own names, and an attached enterprise was formed in some cases to enable some transactions to take place. An example of such an enterprise is the Small Enterprise "Laser Physics," organized in 1990 to market products and technology for the Institute of Laser Physics (formerly a department of the Vavilov Scientific-Technical center) in St. Petersburg.[60] The small enterprise works with a skeleton staff, and hires Institute employees as needed on a contract basis. The manager looks for commercial opportunities wherever he can find them, and has sold equipment made by former suppliers to the Institute, thus becoming a specialist in his particular technology area instead of just a captive of the parent Institute. The employees at the parent Institute are not complaining, since they can supplement their income through contract work with the small enterprise.

Some attached enterprises were formed for the sole purpose of enabling the parent enterprise to turn non-cash bank accounts into cash. In a holdover from the days of strict centralized planning, banks police how much money can be received as cash to pay employees, and to engage in unplanned activities. Under inflationary conditions, enterprises were under great pressure to obtain more money as cash from their bank accounts. An attached company could pretend to perform contract work or hire fictitious employees, and thus help the parent enterprise obtain cash from the bank. Official recognition of the problem serves as evidence that the practice was widespread.[61]

Unfortunately, sometimes the relationship between parent and attached companies goes sour and the conflict can wind up in court. Some court cases show that Russian parent companies do not always understand the formal process involved with forming a new company, and once the new company is formed may continue to treat it as a department subordinate to the general director's office. In a Saratov case, a metallurgical smelting center took one of its shops and decided to form the small enterprise *Volzhskie zori* ("Volga Dawn"). [62] The parent enterprise went to the local government and sought a decree authorizing registration of the small enterprise. The Arbitrazh Court found the registration invalid because fundamental steps in the company formation process had been neglected: no charter had been drafted, and there had been no meeting of founders at which a formal decision to form a company was made.

In a case from Chechnya, obviously before tragic events overtook the region, the managers of a construction materials production combine dissolved the Construction Materials Combine No. 2, an attached enterprise it had formed earlier, for violations of "financial discipline."[63] The Arbitrazh Court found that since Combine No. 2 had been organized as a separate enterprise, dissolution could not take place by fiat of the general director. Rather, it had to take place in accordance with the law on company dissolution which requires a decision from the board of directors and approval of the employees. Managers seek the advantages of forming attached companies, but are not always comfortable with the degree of independence that the new company attains. Not surprisingly, the discomfort with independence is most acutely felt when valuable property is involved.

While attached enterprises often enjoy a relationship of mutual convenience with their parent enterprises, prosecutors investigating defense enterprises in 1994 found that arrangements in many cases were a little too convenient. They found, for example, that a deputy director of the Russian Space Equipment Construction Institute, together with other managers, founded a partnership and transferred major pieces of equipment and buildings to the private partnership.[64] Prosecutors found the transfer unlawful, and insisted on return of the property to the Institute. The

Chief Engineer of the same institute also founded his own firm and sold it an entire computer center for a nominal fee.

Hindering Entrepreneurship: Enterprise Formation in the Privatization Context

The privatization program had the laudable goal of stopping the wholesale stripping of assets that has been dubbed "spontaneous privatization." Unfortunately, by shifting decision-making control to the property committees, and requiring a rather cumbersome procedure for privatization, the program has overcompensated and in many cases serves to hinder legitimate entrepreneurs. For example, the employees of a state-owned elevator enterprise in Nizhegorodskii oblast formed a new company, *Universal*, and decided to purchase the assets of their own elevator enterprise.[65] The local prosecutor challenged the asset purchase, and eventually prevailed. The court found violations of the privatization procedures which required signatures of at least half the employees and the formation of a detailed privatization plan based on the approved options. Were the employees in this case trying to steal state property, as in the many cases of corruption, or were they making a reasonable bid to buy out their company?

In another case, the employees of an automobile and motorcycle repair station in Sergiev Posad (in the Moscow region) requested the separation of the repair station from its parent company, the Izhmash factory.[66] The Izhmash factory sought to block the separation and privatization of the division, arguing that not all of the employees of the repair station and its outlets in neighboring towns had actually approved the plan to separate. The Supreme Arbitrazh Court agreed. This case shows how determined a parent enterprise can be to keep the old structures intact.

In Tyumen, the employees of a state-owned store tried to form a new enterprise to take over operations of the store.[67] Since no attempt was made in this case to obtain permission from the local property committee, the formation of the new enterprise was quite properly declared invalid, and the attempt to take over the store illegal. The overall process of transfer of good retail space to new management in Tyumen has been excruciatingly slow.[68] In many cases across Russia, local bureaucratic resistance plays a role in hindering the entrepreneurial process.

In theory, privatization has provided a valuable one-time opportunity for subdivisions of companies to emancipate themselves from a parent enterprise. A Supreme Arbitrazh Court ruling from 1992 points out that if a subdivision wanted to break off from an existing enterprise, the consent of the employees of the enterprise as a whole (not just the renegade division) was required.[69] If, however, the parent enterprise is changing its legal status via the privatization process, then the employees do not have the power to block new company formation. Theoretically, then, as long as the arduous privatization procedures are followed, the result could be a more rational industrial structure as new viable companies emerge from obsolete structures.

In practice, however, many managers will fight to keep old structures intact (often successfully), as indicated in some of the examples above. Some managers also seek special decrees which prohibit subdivisions from splitting off. These efforts can be successful, even when general legislation would allow subdivisions to split off.[70]

The tendency to favor centralized organization can especially be seen in science and technology. President Yeltsin, in a decree in April 1992, placed restrictions on the separation of "test production, test experimental production and test study production sites" from state scientific and technological organizations.[71] The decree cited the dangers of destroying the technological unity of science, test production, and scholarly endeavors. The "unity of science" appears to be an argument made by institution managers who are resisting the breakup of their institutions. The requirement that test production be married to research is an arbitrary one, and not always conducive to effective R&D.

The Chief Justice of the Supreme Arbitrazh Court of the Russian Federation reiterated this prohibition in August of 1994.[72] The Chief Justice somewhat alarmingly pointed out that the rule applies to proposed separation of test production sites, whether or not the test production site is a separate legal person or a subdivision of the parent enterprise. The letter of the court chairman says that courts should evaluate whether the technological unity of science will be destroyed once a proposed separation takes place. Thus even a separate legal person might find its future growth and flexibility hindered by this requirement for technological integrity, and future new company formation may be restricted at the very institutions that should be the seedbeds of new economic development.

In summary, important underlying problems of new company formation in Russia failed to be addressed in the perestroika period, and policies tended to encourage corruption. The process of properly defining property rights, launched after the failed 1991 coup, is complex, and has led to the temporary result of hindering entrepreneurship. Two positive results are already evident, however. First, the privatization program created capital markets and many private owners, and gave Russia the opportunity to develop a vibrant private economy. New companies need a positive business environment in which to function. Second, the privatization program and improved enterprise legislation set in motion a process, bolstered by a functioning court system, whereby enterprises will eventually clarify their status and improve conditions for launching new companies. Russia will lose many opportunities due to excessive centralization in research and development, but individuals can be relied on to create many more opportunities.

Appendix

Current Legal Options for New Company Formation and the New Civil Code
The types of enterprises contemplated by the Civil Code are as follows: partnerships (general and limited), societies (limited liability, additional liability, joint-stock, daughter, and dependent), production and consumer cooperatives, state and municipal unitary enterprises, social or religious organizations, charitable funds, institutions, associations, and unions. All of the above enterprise types are legal persons. It is also possible for an existing enterprise to form a representation office or a branch (*filial*), but these types of organizations are not separate legal persons (Civil Code, Article 55).

The Civil Code makes a distinction between enterprises in which the participants have limited, defined rights (such as rights of shareholders), and enterprises in which the participating founder has ownership rights (Civil Code, Article 48, para. 2). The participating founder has ownership rights to the property of state and municipal unitary enterprises, including daughter enterprises and other institutions formed by them. Participants in economic partnerships and societies (economic partnerships and societies include all of the types of partnerships and societies listed in the previous paragraph), and in production or consumer cooperatives have only limited rights to the property of the enterprise. The limited rights as shareholders or participants are defined in legislation and in the organization's charter. Founders or participants do not have property rights to social and religious organizations, charitable funds, associations, and unions.

Thus, a potential investor evaluating a newly formed company should try to glean three major pieces of information. First, it is important to know whether the newly formed company is registered as a separate legal person (and is therefore not merely a representation office or branch of another enterprise). Second, it is important to know whether the enterprise under examination owns its property, or whether the property is state property (and therefore subject to separate sets of rules on state property). The property of an enterprise is state property in the case of a state or municipal enterprise, a daughter society of a state or municipal enterprise, or an institution formed by a state or municipal enterprise. A daughter society is defined as any enterprise in which one owner owns a controlling share (Civil Code, Article 105). The property of a daughter enterprise would be considered state property if a state enterprise owned a controlling share. A dependent society is one in which one owner owns more than a 20 percent share (Article 106). Parent enterprises must report the existence of dependent societies, but the status of the parent's property does not affect the status of the property of the dependent society.

Finally, when evaluating a new enterprise, it is necessary to know whether the enterprise would be considered a commercial or non-commercial enterprise

(Civil Code, Article 50). Non-commercial enterprises may engage in commercial activities only if these activities are in accordance with the narrow purpose stated in their charters. Thus, a new enterprise would be a non-commercial enterprise if it is a consumer cooperative, a social or religious society, an institution formed by a property owner, or a charitable or other fund (other forms may be authorized by law). Of these, the most poorly defined is the "institution" category (Civil Code, Article 120). It is possible that a group of computer programmers could function in the form of an "institution" formed by another enterprise, but might operate under some restrictions, both because the parent enterprise might be a state or municipal enterprise (in which case the property is state or municipal property), and because an "institution" is a non-commercial enterprise. It should be possible to understand what restrictions apply to a non-commercial institution by examining the organization's charter.

Among enterprises that are separate legal persons, whose property is not state or municipal property, and that are commercial enterprises, the organizational form might be one of several types. The first is the general partnership (Civil Code, Articles 69–81). The partners in a general partnership may be other economic societies (i.e., enterprises), or individuals. The most striking characteristic of the general partnership is that the partners do not have limited liability. Entities may participate in only one general partnership at a time. An interesting hybrid partnership type is the limited partnership (Articles 82–86). The limited partnership allows for partners who have unlimited liability, and additional investors (participants) who enjoy limited liability. A general partnership may be a participant in a limited partnership.

The next organizational type is the limited liability society. As with many of these organizational types, the Civil Code states that laws will provide more detailed rules in the future. For instance, for a limited liability society, future laws may limit the number of possible participants (Civil Code, Article 87, para. 2). Participants make contributions to the charter capital of a limited liability society, and one-half of all contributions must already be in place at the time of registration (Article 90, para. 3). The remaining capital must be paid in within the first year. If the assets of the society diminish to the point that they are less than the amount of the charter capital, the society must report a reduction in the amount of charter capital, and notify creditors of the reduction.

Future legislation will establish a minimum amount of charter capital which must be declared before an enterprise may be registered. Also, the charter capital should be sufficient to guarantee the interests of the society's creditors (Article 90, para. 1). As with previous enterprise legislation, the charter capital is treated as a static asset. The rules overlook the possible willingness of creditors to loan based on the prospective business activity (rather than on assets), and overlook the possibility that a good manager might need to draw down assets in order to make the business grow, to enter new markets, or to develop new products. Presumably the rules are designed to prevent undercapitalized companies, but they are more appropriate for a stable, ongoing concern than for a robust start-up company. The rules seem to discourage risk-taking in high-tech or other capital intensive businesses, since every step of the start-up has to pay for itself under the rules as described in the Civil Code.

The other remarkable feature of a limited liability society is that the charter may place restrictions on the right of participants to sell their share to third parties. The right to sell one's share to an outside party is presumed if the charter is silent on the issue. Participants have the priority right to purchase the share of any participant wanting to leave the society.

A new type of organizational form is the society with additional liability (Civil Code, Article 95), a variant of the limited liability society. In a society with additional liability the parties agree to assume responsibility for the charter capital contributions of the other parties. This organizational form will require more detailed explanation in future legislation.

The rules concerning open and closed joint-stock societies have not changed significantly (Civil Code, Articles 96–104). The primary difference between a joint-stock society and a limited liability society is that the joint-stock society issues shares of stock. Many other provisions are the same, including the prohibition on reducing the amount of charter capital without notifying registration authorities and creditors.

An open joint-stock society differs from a closed joint-stock society in that shareholders in an open joint-stock society may freely sell their shares to outsiders. An open joint-stock society may also make public offerings of stock, and is required to issue audited financial statements on an annual basis. All joint-stock societies may issue preferred shares of no more that 25 percent of the outstanding shares, and may issue bonds (*obligatsii*) up to the amount of the charter capital, within certain restrictions.

The Civil Code provisions on joint-stock societies make frequent reference to a law on joint-stock societies that is to provide further detail on, for instance, the minimum allowable charter capital, types of questions that may only be decided by the board of directors, and many other detailed issues.

The primary difference between a limited liability society and a closed joint-stock society (both of which usually limit sale of shares to outside parties) appears to be that different legislation in the future will govern these types of organizations. Legislation concerning joint-stock societies can be expected to concern itself with protecting the rights of shareholders, since in the case of open joint-stock societies the shareholders may be ordinary citizens. All joint-stock societies (not just open ones) may have more stringent financial reporting requirements, and they must adhere to separate securities regulations. Joint-stock societies may use obligations as a method of raising capital, a method not explicitly permitted for limited liability societies. It is too early to tell how these two types of organizations will evolve.

A final type of enterprise, or commercial society, is the production cooperative (Civil Code, Articles 107–112). This type of enterprise is sometimes called a brigade. Production cooperatives are now legal persons, and own their property, as noted above. The organizational form is flexible, with no fixed charter capital required. However, the participants do not necessarily have limited liability (Article 107, para. 2), and the operations of the cooperative are subject to disruption when a member leaves. All property of the cooperative must be divided into shares (with the exception of property designated as indivisible funds), and a member who wishes to depart must be paid his share at the end of the accounting year. The limited liability issue, and the problem of the departing member, could be remedied through careful drafting of the charter, but the legal assumptions are not favorable for this organizational type. Another example is that cooperative members receive distributions based on their labor contributions, unless otherwise specified in the charter. This organizational form seems suited to groups of laborers or perhaps computer programmers, but would not work as well for a more complex type of company. As with other types of organizations, subsequent legislation will provide more detail.

The Civil Code provisions on enterprises allow wide diversity in the types of organizations allowed to function in Russia. Property rights are clarified, and in some cases strengthened. There are some rigidities in the system, however, especially in the area of the charter capital of limited liability societies and joint-stock societies.

Notes

1. Decree of the Council of Ministers of the USSR of September 2, 1982 No. 816, *O poriadke sozdaniia, reorganizatsii i likvidatsii predpriiatii, obedinenii, organizatsii i uchrezhdenii* [On the Rules for Founding, Reorganizing and Liquidation of Enterprises, Combines, Organizations and Institutions], 5 *Svod zakonov*, 379–383.

2. Decree of the Council of Ministers of the USSR of August 19, 1982 No. 765 *O zhilishchno-stroitelnoi kooperatsii* [On Housing-Construction Cooperatives] 6 *Svod zakonov*, 233–239. This is merely an example of the line of decrees on consumer cooperatives to provide housing, garage, and summer houses. This discussion leaves out the large area of agricultural cooperatives.

3. Decree of the Council of Ministers of the USSR of February 5, 1987 No. 160, *O sozdanii kooperativov obshchestvennogo pitaniia* [On the Foundation of Public Catering Cooperatives], 10 SP SSSR, item 41 (1987). Item 42 is a companion decree concerning consumer goods cooperatives.

4. Decree of the Council of Ministers of the USSR of March 7, 1988, No. 307 *O sozdanii kooperativov po razrabotke programmnykh sredstv vychislitelnoi tekhniki, a takzhe po okazaniiu informatsionno-vychislitelnykh i posrednicheskikh uslug v oblasti informatiki*

[On the Foundation of Cooperatives for Developing Computer Software, and Also for Providing Computer-Information and Middleman Services in the Area of Informatics], 14 SP SSSR, item 37 (1988).

5. Law of the USSR of May 26, 1988, *O kooperatsii v SSSR* [On Cooperatives in the USSR], 22 Vedomosti SSSR, item 355 (1988) [Hereinafter cited as the Law on Cooperatives]. Article 3, section 1, paragraph 3 states that cooperatives engage in any activity not prohibited by legislation.

6. When you have the status of a legal person in Russia, it means that you are acknowledged as a separate company with the right to conclude contracts in your own name (along with a bundle of other rights). As a practical matter, an organization has to have a charter which says it is a legal person, or banks and government functionaries may refuse to acknowledge that the organization exists. The practical need exists for a charter that says you are a legal person, even if by statute (in the case of cooperatives) you supposedly have all of the rights granted to legal persons.

7. Decree of the Council of Ministers of December 29, 1988, No. 1468 *O regulirovanii otdelnykh vidov deiatelnosti kooperativov v sootvetstvii s zakonom o kooperatsii v SSSR* [On the Regulation of Certain Types of Activities of Cooperatives in Accordance with the Law on Cooperatives of the USSR], 4 SP SSSR, item 12 (1989).

8. Law on Cooperatives *supra* , note 7, article 11, para. 3.

9. *Primernyi ustav kooperativa po proizvodstvu tovarov narodnogo potrebleniia Odobren postanovleniem SM SSSR* [Model Charter of a Cooperative for the Production of Consumer Goods Approved by Decree of the CM of the USSR], 9 *Ekonomicheskaia gazeta,* at 12 (1987), Article 1, paragraph 2. Model charters for public catering and service cooperatives are also included.

10. Law on Cooperatives, *supra* note 6, Article 18 deals with planning, and Article 19 deals with price controls. The law allowed the state some leverage for controlling prices when, for instance, raw materials were obtained at controlled state prices. As a practical matter, cooperative prices were not effectively controlled.

11. Unpopularity of cooperatives is discussed in Janet Lee, "The Evolution of Cooperative Legislation: A Case Study of Reform in the Soviet Union," *Stanford Journal of International Law* 27, No. 1 (1991).

12. Decree of the CM of the RSFSR of 18 July 1991 No. 406, *O merakh po podderzhke i razvitiiu malykh predpriiatii v RSFSR* [On Measures for Support and Development of Small Enterprises in the RSFSR], GARANT, and Edict of the President of the RF of November 30 1992 No. 1485, *Ob organizatsionnykh merakh po razvitiiu malogo i srednego biznesa v RF* [On Organizational Measures for Development of Small and Medium Business in the RF], GARANT.

13. Decree of the Presidium of the Supreme Soviet of the USSR of January 13, 1987 No. 6362-XI *O voprosakh, sviazannykh s sozdaniem na territorii SSSR i deiatelnostiu sovmestnykh predpriiatii, mezhdunarodnykh obedinenii i organizatsii s uchastiem sovetskikh i inostrannykh organizatsii, firm i organov upravleniia.* [On Questions Concerning the Founding and Activities on the Territory of the USSR of Joint Ventures and Foreign Organizations, Firms and Management Organs], official copy; Decree of the Council of Ministers of the USSR of January 13, 1987, No. 49, *O poriadke sozdanii na territorii SSSR i deiatelnosti sovmestnykh predpriiatii s uchastiem sovetskikh organizatsii i firm kapitalisticheskikh i razvivaiushchikhsia stran* [On the Rules of Founding and Activities on the Territory of the

USSR of Joint Ventures with the Participation of Soviet Organizations and Firms of Capitalist and Developing Countries] 9 SP SSSR, item 40 (1987). There were decrees issued at the same time which applied to joint enterprises with enterprises in socialist countries.

14. Decision of the CPSU Central Committee and the USSR Council of Ministers of September 17th, 1987 No. 1074, *O dopolnitelnykh merakh po sovershenstvovaniiu vneshneekonomicheskoi deiatelnosti v novykh usloviiakh khoziaistvovaniia* [On Additional Measures to Improve the Country's External Economic Activity in the New Conditions of Economic Management]. 41 *Ekonomicheskaia gazeta* (1987). The second liberalization came with the Decree of the Council of Ministers of December 2, 1988 No. 1405 *O dalneishem razvitii vneshneekonomicheskoi diiatelnosti gosudarstvennykh, kooperativnykh i inykh obshchestvennykh predpriiatii, obedinenii i organizatsii* [On the Further Development of Foreign Economic Activities of State, Cooperative and Other Social Enterprises, Combines and Organizations], 2 SP SSSR, item 7 (1989).

15. Decree of the Council of Ministers of January 13, 1987, *supra*, Note 13, article 11.

16. Oil and Gas License No. 0004 issued by the Russian Federation Committee on Geology and the Use of Underground Resources. License was issued to the Siberian-American Oil Company (50 percent owned by Americans) on December 1, 1992. A copy of this license is on file with Emily Silliman.

17. Decree of the Council of Ministers of the USSR of April 7, 1989 No. 294, *Ob ekonomicheskikh i organizatsionnykh osnovakh arendykh otnoshenii v SSSR* 20 SP SSSR, item 63 (1989). The option of becoming a leasehold enterprise was reiterated in *Osnovy zakonodatelstva SSSR ob arende ot 23 noiabria 1989g. (c izmeneniiami ot 7 marta 1991g)*, [Fundamentals of USSR Legislation on Leasing of November 23 1989 (with changes of March 7, 1991), Article 16, [hereinafter referred to as the Fundamentals on Leasing], GARANT.

18. One example of such an enterprise is Lesenergo in Moscow, which supplies equipment to the paper industry. In the case of Lesenergo, a group of managers, not the employees as a whole, gained control of the enterprise.

19. All references to the Arbitrazh Court in this paper refer to the Russian system of economic courts—the courts that have jurisdiction over commercial, financial, and privatization disputes.

20. Instructional Letter of the Supreme Arbitration Court of the RF of April 11, 1994, N S5-7/sz-235.

21. Decision of the CPSU and the CM of September 1987 and 1988 Decree (article 2) *supra*, note 14.

22. Law of the USSR of June 4, 1990 *O predpriiatiiakh v SSSR* [On Enterprises in the USSR], 25 Vedomosti SSSR, item 460 (1990). The Russian law is Law of the RSFSR of December 25, 1990 *O predpriiatiiakh i predprinimatelskoi deiatelnosti* [On Enterprises and Entrepreneurial Activity], 30 Vedomosti RSFSR, item 418 (1990).

23. Law of the RSFSR of December 25, 1990 *O predpriiatiiakh i predprinimatelskoi deiatelnosti* [On Enterprises and Entrepreneurial Activity], 30 Vedomosti RSFSR, item 418 (1990), [Hereinafter cited as the Russian Law on Enterprises]. Planning and prices are governed in articles 22 and 23.

24. Esther Dyson, editor of the software industry journal *Release 1.0*, noted that there was a strong entrepreneurial spirit and a lot of activity among Russian programmers

in 1989 when she began visiting Russia. Personal interview with Emily Silliman, May 1995. Various reports of Dyson's Russian travels appear in *Release 1.0*.

25. Edict of the President of the RSFSR of December 3, 1991 No. 297 *O merakh po liberalizatsii tsen* [On Measures for Liberalization of Prices]. See also: Edict of the President of the Russian Federation of January 29, 1992, *O svobode torgovli* [On the Freedom of Trade], GARANT.

26. A major exception is energy prices, where state price controls remain in effect. Whereas chaos in the energy sector threatens Russia in the longer term, technology enterprises are not impacted directly.

27. Civil Code of the RF of November 30, 1994, N 51-F3 *Grazhdanskii kodeks Rossiiskoi Federatsii (chast pervaia)* [Civil Code of the Russian Federation, Part One], GARANT [Hereinafter cited as the Civil Code]. It should be noted that prior legislation remains in effect until specific laws implementing Civil Code provisions are passed. Most notably, this includes the Decree of the CM of the RSFSR of December 25, 1990, No. 601, *Ob utverzhdenii Polozheniia ob aktsionernykh obshchestvakh* [On Confirming the Regulations on Joint-Stock Societies], GARANT.

28. Federal Law of November 30, 1994 No. 52–F3, *O vvedenii v deistvie chasti pervoi Grazhdanskogo kodeksa Rossiiskoi Federatsii* [On Rendering the Civil Code of the Russian Federation Effective Law), GARANT.

29. Russian Law on enterprises, *supra*, note 22, Article 11, section 3.

30. Law of the RSFSR of December 24, 1990 *O sobstvennosti v RSFSR* [On Property in the RSFSR], Vedomosti RSFSR, Article 14, [hereinafter cited as the Law on Property]. The problem with this provision is that it is circular, stating in part ". . . enterprises which are founded as the owners of property . . . have the rights of ownership over property contributed to them in the form of contributions or other payments from their participants . . ."

31. Edict of the President of the RSFSR of December 3, 1991, No. 256, *O merakh po stabilizatsii raboty promyshlennogo kompleksa RSFSR v usloviiakh ekonomicheskoi reformy* [On Measures for Stabilizing the Work of the Industrial Complex of the RSFSR under the Conditions of Economic Reform], GARANT.

32. Law of the RF of July 3, 1991, with changes of June 5, 1992 and December 24, 1993, *O privatizatsii gosudarstvennykh i munitsipalnykh predpriiatii v RF* [On the Privatization of State and Municipal Enterprises in the RF], GARANT.

33. Edict of the President of the RF of October 28, 1994 No. 2026, *O gosudarstvennom komitete RF po upravleniiu gosudarstvennym imushchestvom* [On the State Committee of the RF on the Management of State Property], GARANT.

34. Vladimir Sanko, *Za chto zhe vy Vovku Polevanova?* [Vovka Polevanov is Not to Blame], *Nezavisimaya Gazeta*, February 11 at 2.

35. Decree of the Supreme Soviet of the RF of June 11, 1992, *O vvedenii v deistvie gosudarstvennoi programmy privatizatsii gosudarstvennykh i munitsipalnykh predpriiatii v RF na 1992 god*, [On the Entrance Into Force of the State Program for Privatizing State and Municipal Enterprises in the Russian Federation for 1992 of June 11, 1992], 28 Vedomosti SND RF i VS RF, item 1617 (1992).

36. State Program for Privatizing State and Municipal Enterprises for the Russian Federation for 1992, *supra* note 34, Preamble.

37. Resolution of the Russian Duma of December 9, 1994, On the results of the first (voucher) stage of privatization, *Ekonomika i Zhizn*, #52, 1994 at 11.

38. See generally Nesterov and Bukhvald *Vauchernaia privatizatsiia: v chiu polzu schet?* [Voucher Privatization: In Whose Favor?]; *Ekonomika i zhizn*, #6, 1995 at 9; E. Prokova, *Pobeda ili porazhenie* [Victory or Defeat] *Ekonomika i zhizn*, #3, 1995 at 10.

39. Ibid.

40. Russia: Special Survey, *The Economist*, April 8, 1995 at 6.

41. Ibid.

42. See generally J. Robert Brown, "Order from Disorder: The Development of the Russian Securities Markets," *University of Pennsylvania Journal of International Business Law* 15, p. 509.

43. Ibid.

44. See *Order from Disorder*, *supra*, note 41.

45. Interview with Burkov, Chairman of the Duma Committee on Property, Privatization and Economy, *Ekonomika i zhizn*, #52, 1994 at 10.

46. Decree of the President of the RF of July 22, 1994, No. 1535, *Ob osnovnykh polozheniiakh Gosudarstvennoi programmy privatizatsii gosudarstvennykh i munitsipalnykh predpriiatii v RF posle 1 iiulia 1994 goda* [On the Fundamental Provisions of the State Program for Privatizing State and Municipal Enterprises in the RF After July 1994], GARANT.

47. *Supra*, note 34.

48. See State Privatization Program of 1992, 1994, Article 2.1.

49. 1994 State privatization program, *supra*, note 45, article 3.

50. Decree of the Government of the RF of July 26, 1994 No. 870 *O privatizatsii obektov nauchno-teshnicheskoi sfery* [On Privatizing of Scientific-Technical Facilities], GARANT.

51. Decree of the Government of the RF of December 19, 1994, No. 1399 *O merakh po stabilizatsii ekonomicheskogo polozheniia predpriiatii i organizatsii oboronnogo kompleksa* [On Measures for Stabilizing the Economic Condition of Defense Complex Enterprises and Organizations], GARANT.

52. Established by Edict of the President of the Russian Federation on July 8, 1994 No. 1482, *Polozhenie o poriadke gosudarstvennoi registratsii subektov predprinimatelskoi deiatelnosti*, [Statute on the Rules for State Registration of Subjects of Entrepreneurial Activity], GARANT.

53. See commentary on the registration rules in William G. Frenkel, "Overview of the Revised Procedure for Company Registration in the Russian Federation," 6 SEEL 1 (1995) at 1.

54. This is a very rough calculation based on a hypothetical profit and loss statement, using current tax rates. Accurate calculations may only be made for a specific business, using real numbers, and with the help of a Russian accountant who knows exactly (for example) how to calculate self-cost.

55. Ruling of the State Committee on Property of the RF of February 5, 1993 No. 217-r (with changes from February 26, 1993 and May 17, 1993), *Ob uporiadochenii protsessa ucheta i razgranicheniia prav sobstvennosti na nezhilye pomeshcheniia* [On Bringing Order to the Process of Accounting for and Assigning the Rights of Ownership of Non-Residential Space], GARANT.

56. *Kommersant*, October 6, 1994, p.2, available on LEXIS.

57. See V. Iakhontov, E. Novomlinskaia, and A. Schukin, *Khoroshii gorod! Otlichnaia nedvizhimost*, [Good City! Wonderful Property] *Kommersant*, #2, January 24, 1994, at 18.

58. Personal interviews by Emily Silliman with the managers of the parent company, Khokkei, and the attached company in 1992 and in March 1993.

59. Labor Code of Russia, Articles 33 and 35, list possible causes for firing an employee, but also require trade union approval of firings, which approval is often not forthcoming. Law of the RSFSR of December 9, 1971, as updated through September 25, 1992 *Kodeks zakonov o trude RF* [Labor Code of the Russian Federation], "Respublika" (1992).

60. Interview with a scientist who is both an employee of the Institute and does contract work for the Small Enterprise, 1994.

61. Decree of the Council of Ministers of November 2, 1990, No. 1132 *O vzaimootnosheniiakh gosudarstvennykh predpriiatii s sozdannymi pri nikh kooperativami*" [On the Relationship between State Enterprises and the Enterprises Which are Founded Attached to Them], GARANT. The decree states in part, "We have numerous facts that state enterprises are using cooperatives founded by them as a channel for turning non-cash financial resources to cash resources."

62. Letter of the State Arbitrazh of the RSFSR of February 11, 1992 N N-10/1, section 5, GARANT.

63. Ibid., section 4.

64. Iu. Tolokonnikov, *Sokhranits federalnuiu sobstvennost* [To Preserve Federal Property], 8 Zakonnost (1994), 16–20.

65. Decree of the Plenum of the Supreme Arbitrazh Court of the RF of October 21, 1993, No. 27, GARANT.

66. Decree of the Plenum of the Supreme Arbitrazh Court of the RF of February 23, 1994, No. 4, GARANT.

67. Letter of the Supreme Arbitrazh Court of December 10, 1992 N S-13/OP-357 *O razreshenii sporov, sviazannykh s uchrezhdeniem, reorganizatsiei i likvidatsiei predpriiatii* [On the Resolution of Disputes Concerning the Founding, Reorganization, and Liquidation of Enterprises], Section 8, GARANT.

68. Personal observations of the authors in Tyumen in 1992 and 1993, and reports from Alexander Gorshkalev, an entrepreneur in Tyumen.

69. Letter of the Supreme Arbitrazh Court, *supra*, note 66, Section 6. Article 32, section 3 of the Russian Law on Enterprises, *supra*. Note 22 contains the restriction of separation of a division without employee approval. This law has been superseded by the Civil Code, *supra* note 26. Employees no longer have the authority to block separation of a division of the company, although they may have votes on the matter as shareholders.

70. Decree of the Plenum of the Supreme Arbitrazh Court of the RF of February 23, 1994, No. 3. See also the special decree, Edict of the President of the Russian Federation of February 16, 1992 No. 152-s. GARANT. This is one example of a special decree prohibiting separation of subdivisions from a parent company. The decree was upheld in court.

71. Edict of the President of the Russian Federation of April 27, 1992 No. 426, *O neotlozhnykh merakh po sokhraneniiu nauchno-tekhnicheskogo potentsiala RF*, [On the Urgent Measures for Preserving the Scientific-Technical Potential of the RF], 18 Vedomosti SND RF i VS RF, item 1028.

72. Letter of the Chairman of the Supreme Arbitrazh Court of the RF of August 10, 1994, No. 01-7/OP-555 *Ob otdelnykh rekomendatsiiakh, priniatykh na soveshchaniiakh po*

sudebno-arbitrazhnoi praktike, [On Specific Recommendations Made at Meetings on Judicial-Arbitrazh Practice], section 3, GARANT.

Index

About the Editor

Judith Sedaitis is a Research Associate in the Program on Industrial Restructuring and the Political Economy of Russia at the Center for International Security and Arms Control (CISAC) at Stanford University. She received her B.A. and M.A. degrees from the University of Chicago. In 1994, she received her Ph.D. in sociology from Columbia University with a concentration in organizational sociology and Russian area studies. She was also a Fellow at the W. Averell Harriman Institute for Advanced Studies of the Soviet Union.

Dr. Sedaitis applies organizational theory and network analysis to examine organizational change in transitional economies. Her most recent research focuses on technology transfer in the Russian military R&D sector and the role of strategic high technology alliances in facilitating technology transfer between developed and developing nations. Her most recent publications appear in *Technology Forecasting and Social Change*, *International Studies of Management and Organization*, and in *Legacies, Linkages and Localities: The Social Embeddedness of the Economic Transformation in Central and Eastern Europe* (Oxford University Press, 1996).

About the Contributors

ANDREW J. ALDRIN is a researcher at the RAND Corporation.

JOHN A. ALIC lectures on technology policy at the Nitze School of Advanced International Studies (SAIS) of the Johns Hopkins University and was formerly a senior researcher at the Office of Technology Assessment of the U.S. Congress.

HARLEY BALZER is director of Russian Area Studies at Georgetown University.

LORI A. COAKLEY is Assistant Professor of Management, Bryant College, Smithfield, Rhode Island.

LINDA R. COHEN is Professor of Economics at the University of California, Irvine.

DAVID ELLERMAN is an economist at the World Bank.

CORINNA-BARBARA FRANCIS is Professor of Political Science at Brown University.

JACQUES S. GANSLER is senior vice-president at The Analytic Sciences Corporation and former U.S. deputy assistant secretary of defense.

DAVID V. GIBSON is a senior research fellow at the IC^2 Institute at the University of Texas, Austin.

STEPHEN L. GOMES is president of Venture One Associates and former chief executive officer of American Technology Initiative.

JOHN HAGEDOORN is Professor of International Business Studies and Research Fellow at the Maastricht Economic Research Institute on Innovation and Technology (MERIT), University of Limburg, Maastricht, the Netherlands.

EDWARD KAYUKOV is a student at the Stanford University School of Law.

YEVGENY KUZNETSOV is an economist at the World Bank.

ROGER NOLL is Morris M. Doyle Professor of Public Policy, Department of Economics, Stanford University.

RAYMOND RADOSEVICH is Professor of Technology Management at the Robert O. Anderson School and Graduate School of Management, University of New Mexico.

LINDA M. RANDALL is Assistant Professor of Management at the University of Rhode Island.

BERT SADOWSKI is Assistant Professor of International Business and a Research Fellow at MERIT, University of Limburg.

JACQUES SAPIR is Assistant Professor, Centre d'études des modes industrialisation & Groupe de Sociologie de la Défense, École des Hautes Études en Sciences Sociales, Paris.

LARRY W. SCHWARTZ is a lawyer with the firm Brobeck, Phleger & Harrison in San Francisco.

JUDITH B. SEDAITIS is a Research Associate at the Center for International Security and Arms Control, Stanford University.

EMILY SILLIMAN is an attorney and a private consultant on American ventures in Russia.

GARY S. SMITH is a consultant to the Technology Innovation Project at Anderson Schools of Management, University of New Mexico.